# THE SPECTRUM™
# CHEMISTRY

GARY M. WILEMON, PH.D. AND
DURELL C. DOBBINS, PH.D.

*Beginnings Publishing House, Inc.*
328 Shady Lane
Alvaton, KY 42122

ISBN 0-9666578-6-1



Direct all inquiries to Beginnings Publishing House, Inc., 328 Shady Lane, Alvaton, KY 42122.

Printed in the United States of America

## Contents

Dedicated with humble appreciation
to the world's most delightful parents,
and to their children—
tomorrow's finest scientists.
To God be the glory.

Getting Ready

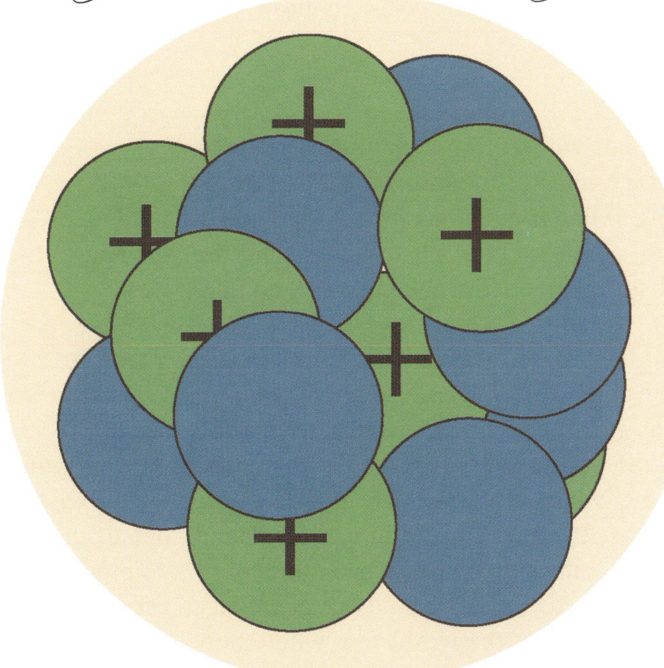

## 1: "Can We Talk?"

Welcome to Chemistry! How do you feel? People tend to feel one of two ways when embarking on a journey through chemistry; some tend to feel a little intimidated, but others approach the subject with a sense of anticipation.

Why would someone be intimidated by chemistry? Well, some folks feel that way because they don't really know what to expect. It's pretty natural to be a little uncomfortable with the unknown. Some may be anxious because they have heard others talk about chemistry as though it were a student-eating monster from the pit. (Can't you see some mother somewhere say to her finicky eater, "Eat your English peas or I'll make you take chemistry!" Brrrr.) You see, some folks have the opinion that chemistry is difficult and is, therefore, something to be avoided at all costs. To those folks we say, "Get a life."

Now don't get me wrong—chemistry is a challenging subject and will require some effort on your part. It contains a number of new concepts. It requires problem-solving skills. It even has its own language. But if you make a consistent effort to understand the material and do lots of practice problems you'll do fine.

Why would someone look forward to chemistry with a sense of anticipation? Because chemistry can be fun! Frankly, it can be very interesting to explore what's in the stuff we use everyday. It's fascinating to see what happens when substances are mixed under various conditions. It's also satisfying to understand *why* things happen rather than sitting back and wondering "What happened?" A basic study of chemistry will open your eyes to see the world in a whole new way. So, don't be intimidated. Instead, expect to be challenged, expect to learn more than you can now imagine, and anticipate that you're going to have a lot of fun.

Now, let me introduce you to chemistry. **Chemistry** is the study of the composition and structure of matter (you might call it "stuff"), and the changes that matter undergoes. That's a pretty broad definition, so you might expect that chemistry has pretty broad applications. And so it does. Chemistry touches every aspect of our lives, from the food we eat to the air we breathe. Our culture and environment are shaped in large part by the limitations of chemistry. Let me give you a couple of examples:

- Why can't you drive your car 1,000 miles on one gallon of gasoline? Is it a conspiracy fostered by the oil companies in order to sell more gas? No, it is because the chemical reaction that produces the energy from the gas just doesn't produce enough energy to send something as heavy as a car over a distance of 1,000 miles.
- Why do we have to paint houses, bridges and other things if we want them to last? Because the paint affords some protection from the detrimental chemical reactions that would occur if these structures were exposed to the elements. (Why do you think the Statue of Liberty is green?!)

Everything is the way it is because of its chemistry. Once you get a handle on the basics of chemistry the rest of the world starts to make a lot more sense.

### You Are Here

After you get your driver's license your mother might ask you to run an errand to a shopping mall. She will tell you which store she wants you to go to and what to get. To find the right store, you will approach one of those maps that shows the mall's layout. When you look at a map, you always find two points of reference: your current location and your destination. The people who make those maps usually help us by placing a dot on the map labeled "You Are Here." The next few paragraphs should help you figure out where *you* are among the sciences.

Some of you have come to us having completed ***The Rainbow*** curriculum. If so, you have already been exposed to the following information. This will serve as a brief review.

## One Door Up from Physics

In the science courses of your younger years you might have learned that there are three **basic sciences** or "pure" sciences.. They are **physics**, **chemistry** and **biology**. Other sciences (geology, astronomy, medicine, environmental science, etc.) are referred to as "**applied sciences**."

The three basic sciences are listed in order of complexity. Physics is the simplest (most fundamental) of all sciences. In physics we learn about the aspects of the universe that have been reduced to their absolute simplest forms. We might call them "rudiments," "fundamentals" or "first principles" of the universe. There we learn about the basic **forces** of nature. We learn about **energy**—the price that is paid to do **work**. We also learn about **matter**. Matter is the *stuff* of the universe. Everything that is *material* is made of it.

As we said, chemistry may be viewed as a focus on matter. Chemistry does not dote on large, complex matter like planets, oceans or humans (although we will gain greater *understanding* of these complex things through chemistry). Instead, it focuses on the atoms that make up the elements, and the combinations of these atoms to form molecules. It applies what we have learned from physics, so we will learn how (for instance) atoms interact with energy. (See Figure 1.) We will also learn to describe the details of how they combine with other atoms to form things of a higher order. If you've completed ***The Rainbow*** you've covered sufficient physics to master this course. If you've taken a course in "physical science" (which is some combination of physics and chemistry) it should also have provided you with sufficient background.

**Figure 1.** What does chemistry have to do with physics? Physics is the science that addresses the fundamentals of the universe. Since matter is physical and obeys physical principles, you can't fully understand chemistry without having a basis in those physical principles. Here are some of the principles that we will introduce you to in the weeks to come. Don't try to learn them all now. Just read them over to get a sampling of the kind of physical knowledge that goes into understanding chemistry. We'll teach it to you as we go.

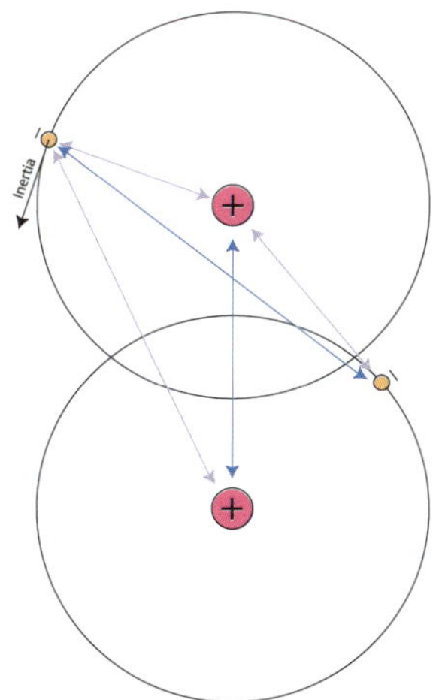


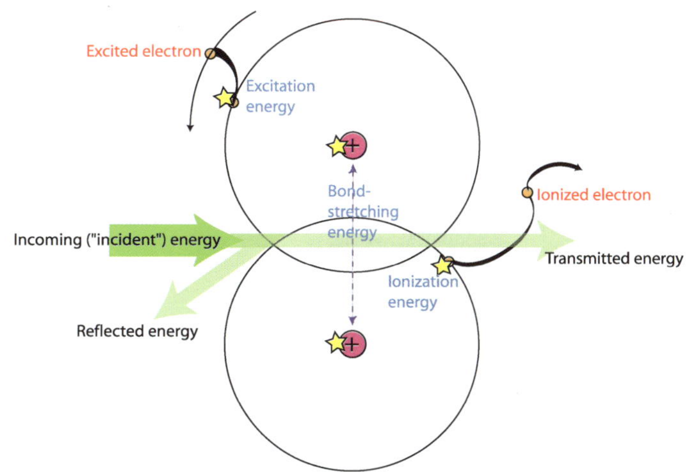


**1A.** Protons and electrons are attracted to one another. Those attractions are responsible for the "chemical bonds" that hold atoms together. Like particles repel.

**1B.** Atoms absorb energy. You may have learned this in days gone by. That absorption of energy is responsible for many of the things that atoms do, like becoming more reactive and losing electrons to become ions.

1C. Atoms and molecules have both potential and kinetic energy. Without kinetic energy their electrons could not move within atoms, and the atoms and molecules themselves would be stationary. When they combine with other atoms or molecules, their potential energy level either increases or decreases, and they either absorb or give off heat. Of course, potential and kinetic energy are among the topics of physics.

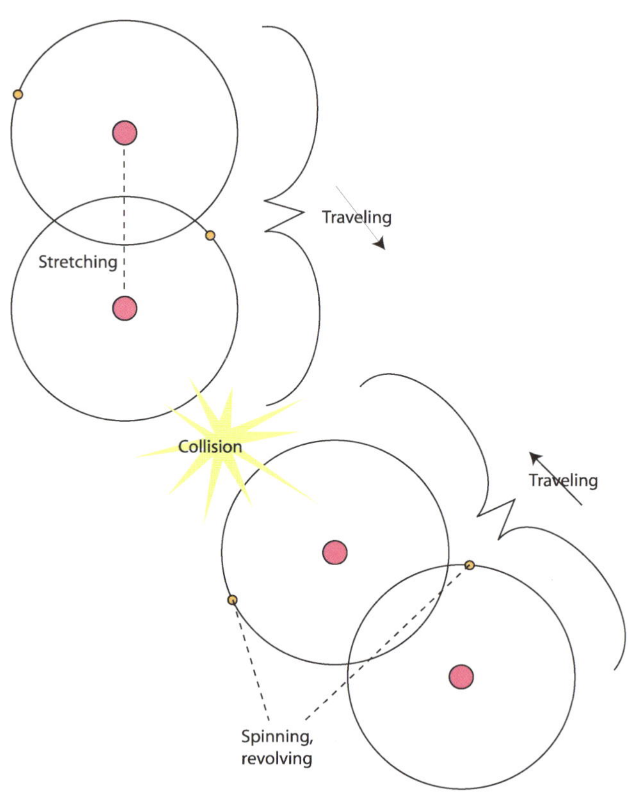


All interactions among chemicals will be seen to arise from natural forces and all these interactions obey the physical laws. Since gravity has practically no effect on the small masses of atoms, the only forces of any concern in the study of the atom will be electromagnetic. In chemistry, we focus on the interactions between light (and the other forms of electromagnetic radiation) and matter. In a special topic on the subject of chemistry, we will dive deeply into an atom to learn about its nucleus. (See Figure 2.) That is the domain of the nuclear forces.

## One Door Down from Biology

Chemicals may be classified into two groups. First, there is a host of simple chemicals such as water, oxygen, carbon dioxide and ammonia. And there are thousands of other, less-well-known molecules. Each has its own place in the universe. However, there are other molecules that are large by comparison to the simple chemicals and display an extremely high level of organization. This second class of substances is unique to living things. They are called **biochemicals** and each of these molecules has a specific *purpose* in the operation of living things. (See Figure 3.) Chemicals, then, serve as the stuff of biology. And, as you will soon learn, the interactions of these chemicals give living things the energy and the machinery to do what they do.

**Figure 2.** The nuclear forces that hold together the nuclei of atoms are among the fundamental forces of nature. We learn about these forces in physics.

## All One

When God created the universe, He didn't create physics or chemistry or biology. These are *our own* words that *we* have made up to describe that creation. The line separating physics and chemistry is not clear. Neither (in a physical sense) is the line that divides the living from the nonliving. We have classified nature into separate disciplines so that we can understand it. We may forget that these disciplines are "creations" of our own and cannot be cleanly separated. We may often speak of the physical parts of chemistry or the chemical parts of biology. There are even disciplines that are named "physical chemistry" and "biochemistry" showing the connectedness among the larger disciplines.

This is why your course is called ***The Spectrum***. You may think of red light as being distinct from yellow light. However, each is part of a spectrum of wavelengths of light that are continuous from red to yellow-red to yellow. Likewise, you may think of chemistry as a unique subject, but it is not clearly distinct from physics or biology. All three are simply samples from an entire spectrum of information. You will not be an adequate chemist until you understand that molecules do as they do because they obey the laws of physics. So, while you study through this book that focuses on the large subject of chemistry, we'll remind you of the physical forces that make chemistry work. We will also remind you of the impact that chemistry has on the operation of your own body and of the world around you.

### How To Take this Course

Well, I guess it's time I gave you the secret of how to succeed in a study of chemistry. Please read this carefully. Your knowledge and implementation of

**Figure 3.** What does chemistry have to do with biology? Biology is the science of living things. Living things, like everything else in the world, are made of chemicals. Living things are limited in what they can do by the restrictions inherent in their chemicals. Because of the marvelous versatility of these tiny structures, though, our freedom of activity is pretty remarkable. Once again, don't try to learn it all now.

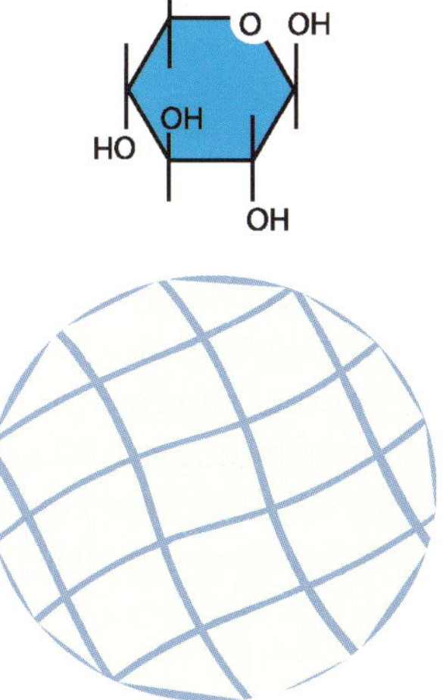


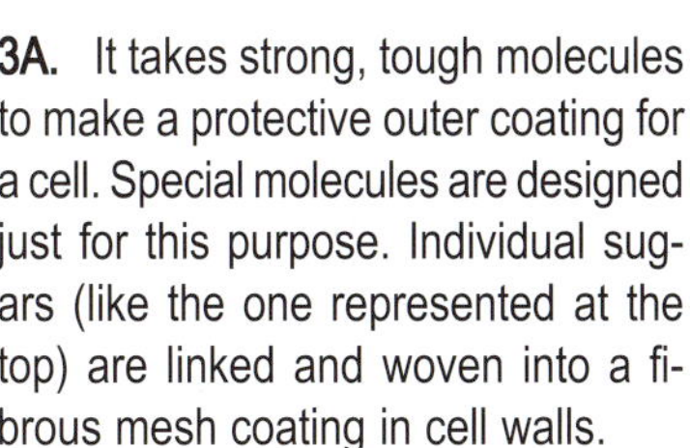

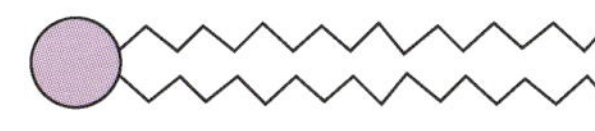

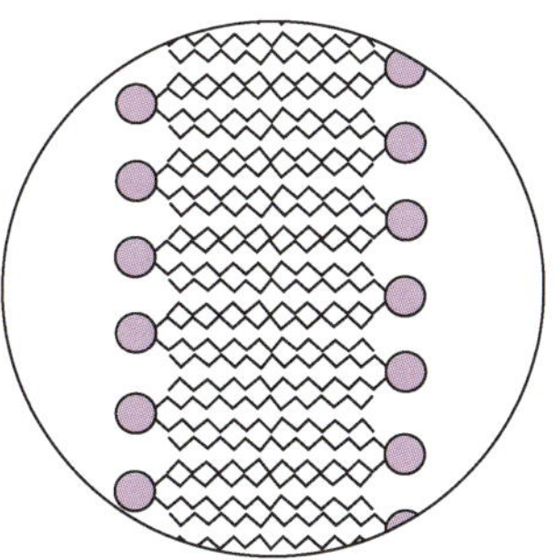

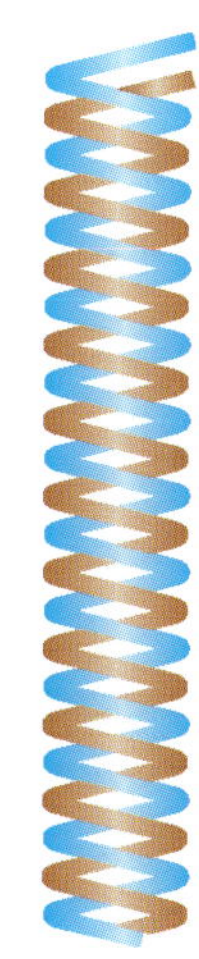

**3A.** It takes strong, tough molecules to make a protective outer coating for a cell. Special molecules are designed just for this purpose. Individual sugars (like the one represented at the top) are linked and woven into a fibrous mesh coating in cell walls.

**3B.** Lipids are highly reduced, water-fearing chemicals. These properties make them excellent barriers against the passage of liquids. It is no surprise to find them in membranes that surround cells and that divide the insides of cells into compartments.

**3C.** Nucleic acid bases are linked together into long, threadlike strands. Two strands are wrapped around one another into a long helix called DNA. The remarkable feature of DNA is that the sequence of molecules encodes a story--the story of life.

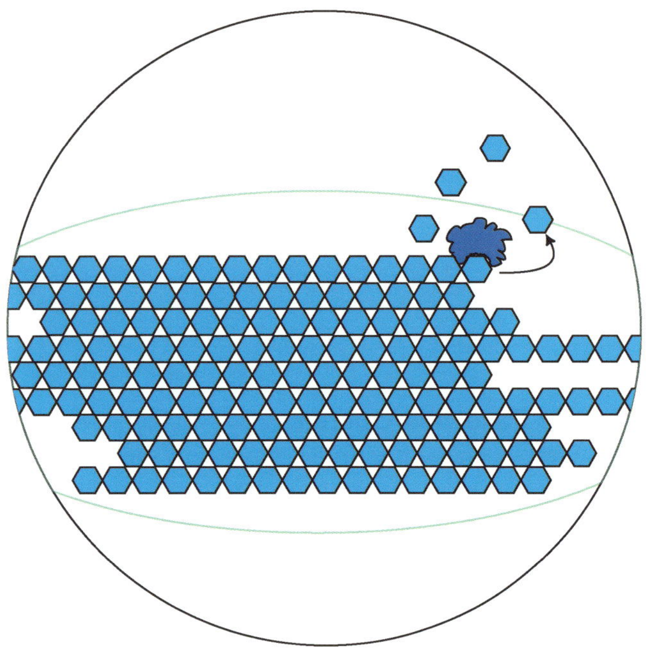

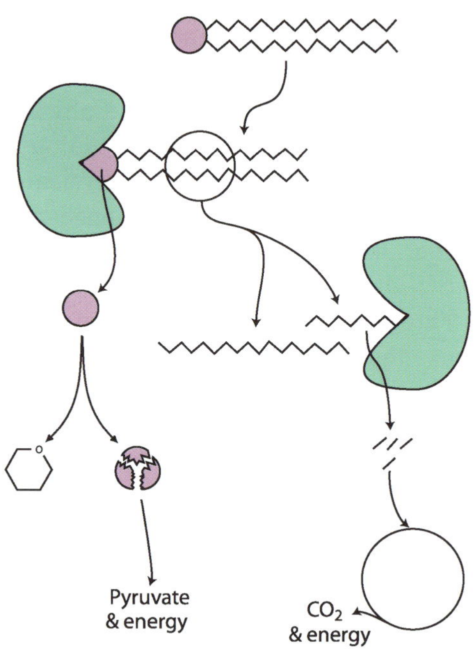


**3D.** In addition to their duty as structural molecules, sugars may be linked together to form storage polymers--glycogen and starch. Nearly all of the chemicals that make up living cells serve both structural and functional purposes—an extremely efficient design feature.

**3E.** Likewise, the lipids that make such nice membranes inside cells also function as deep energy reserves. Those long fatty acid tails are broken down to produce lots of energy. The phosphate-linked head can also be recycled. Just imagine thousands of chemical processes constantly going on inside us.

this secret may well determine the fate of the free world as we know it. Okay, maybe it's not that important, but it could make the difference between your success and failure in this course.

- **Read with a pencil in hand and with a pad of paper and a calculator next to you.** Keep them with your textbook and *never sit down to read without them*. When you encounter a question in the text, don't continue without answering it on paper. When you see a problem, solve it (*even if it has been solved for you in the book*). Be sure you understand the solution before you leave it. If necessary, return to it after completing the lesson and solve it again from memory. You *will* be held accountable for it on the next quiz.
- **Don't read around the equations, but read through them.** Part of the fear of science will go away, and you will begin to think like a scientist when you learn that equations make sense. When you are reading words and you come across an equation, don't skip the equation and continue reading words. Stop reading words and learn what the equation is saying to you. *Then* continue with the words.
- **Work problems.** The best way to make sure that you are understanding the concepts you are learning in chemistry is to solve as many problems as you can on each topic. I have seen many a chemistry student try to study for a test by reading and rereading the chapters and skimming over the examples. Warning! Warning! Do not do this! Yes, you should read the chapters, but chemistry stu-

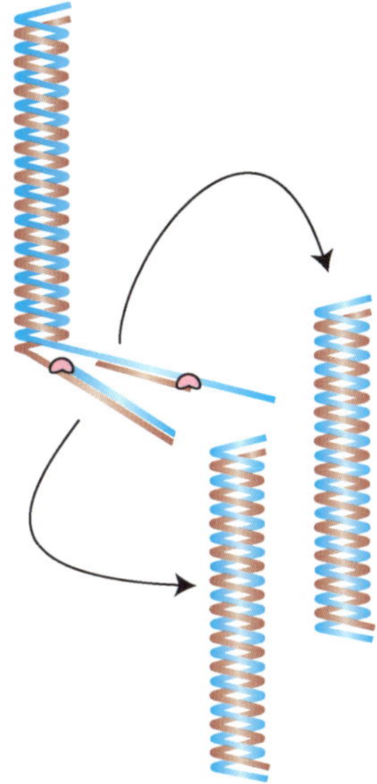

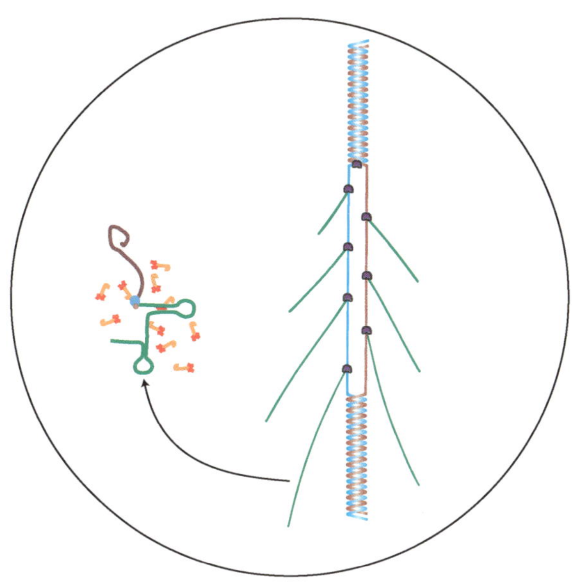

**3F.** When you have a Ph.D. in chemistry, we'll ask you to design a molecule that carries an encyclopedia's worth of information and that will reproduce itself. That's DNA. In *E. coli* bacteria, a molecule of DNA can make a copy of itself in 15 minutes. The copies will repeat this action every 15 minutes thereafter.

**3G.** From biochemistry you may also learn the details of how the information in DNA is translated by the cell. From this code, the cell gets instructions to do everything it does, including the building of all the right equipment to do it. All of this ability is programmed right into the molecules from which DNA is built!

dents are judged more on their ability to figure things out than on their ability to remember what they have read.

- **Become practiced at converting words into math problems.** You will encounter a lot of what we call *word problems* in chemistry. If those make you a little nervous, don't worry. Perhaps the most lasting benefit of this study will not be the chemistry you absorb, but the problem-solving method that you learn. There is a structured way to go about solving word problems. Those of you who took our course called ***Bridge Math*** know it well. We'll talk about it more in Lesson 3.

Okay. Now you're ready. Go ahead and get your calculator, pencil and notebook now, and keep them with your book when you put it away. When you come back on your next science day, please have them ready. You'll need them right away. Oh, yes, and *please have fun*!

## 2: The Language of Chemistry

One of the things that trips folks up when they study chemistry is that there is so much new terminology with which to become familiar. Each of these lessons will likely contain some terms that will be new to you. So let's begin by laying some groundwork in the area of definitions and symbols. You have probably already learned these in an earlier science course. Don't fret about having to go over it again. It'll be good review. It's so important to the study of chemistry that we must be sure you don't miss a single word. If you have been over this material before, it should be really easy for you. Right?

### Elements

If you recall, all matter is made up of elements. **Elements** are defined as fundamental ("elementary") substances which cannot be broken down into simpler ones by ordinary physical or chemical means. As an example, suppose we have a piece of copper wire, and we cut it in half. Then we take one half, and slice it in half again. Suppose we continue this process of cutting the copper down by halves; assume we are even able, somehow, to divide the copper down to a submicroscopic level. Eventually we would get to the point where what is left is a single copper atom. While it is true that a copper atom is composed of still smaller subatomic particles (**protons**, **neutrons** and **electrons**), if we divide the copper atom any further, we won't have copper anymore. Copper is one of those fundamental substances—an element—*composed of a single type of atom*. It is distinguishable from every other type of atom by how many protons are in its nucleus.

As of the time this book went to press, there were 113 elements. (New ones are made in laboratories from time to time.) We will deal with only the most common ones, which amount to about 40% of them. Since atoms of one element combine with atoms of another element to form different compounds, it is convenient to have a shorthand method for referring to them. For this reason, each element has been given its own symbol. Table 1 shows the most common elements along with their symbols. The names of all the elements (and the tentative names of the elements that are as yet unnamed) are provided inside the front cover of this book in the **periodic table of the elements**—an orderly arrangement of the elements' symbols. We will discuss this table in great detail over time.

**Table 1.** Common elements and their symbols.

| | | | | | |
|---|---|---|---|---|---|
| Aluminum | Al | Copper | Cu | Oxygen | O |
| Antimony | Sb | Fluorine | F | Phosphorus | P |
| Argon | Ar | Gold | Au | Platinum | Pt |
| Arsenic | As | Helium | He | Potassium | K |
| Barium | Ba | Hydrogen | H | Radium | Ra |
| Beryllium | Be | Iodine | I | Silicon | Si |
| Bismuth | Bi | Iron | Fe | Silver | Ag |
| Boron | B | Lead | Pb | Sodium | Na |
| Bromine | Br | Lithium | Li | Strontium | Sr |
| Cadmium | Cd | Magnesium | Mg | Sulfur | S |
| Calcium | Ca | Manganese | Mn | Tin | Sn |
| Carbon | C | Mercury | Hg | Titanium | Ti |
| Chlorine | Cl | Neon | Ne | Tungsten | W |
| Chromium | Cr | Nickel | Ni | Uranium | U |
| Cobalt | Co | Nitrogen | N | Zinc | Zn |

Hint: Memorize the symbols for the elements listed in Table 1. It will take you a little extra time now, but it will save you scads of time later.

Please notice that each of the element symbols starts with a capital letter. Some them have a lower case letter following. To avoid confusion, it is very important to reproduce these symbols as written. For example, Co is the symbol for Cobalt, but CO is the formula for carbon monoxide. Also notice that some of the symbols appear to have no relation to the name of the element. That's because some of the symbols were derived from their Latin names. Tungsten was once known by its Latin name *wolfram*, and lead was known as *plumbum*. Thus, they have the symbols W and Pb, respectively. Table 2 shows the names of these elements and the Latin names from which they are derived. You needn't memorize these, but seeing them will help you to remember their correct symbols.

**Table 2.** Elements having symbols that are taken from their Latin names.

| Name | Symbol | Latin Name |
|---|---|---|
| Antimony | Sb | *Stibium* |
| Copper | Cu | *Cuprum* |
| Gold | Au | *Aurum* |
| Iron | Fe | *Ferrum* |
| Lead | Pb | *Plumbum* |
| Mercury | Hg | *Hydragyrum* |
| Potassium | K | *Kalium* |
| Silver | Ag | *Argentum* |
| Sodium | Na | *Natrium* |
| Tin | Sn | *Stannum* |
| Tungsten | W | *Wolfram* |

## Compounds

Remember, we observed earlier that elements combine to form other substances. **Compounds** are defined as substances that contain two or more elements chemically combined in definite proportions by mass. For example, water is a compound made up of the combination of hydrogen and oxygen; and it always contains one oxygen atom combined with two hydrogen atoms. This consistently gives it a definite composition of 11.2% H (two light atoms) and 88.8% O (one somewhat heavier atom) by mass. (Notice how I've started to sneak element symbols in on you?) That means if we had 100 g of water and we were able to separate it into its individual elements, we'd have 11.2 g of H and 88.8 g of O. The fact that specific compounds are comprised of specific proportions of elements by mass is called the **Law of Definite Composition**. This law basically says that water is water is water is water and that water will always have the composition of water. If the composition changes, it's no longer water. Do I make my point?

Law of Definite Composition: A specific compound is comprised of a specific proportion (by mass) of elements.

Law of Multiple Proportions: A particular set of elements can combine in different ratios to produce different compounds.

Hydrogen peroxide is another compound of H and O; but in this case two atoms of hydrogen and two atoms of oxygen are involved—different composition, different compound! The relative amounts of the two elements in hydrogen peroxide are 5.9% H and 94.1% O by mass. The ability of two or more elements to combine in *different* ratios to produce a variety of *different* compounds (such as water and hydrogen peroxide) is the called the **Law of Multiple Proportions**. (See Figure 1.)

The most important thing to remember is that every specific compound has a very specific composition of elements. Compounds do not vary in this respect. One compound has a single, definite composition. If the composition changes, it's no longer the same compound. Likewise, every different compound has its own chemical formula that shows the ratio of the various atoms in the compound. Figure 2 shows the chemical formulas of water and hydrogen peroxide next to those molecules and explains how to understand those formulas. Figure 3 illustrates a variety of molecules and their chemical formulas. Studying them carefully will help you understand how to write chemical formulas of your own. Figure 4 shows how to name molecules that have repeating clusters of identical atoms.

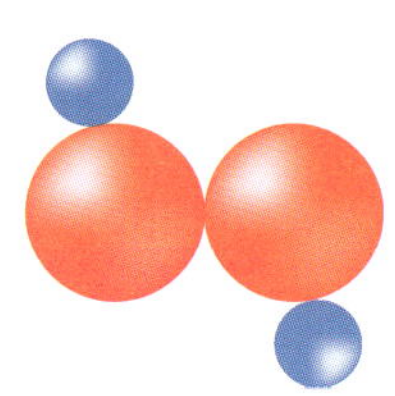

**Figure 1.** (Left) Water always consists of one atom of oxygen and two atoms of hydrogen. This gives all water a constant proportion by weight of hydrogen to oxygen. Likewise, hydrogen peroxide consists of two atoms of oxygen and two of hydrogen. This gives all hydrogen peroxide a constant proportion by weight of hydrogen to oxygen. Both are demonstrations of the law of constant composition. The fact that each of these compounds consists of only hydrogen and oxygen, but that the two have different proportions of the two elements, illustrates the law of multiple proportions.

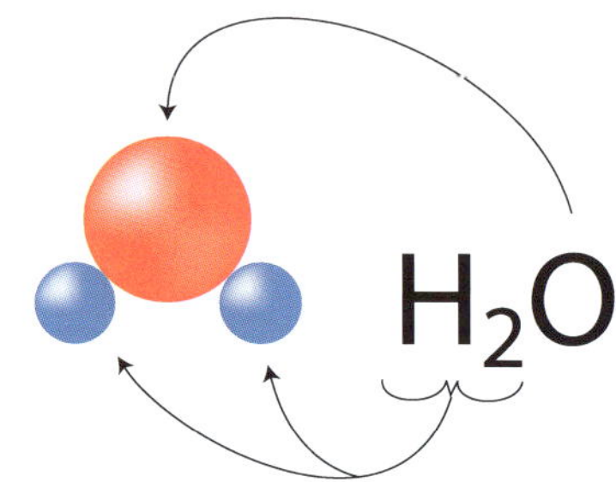


**Figure 2.** (Right) The chemical formula of a substance provides the number of atoms of each element. The number of atoms is always subscripted to the right of the element's symbol. If no number is written next to an element's symbol, the number of atoms of that element is assumed to be one.

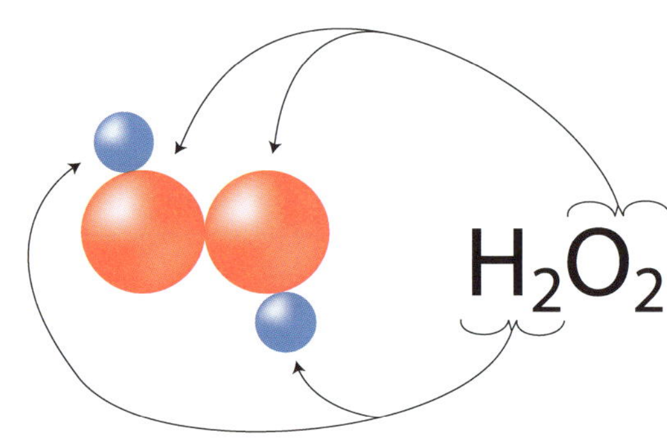


**Figure 3.** (Below) The figure shows one molecule of each of several substances with their chemical formulas. One of the molecules is not a compound. Do you know which?

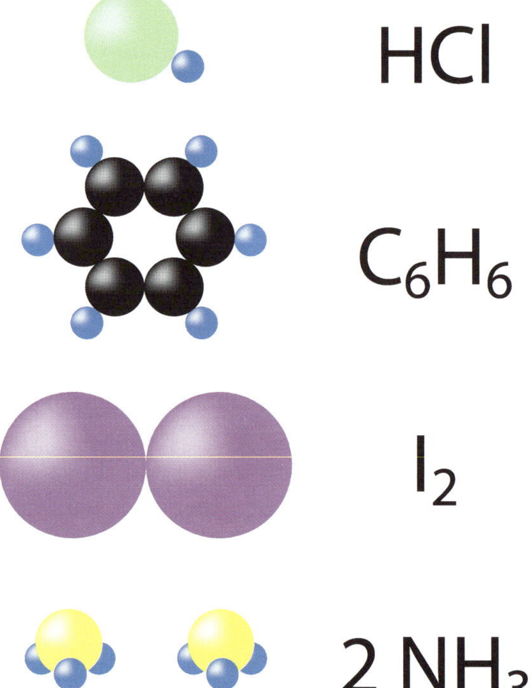


### Molecules

The most fundamental uncharged unit of a compound is called a **molecule**. Suppose we took a drop of water and started dividing it like we did earlier to our piece of copper. Eventually, we'd get down to a single molecule of water. If we divided any further, we wouldn't have water anymore. Figure 5 illustrates this principle using an unknown compound made up of two atoms. When you have divided any substance until the remaining quantity is a molecule, you would have to affect the structure of the molecule to divide that substance any further.

We write the formulas for compounds using the elements' symbols and the relative amounts of the various atoms. In the case of water, the **chemical formula** is $H_2O$. That means two H atoms combine with one O atom to form a water molecule.

While a compound, by definition, consists of more than one kind of atom, molecules may sometimes be composed of only one kind of atom. In fact, there are seven elements that exist primarily as **diatomic** molecules—molecules that contain two identical atoms. (Although the word *diatomic* literally means "two atoms," not just any molecule containing two atoms is thought of as diatomic. The two atoms must be identical.) These seven elements are hydrogen, nitrogen, oxygen, fluorine, chlorine, bromine and iodine. Therefore, when these elements are written to show the way they occur in nature, they are expressed as $H_2$, $N_2$, $O_2$, $F_2$, $Cl_2$, $Br_2$ and $I_2$.

$Mg(NO_3)_2$

**Figure 4.** Notice that there are two identical groups of atoms in this molecule. Each consists of a single nitrogen atom (yellow) and three oxygen atoms (red). Each of those two groups is designated as $NO_3$, so they show up in the formula as $(NO_3)_2$.

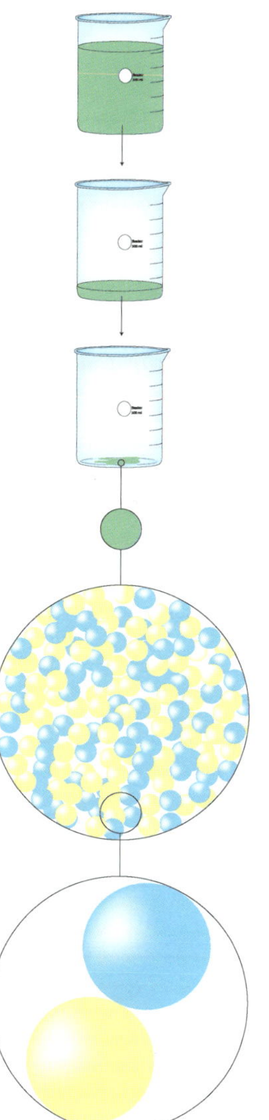

**Exercises**

1) Make a set of flashcards to help you in learning the element symbols. (A pack of colored flashcards and a marker are provided in your lab set.) Put the name of the element on one side of a card and its corresponding symbol on the other side. Drill with these cards until you have them down cold.

2) How many total atoms are present in each of the following molecules?

   a) $(NH_4)_3PO_4$
   b) $Ca(HSO_4)_2$
   c) $Li_2(C_2H_3O_2)_2$

3) What is the difference in meaning in each of these pairs?

   a) Si and SI
   b) Pb and PB
   c) 2 H and $H_2$

4) Which of the following molecules are termed *diatomic*?

   $Cl_2$, $Br_2$, CO, HI, $H_2O$, $O_2$, $F_2$

**Figure 5.** Any substance may be divided into smaller and smaller portions until the tiniest portion of that substance is left. That tiniest, irreducible portion will be a single molecule. In this figure, the amount of a substance is being further and further reduced until a single molecule is all that is left. Notice that this molecule is composed of two atoms. They are bound together by the forces that attract them. If they were to be separated the result would no longer be the same substance.

## 3: How Do You Measure "Up"?

And if you did measure "up," how good would your measurement be? Today's lesson is all about measurements: how to choose a device for making them, how to make them, how to express them and how to judge their usefulness. If you are to learn anything about an experimental science such as chemistry, you might as well get used to lots of measurements. If you have been through our course called ***Bridge Math,*** this lesson will be a review for you.

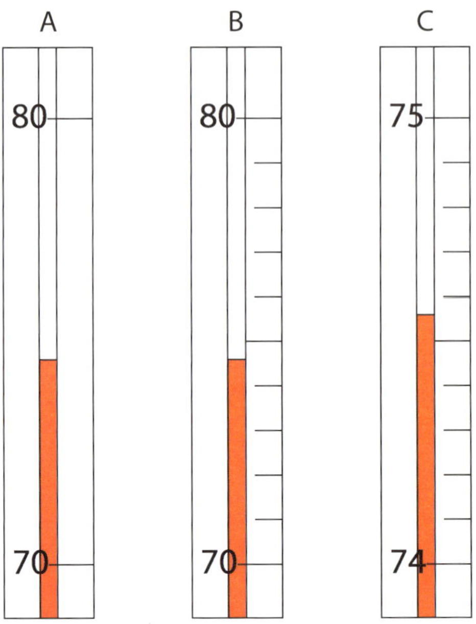


**Figure 1.** These three thermometers are graduated differently. Which thermometer is calibrated with (and allows you to read it with) the greatest precision?

### Significant Figures

Basically, there are two kinds of numbers: exact and inexact. Exact numbers are those obtained either by counting or by definition. If we have a dozen eggs, then, by definition, we have 12 eggs—we don't have 11, or 12.3, or 13. Alternately, we might have several loose eggs in our refrigerator and when we count them, we find that we have 16 eggs. In either of these two situations there is no question as to *exactly* how many eggs we have.

When we make measurements, though, the situation is not quite so clear cut. Measurements are, to some extent, inexact. Use Figure 1 to assist you in the following example. Suppose you want to know how hot it is outside, so you go to a thermometer that is hanging on the side of the house. The red line on the thermometer is not quite halfway between the 70°F mark and the 80° mark, just like Thermometer A in the Figure. What is the temperature? It's not 70°, because the line is above 70°. But it's not 80° either, because the line is not up to the 80° mark. It's somewhere in between, and there is not an exact way of deciding. You have to estimate. Since the red line is not quite halfway between 70° and 80°, you estimate that the temperature is 74°. You may feel that your estimate is fairly close, but you recognize that there is a "degree of uncertainty." That's what we mean by "inexact".

What is it that determines the degree of uncertainty in your measurements? There are three things. First, it has something to do with your skill in making the measurement. You are probably pretty skilled at reading a thermometer. But there are other measuring devices that require considerably more skill. Some require months or years of training and practice.

**Table.** Five measurements from a single thermometer of the temperature of boiling water.

| Thermometer 1 | Thermometer 2 |
|---|---|
| 100.12 | 99.9 |
| 100.11 | 100.4 |
| 100.11 | 101.0 |
| 100.11 | 99.2 |
| 100.13 | 100.1 |

The second factor that affects the uncertainty of measurements is the accuracy of the measuring device. **Accuracy** is the extent to which a measurement gives a true value of the quantity being measured. For example, look at the five measurements of the temperature of boiling water listed under "Thermometer 1" in the first column of the Table. Each of these measurements was made by the same thermometer. By definition, the temperature is 100°C, but that's not the value being shown on this thermometer. It has a tendency to

**Figure 2.** Here are two different analytical balances measuring with very different levels of precision. The one on the left measures to the nearest 0.00001 g (It is contained in a case to prevent air movement from affecting measurements.) The weight shown on its display is that of a single hair approximately 5 cm in length. The balance on the right measures to the nearest 0.1 g.

register a little higher. All five measurements are very close to the same value, but that value is not 100.00°C. This is an example of inaccuracy.

When it comes to accuracy, some instruments may be well calibrated, but none are exactly accurate. The question is *how* accurate (or inaccurate) is a given instrument. One thermometer may measure the boiling point of water as 100.001°. Most people wouldn't pay the money it costs to purchase a thermometer that delivers that kind of accuracy. Another might measure it as 90.4°C. Not many people would pay a nickel for that thermometer! In order to measure with confidence, every instrument (including every thermometer) must be **calibrated** by comparing it to some known **standard**. In our example, the standard is boiling water.

The third factor that affects uncertainty in our measurements is the **precision** of the measuring system. In the sciences, precision is the extent to which repeated measurements agree with one another. For example, five measurements of the temperature of boiling water were taken using Thermometer 2, as shown in the Table. The temperature values hit all around 100°. If you average the numbers, you get a value that is very close to 100°, so it's not an accuracy problem. The thermometer just can't seem to settle on the same reading twice. Compare this with the accuracy problem you saw with Thermometer 1. It's a different problem.

The precision of an instrument is its own unique quality. Most manufacturers make their instruments readable at the instrument's precision limit. That way you don't lose precision because the manufacturer failed to add enough lines on the scale. So then, how much precision you get out of an instrument will be determined by how finely graduated it is. (See Figure 2.)

Let's go back to our thermometers in Figure 1. We have already concluded from Thermometer A that the temperature is about 75°F. We did so by estimating that the red fluid in the thermometer comes up to a point that was roughly halfway between the 70° and the 80° marks. Now take a look at Thermometer B. What is the temperature? The red line is about halfway between the 74°F and 75° marks. So you might confidently conclude the temperature is near 74.5°. Notice that this time you are able to estimate the temperature to the first place to the right of the decimal point because of the finer graduations on the thermometer. If the thermometer were even more finely graduated, as is Thermometer C, then there would be even less uncertainty in your measurement. The temperature can now be confidently estimated in the second decimal place. So then, the precision of your reading clearly depends on the graduation of your thermometer.

In every discipline of science, the term **significant figures** is used to describe the number of digits we can use with confidence. We define significant figures to include **all the digits in a measurement that can be determined without doubt, plus one estimated digit**. Therefore, for thermometer A, where we estimated the temperature at around 75°F, there are two significant figures. The seven and the five are considered to be significant in that each offers some useful information. Thermometer B adds to our reading precision. From it we can get *three* significant figures. The seven, the four and the five are all "significant." The temperature was clearly above 74°, and we estimated the ".5". From thermometer C we determined the temperature to be 74.56°. Again, the temperature was clearly above 74.5°, and we estimated the "6" in the hundredths place. So, in this instance, there are *four* significant figures.

Well, what's the big deal? What's so "significant" about significant figures? It all goes back to that degree of uncertainty. When we make a measurement and there is a degree of uncertainty, that's one thing. But when we take that measurement and use it in calculations involving other measurements, each having its own degree of uncertainty, then the degree of uncertainty is compounded. We may wonder how much confidence we can have in our calculations. There is one thing to keep in mind: **the results of a calculation cannot be more precise than the least precise measurement used within that calculation**. Since that is so, there are rules that we apply to calculations involving significant figures. When applied to addition and subtraction, here is the rule you follow:

Rule 1: When adding or subtracting, the answer must be expressed to the same precision as the least precise measurement.

For example, if I saw a small basket of tomatoes, I might estimate the number of tomatoes in the basket at "10." There might be 12 or there might be 8, but I guessed the number at 10. Now suppose I counted precisely 5 tomatoes on the table beside the basket. How many tomatoes are there altogether? Can I say with assurance that there are exactly 15? Certainly not! If the 10 was an estimate, then the 10 plus 5 is subject to the same amount of error as the 10! So the error associated with two added numbers is potentially as great as the greatest error associated with either of them. We can't show more certainty in the answer than either of those numbers shows.

Let's put this principle to work by solving a real problem. Let's say we had a pile of sulfur and a pile of carbon. Suppose we wanted to know, without mixing the two piles together, what would be their combined mass. We know that the masses of the individual piles were determined using two different well-calibrated analytical balances as follows:

Mass Sulfur = 57.24 g
Mass Carbon = 10.4 g

There are two answers to every problem. There is the answer that the calculator displays containing as many digits as it can generate from the numbers you give it. Then there is the correct answer stated to the proper precision. The calculator gives you:

57.24 g + 10.4 g = 67.64 g

But the least precise measurement was known only to the first decimal place. Therefore, we can know the answer of this calculation only to the first decimal place. If we are to state the answer to its proper level of precision, the answer is **67.6 g**.

The same rule applies to subtraction as to addition. If we were to subtract 41.2197 lb from 58.43 lb, the calculator would give us the following:

58.43 lb – 41.2197 = 17.2103 (calculator answer)

But the least precise measurement was known only to 2 decimal places. Therefore, we can know the answer only to 2 decimal places; the answer is **17.21 lb**. Now let's move on to multiplication and division.

Rule 2: When multiplying or dividing, the result must have no more significant figures than the measurement with the fewest significant figures.

To illustrate this rule, suppose I wanted to estimate the area of a rectangle, so I measured the width of the rectangle to be 4.1 centimeters (cm). Based on the two significant digits in my measurement, the measuring device must have had 1-cm graduations. As always, I estimated the last significant digit. The real value could be anywhere between 4.0 and 4.2, but I believed it was closer to 4.1 than to either of these. Now let's suppose I measured the height of the rectangle using a device that allowed much greater precision and found it to be 2.000 cm. The area of a rectangle is calculated by multiplying 4.0 cm by 2.000 cm. Can I say with confidence that the area of the rectangle is 8.000 $cm^2$? No way! It's impossible to know the area with more precision than I know the length of either of the sides! (See Figure 3.) The area can be anywhere from:

$$4.0 \text{ cm} \times 2.000 \text{ cm} = 8.0 \text{ cm}^2$$

to:

$$4.2 \text{ cm} \times 2.000 \text{ cm} = 8.4 \text{ cm}^2$$

depending on the *actual* length of the long side. To reflect the correct amount of precision in my measurement, I would have to restrict the number of significant digits to the number of significant digits in the least precise measurement (4.1 cm has only two significant digits). The answer, correctly expressed, is 8.2 $cm^2$.

Let's put what we have learned to work. What is the area of a rectangle having a length of 11.39 in and a width of 2.38 in?

$$A = \text{length} \times \text{width}$$
$$(11.39)(2.38) = 27.1082 \text{ (calculator answer)}$$

But the measurement that had the smallest number of significant figures was 2.38 inches (only 3 significant figures). Therefore, we can know the answer only to the same number of significant figures; the answer is 27.1 $in^2$. (You didn't really think you knew the area to the fourth decimal place, did you?)

Division follows the same rule as multiplication. Calculate the answer to (24.6 cm)(3.00 g) / 9.3 cm

$$(24.6)(3.00) / (9.3) = 7.9354839 \text{ (calculator answer)}$$

But the measurement that had the smallest number of significant figures was 9.3 (only 2 significant figures). Therefore, the answer is 7.9 g.

You might have wondered about that last example. How many significant figures are present in the number 3.00? Actually, whether a zero is significant or not can be a little confusing; so naturally we have some more rules to help us out.

1) All zeroes that fall between nonzero digits are significant. For example:

**Figure 3.** The height of this rectangle is measured very precisely (2.000 cm), but the width of it is not (4.0 cm). The figure shows the effect that the uncertainty has on the calculated area ("A"). The shaded areas on the right side of the rectangle show the change in area that would come about with the smallest possible change in the measured width. Notice the uncertainty in the width is multiplied by the height. Even though we don't like uncertainty, we must reflect it in our answers.

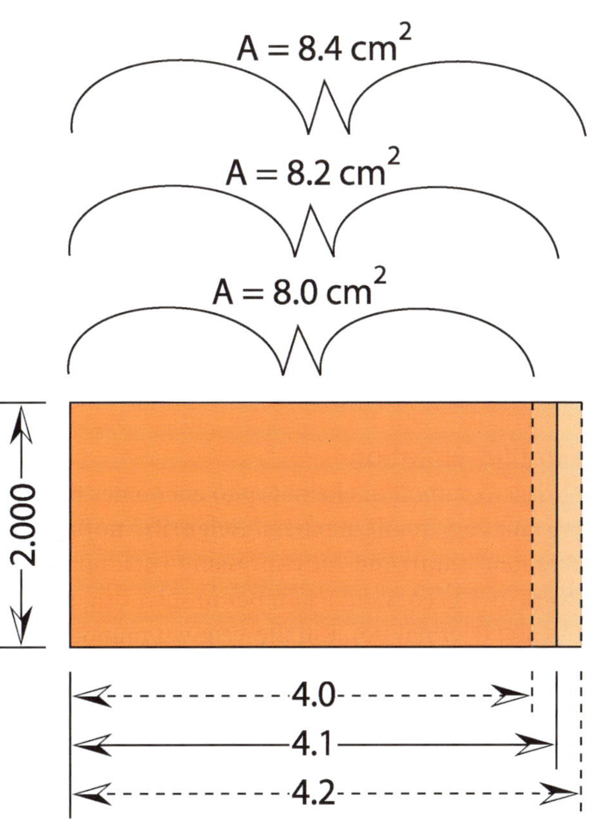

| | |
|---|---|
| 404 | 3 significant figures |
| 105 | 3 significant figures |
| 76.02 | 4 significant figures |

2) Zeroes that precede the first nonzero digit are **not** significant. For example:

| | |
|---|---|
| 0.03 | 1 significant figure |
| 0.0404 | 3 significant figures |

3) Zeroes at the end of a number that has a decimal point are significant. For example:

| | |
|---|---|
| 45.0 | 3 significant figures |
| 450. | 3 significant figures |
| 450.0 | 4 significant figures |
| 3.00 | 3 significant figures |

4) Zeroes at the end of a number without a decimal point are not significant unless otherwise stated. For example:

| | |
|---|---|
| 30 | 1 significant figure |
| 480 | 2 significant figures |

If you were to think your way through each of these rules, you would find that they all make sense in preserving the right amount of precision in your answers. A full treatment of the logic behind each of these rules is given in ***Bridge Math***. While this is a very short introduction to these concepts, they are concepts that any science major will face time and time again. (Not to mention on the next quiz.) Spend whatever time it takes to be comfortable with them.

## Scientific Notation

It is common in chemistry to encounter both very large and very small numbers. **Scientific notation** is a convenient shorthand for expressing such numbers as powers of 10. To write a number in scientific notation, move the decimal point in the original number until it is just to the right of the first nonzero digit. Then multiply this new number by 10 raised to an exponent ("power") equal to the number of places the decimal was moved. If the decimal is moved to the right (making the number larger), then the exponent must be decreased to compensate. If the decimal is moved to the left (making the number smaller), then the exponent must be increased to compensate. Here are some examples:

0.0004503 → $4.503 \times 10^{-4}$
The decimal point was moved 4 spaces to the right.

7,100,000,000 → $7.1 \times 10^{9}$
The decimal point was moved 9 spaces to the left.

4503 → $4.503 \times 10^{3}$
The decimal point was moved 3 spaces to the left.

0.034 → $3.400 \times 10^{-2}$
The decimal point was moved 2 spaces to the right.

2,000,000,000 → $2 \times 10^{9}$
The decimal point was moved 9 spaces to the left.

Notice that, in the last example, only one significant digit is reported. The scientific notation must be modified by losing the decimal point to preserve the correct number of significant digits. The subject of scientific notation is covered at length in ***Bridge Math*** as well. If this concept causes you grief, you may want to take that course. As is true with significant figures, scientific notation will be encountered frequently throughout your science education.

## Standard Units of Measurement

Once upon a time, at the beginning of this lesson, we spoke of measurements. (Remember?) Please keep in mind that every measurement results in both a number (or value) and a unit that tells what is being measured. A common pitfall for chemistry students is a failure to keep up with the units in problems. In chemistry, we use metric units. The **metric** system involves a series of standard units and uses factor-of-ten prefixes to express larger or smaller quantities. Inside the back cover of your textbook you will find a Table of these prefixes. You should memorize the ones from micro- to kilo-. If you are planning to be a professional scientist, or any unusually knowledgeable person, you might as well get to know all of them.

From time to time, you may also hear someone speak of the **International System** of units. This is a select group of units that have been adopted by a scientific body to provide a standard for reporting in familiar units. That way scientists meeting with those from other nations won't confuse them with the use of some goofy system of measurements that is unfamiliar. This International System of units (*Le Système International des Unites*)—or **SI** units—is also called the **mks** (meter-kilogram-second) system. Its other units of measure are the ampere (A) for electrical current, the Kelvin (K) for temperature, the mole (mol) for an amount of a substance, and the candela (cd) for luminous intensity. While, as I said, you may hear of this system or see it on a standardized test, it is largely ignored in practice for all but special reporting purposes.

The unit of mass you will probably use most often is the **gram (g)**. As you will learn in physics, there is a difference between mass and weight. Mass is a measure of an amount of matter, while weight is a measure of the influence of gravity on that matter. At the surface of the Earth, the two are the same (in metric units). In outer space or on another planet, these values are different. Since most chemists operate somewhere near the surface of the Earth (a point to be argued), we tend to use the two terms interchangeably.

The units used most frequently for measuring length are the **meter (m),** the **centimeter (cm)** (see Figure 4.), the **millimeter (mm)** and the **nanometer (nm)**. The units used most frequently for measuring volume are the **liter (L)**, the **milliliter (mL)** and the **cubic centimeter ($cm^3$)** (equal to the milliliter).

Finally, the units most commonly used for measuring temperature are the **Kelvin (K)** and the **Celsius scales (°C)**. Occasionally the **Fahrenheit scale (°F)** will be used, but we don't really like it. The Fahrenheit scale separates the freezing point and boiling point of water by 180 degrees (32° to 212°), whereas the Celsius scale separates them by 100 degrees (0° to 100°). Thus, a Celsius degree is larger, or represents a wider range, than a Fahrenheit degree. A Kelvin degree is the same size as a Celsius degree, but the degree sign is not used with Kelvin temperatures. The Kelvin scale starts at **absolute zero**—the coldest temperature theoretically attainable, and the point at which all particle motion stops). To convert from one temperature scale to another, use the following formulas:

1) K = °C + 273.15 (if your expressions of temperature don't have to be really precise, you can drop the ".15")
2) °C = (°F – 32°)/1.8
3) °F = (°C × 1.8) + 32°

**Figure 4.** The most commonly used unit of length in chemistry is the centimeter. An inch is equal to approximately 2.54 cm. A centimeter is precisely equal to the width of a Popsicle stick.

Later, we'll introduce standard units and conversion factors for units of other measurements, including energy and pressure. For now, just have fun with the exercises.

**Exercises**

1) How many significant figures are there in the following?

a) 24 inches
b) 135,000 ants
c) 0.0045 meters
d) 3.0808 pounds
e) 4.00 liters

2) Calculate the following to the appropriate number of significant digits.

a) $3485.0 + 5.78 =$
b) $0.4329 \times 9.3 =$
c) $459.3 - 5.211 =$
d) $(95.3)(521.4)/(2.11) =$

3) Write the following in scientific notation. (Be sure to preserve the number of significant figures.)

a) 3,600
b) 0.0857
c) 4,703,000
d) 0.00000005

4) A temperature of 4,500°C is equivalent to what Fahrenheit temperature?

5) What is the temperature in Kelvin of an iron bar having a temperature of 350°C?

## 4: Anything Times One Is...

Perhaps the most useful skill you will learn in studying chemistry is not anything "chemical" but rather the ability to solve problems in a structured, organized way. There are actually several methods that might be employed in problem solving, but in our study we will emphasize only one, the **unit factor method**, also sometimes called the factor-label method.

The unit factor method gets its name from the fact that the necessary mathematical transformations are accomplished through the use of "units." For example, suppose you are given the following problem: How many inches are there in a length of 4 feet? You can do this intuitively. You know there are 12 inches in a foot, and if you have 4 feet then each of them has 12 inches, so the answer must be $4 \times 12 = 48$ inches. And it wouldn't have mattered if you had been given 3.4 feet, 56.9 feet, or $4.32 \times 10^5$ feet, you would know to multiply the number of feet by 12 to get the number of inches. But why does this work? Let's think about what we actually do to determine the answers.

We don't really multiply 4 ft by 12; if we do we get an answer of 48 ft. And we don't multiply 4 ft by 12 in; if we do that we get "48 foot-inches" (whatever these are). What we do is multiply 4 ft by 12 inches per foot. If we set up the problem in a formula, it looks like this:

$$? \text{ in} = 4 \cancel{\text{ft}} \times \frac{12 \text{ in}}{\cancel{\text{ft}}} = 48 \text{ in}$$

The reason the "ft" units cancel out is because we have "ft" in the numerator and "ft" in the denominator. And ft/ft = 1. Therefore, all we are left with is an answer in inches.

In essence, what we have done is multiply 4 ft by a "unit factor," that is, a factor of 1. Remember, when we multiply any number by 1, we do not change its value. Well, 12 inches and 1 foot are the exact same length, aren't they? Therefore, 12 in/1 ft is an expression in which a length is divided by itself; it is equal to 1. So by multiplying 4 ft by 12 in/1 ft we have not changed the value of the 4 ft. Yes, the number is different, but that is because it is now expressed in different units. In fact, 4 ft and 48 inches **are** the same length. The premise of unit factors is this: Any value divided by any equivalent value is equal to 1 *regardless of how those values are expressed.* (See the Figure.)

Well, suppose the problem had been given a little differently. Suppose you had been given a length of 48 inches and were asked for that same length in feet. Again, intuitively you know to divide the 48 inches by 12, to get an answer of 4 ft. But what you're really doing is dividing 48 inches by 12 inches per 1 foot, like this:

$$? \text{ ft} = 48 \text{ in} \div \frac{12 \text{ in}}{\text{ft}} =$$

As you know, dividing by 12 in/1 ft is the same thing as multiplying by its reciprocal, 1 ft/12 in, as follows:

$$? \text{ ft} = 48 \text{ in} \times \frac{\text{ft}}{12 \text{ in}} = 4.0 \text{ ft}$$

In this way the "inches" unit cancels, and you are left with an answer in feet.

(Note: In this problem, the values for 1 foot and 12 inches are known by definition. That is, by definition there are *exactly* 12 inches in 1 foot. There are not 12.1 inches in a foot, neither are there 12.01 inches in a foot, nor 12.001, nor 12.0001. The precision of such values that are known by definition, and the number of significant digits associated with them is infinite. In other words, it's as if the number 12 had an infinite number of zeros following its decimal point. If you added anything to the 12—anything at all—or took anything away from it, it would no longer be precisely 1 foot.)

We have also specified 48 inches in the problem. We wish this to represent a fixed number rather than a measured value. Once again, its precision is infinite. When we write the answer of 4 ft, it has been determined using only numbers having absolute precision, so its precision will also be known absolutely. There is no limit to the number of significant digits associated with this result. It's value is precisely 4.

Again, we have multiplied by a unit factor because 1 ft is the same length as 12 in. And *multiplying a quantity by a unit factor does not change the quantity itself, even though it is expressed in different units.*

Now that you understand the concept of unit factors, undoubtedly you realize that you know a large number of them already. Here are some examples:

$$\frac{5280\text{ ft}}{\text{mi}}, \quad \frac{1\text{ ton}}{2000\text{ lb}}, \quad \frac{1\text{ yd}}{3\text{ ft}} \text{ and } \frac{1\text{ g}}{1000\text{ mg}}$$

By now you probably also realize that you can make a unit factor for any equivalency you are given. For example, if A = B, then A/B is a unit factor, and B/A is also a unit factor. (The reciprocal of 1 is also 1.)

Therefore, what you need to do is to train yourself to look at problems in this specific way. When you are given a problem, just remember:

**I**'m **A** **F**ine **O**ld **C**hemist

- **I** Write down what **I**nformation you've been given in the problem. (Don't forget the units!)
- **A** Decide and write down what the problem is **A**sking for, again making a special note of what units are requested.
- **F** Search for and write down one or more unit **F**actors that will take you from where you are to where you want to end up. A list of common factors is provided for you inside the back cover.
- **O** **O**rganize the information you have written down in the form of an equation, including all unit factors (with units) so that all units will cancel except those desired in the answer.
- **C** **C**alculate.

Following are some examples that will further illustrate how to use the unit factor method.

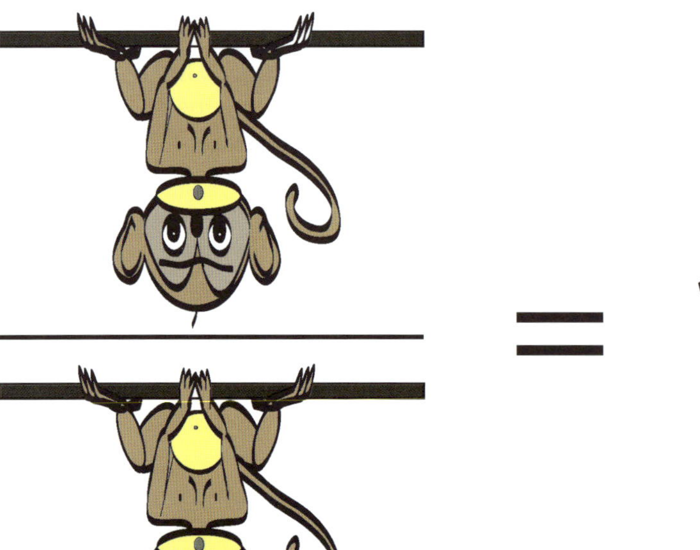

**Figure.** Anything divided by itself is equal to 1. Furthermore, any value divided by its own equivalent expressed in other units is also equivalent to 1.

### Example 1

How many centimeters long is a copper wire that measures 4.71 inches in length?

**I** What **I**nformation is given? A length of 4.71 inches. Write it down:

$$4.71 \text{ in}$$

**A** What is **A**sked for? The length expressed in cm. Write it:

$$? \text{ cm} = 4.71 \text{ in}$$

**F** What unit-**F**actor(s) will "take us" from inches to centimeters? Well, looking at our table of conversion factors inside the back cover we find the following:

$$1 \text{ in} = 2.540 \text{ cm}$$

Notice that this factor contains inches (where we are now) and centimeters (where we want to go). We can write this as a unit factor in either of two ways:

$$\frac{1 \text{ in}}{2.540 \text{ cm}} = 1 \quad \text{or} \quad \frac{2.540 \text{ cm}}{1 \text{ in}} = 1$$

We'll have to use the form that will cancel with the units we want to get rid of. In this case, we have inches in the numerator, ("4.71 in" is considered a numerator; "1" is its implied denominator), so we need the form that has inches in the denominator, then we will organize the problem accordingly.

**O** **O**rganize the problem in this way:

$$? \text{ cm} = 4.71 \cancel{\text{in}} \times \frac{2.54 \text{ cm}}{1 \cancel{\text{in}}}$$

**C** **C**alculate:

$$4.71 \times 2.54 \text{ cm} = 12.0 \text{ cm}$$

### Example 2:

Convert $4.6 \times 10^7$ milliliters to liters.

1) **I**nformation?

$$4.6 \times 10^7 \text{ mL}$$

2) **A**sked for?

$$? \text{ L} = 4.6 \times 10^7 \text{ mL}$$

3) Do we have a unit **F**actor?

From our knowledge of the metric prefixes we know that there are 1,000 mL in 1 liter. So we organize and calculate this way:

$$? \text{ L} = 4.6 \times 10^7 \text{ mL} \times \frac{1 \text{ L}}{1{,}000 \text{ mL}} = 4.6 \times 10^4 \text{ L}$$

### Example 3:

Sometimes we need more than one unit factor to bring about the desired change. How many inches are there in 5.28 miles? Unless we have a very detailed conversion table, we won't know. That's okay. We can set up the problem like this:

$$? \text{ in} = 5.28 \text{ mi} \times \frac{5{,}280 \text{ ft}}{1 \text{ mi}} \times \frac{12 \text{ in}}{1 \text{ ft}} = 3.3 \times 10^5 \text{ in}$$

I find it easier to write out such problems in a slightly different way. (In fact, this is how I'll do them from now on):

$$? \text{ in} = \frac{5.28 \text{ mi}}{} \left| \frac{5{,}280 \text{ ft}}{1 \text{ mi}} \right| \frac{12 \text{ in}}{1 \text{ ft}} = 3.3 \times 10^5 \text{ in}$$

**Example 4:**

Suppose a 1.75 lb bag of metal chips costs $30.98. What is the cost per pound? What is the cost per gram?

1) What is given? 1.75 lb of chips, $30.98 worth of chips.
2) What is asked for? $/lb, $/g.
3) By definition, the cost per pound will be the cost divided by the number of pounds. Therefore:

$$\frac{\$?}{\text{lb}} = \frac{\$30.98}{1.75\ \text{lb}} = \frac{\$17.7}{\text{lb}}$$

4) But what about the cost per gram? Here we need a unit factor to convert from pounds to grams. In our Table of Conversion Factors we find that there are 453.59 g in one pound. Therefore, we organize and calculate this way:

$$\frac{\$?}{\text{g}} = \frac{\$30.98}{1.75\ \text{lb}} \left| \frac{1\ \text{lb}}{453.6\ \text{g}} \right. = \frac{\$0.0390}{\text{g}}$$

One more thing about solving problems. I'm not asking you to "check your brain at the door." When you see an intuitive way to solve a problem, then, by all means, use it. Just be careful! It is very easy to think we see the solution to a problem when we don't. The advantage of the unit factor method of solving problems is that it is very structured. If you follow the rules, making sure all the appropriate units cancel out, then it is hard to make a mistake ("hard," of course, being a relative term). Certainly, when you are confronted with a problem that is too difficult to solve intuitively, the unit factor method is ideal. This is not a babyish way to solve problems. I know excellent mathematicians with 30 + years of experience who still use it faithfully.

**Exercises**

1) A bag of chips has a mass of 193.6 g and costs $1.19. What is the cost of a pound of chips?

2) Please convert $1.6 \times 10^{-2}$ km to centimeters.

3) Suppose 78 kg of mercury costs $420. What would be the price of the Hg in terms of dollars per lb?

4) A rectangular metal container is 12.0 cm long, 4.85 cm wide and 2.60 cm deep. What is its volume in mL?

5) How many centimeters long is a copper wire that measures 4.71 yards in length?

6) Please convert $4.6 \times 10^{7}$ L to milliliters.

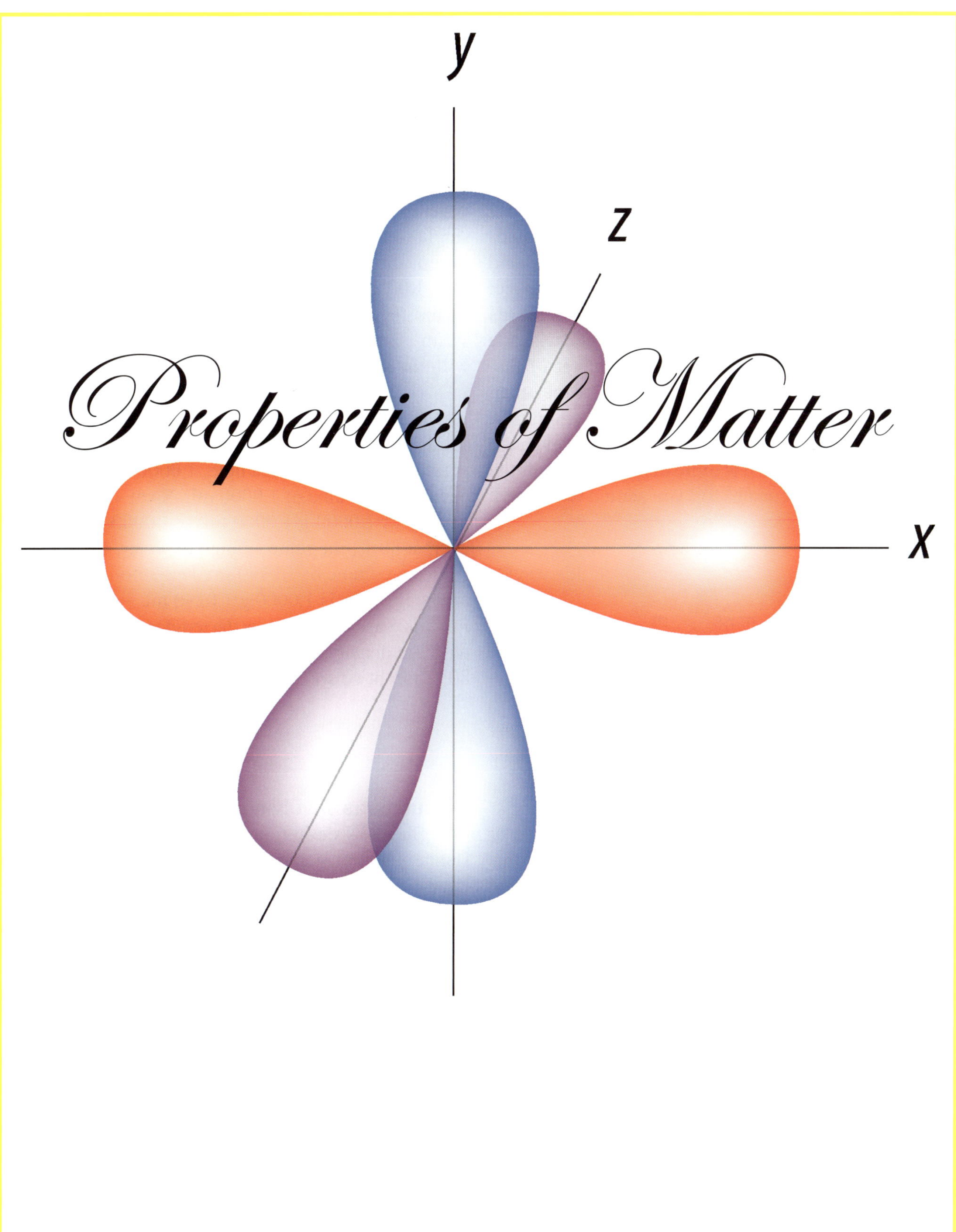

y
z
Properties of Matter
x

## 5: Bony People Tend to Be Dense

**Density** is defined as **mass per unit volume**. In other words, density is a ratio of the *amount* of a material (i.e., its mass) to the *volume* occupied by that amount of material. For example, suppose you cut out two identical pieces of aluminum foil. Since they are the same size, they have the same mass. Then suppose you loosely crumple them up into the shape of equal-sized balls. At this point they have the same mass and each one occupies about the same amount of space; therefore, the density of each ball, whatever it is, is approximately the same. But then suppose you take one of the balls and squeeze it until the diameter of that ball is about half as great as the diameter of the other. Each ball still contains the same mass of aluminum—no material has been lost or gained. Now, however, the ball that was compressed is more **dense**; it contains the same amount of aluminum foil you started with, but it takes up less space.

Mathematically, density is expressed this way:

$$d = \frac{\text{mass}}{\text{volume}} = \frac{m}{v}$$

For solids and liquids, density is usually expressed in units of g/mL (or $g/cm^3$). Since gases have such low densities, they are typically expressed in g/L. The Table shows the densities of some selected substances.

Sometimes folks confuse density with weight. For example, when we go fishing, we might take along corks and lead sinkers. Since the sinkers sink in water and the corks float, we may think of the sinkers as "heavy" and the corks as "light." But suppose we used a laboratory balance and found a large cork and a small sinker that actually had the same mass. Would the sinker float? Would the cork sink? Not on your life. It is not the weight of the sinker or cork that is responsible for its buoyancy, it is its density. *When an insoluble object is placed in a liquid, if it is more dense than the liquid, it will sink. If it is less dense than the liquid, it will float.* Lead is more dense than water, so lead sinkers always sink in water, no matter how small they are. Cork is less dense than water, so corks float.

This principle also holds true when liquids are placed together. When vegetable oil and water are mixed, a two-layer system results. The water is more dense than the oil and forms the bottom layer, and the less dense vegetable oil forms the top layer. If water, vegetable oil and Karo Syrup are added together in a single container, a three-level system results. The Karo Syrup (d = 1.37 g/mL) forms the bottom layer, water (d = 1.00 g/mL) is next, and vegetable oil (d = 0.91) forms the top layer. (Note that Karo syrup—corn syrup—is soluble in water. Given time, the syrup layer will dissolve in the water leaving a two-level system.)

It should also be mentioned that the volumes of solids and liquids change slightly with changes in temperature. At 4°C, for example, 1.0000 g of pure water occupies a volume of 1.0000 mL. So at 4°, water has a density of 1.0000 g/1.0000 mL = 1.0000 g/mL. However, at 80°C, the same amount of water, 1.0000 g, occupies a volume of 1.0290 mL. Therefore, the density of water at 80° is 1.0000 g/1.0290 mL = 0.9718 g/mL. For this reason, sometimes the temperature is also given with the density. For example, the density of water at 80°C might be written as:

$$d^{80} = 0.9718.$$

### Calculations Involving Density

Since, by definition, density is the mass of a substance per unit volume, density can be easily calculated if both the mass and volume of a substance are given. For example, a sample of an unknown liquid has a mass of 62.7 g and a volume of 79.5 mL. What is its density?

$$d = \frac{\text{mass}}{\text{volume}} = \frac{62.7\text{ g}}{79.5\text{ mL}} = 0.789\text{ g/mL}$$

From the data given in the Table, what would we suspect the liquid to be if we knew the liquid was pure? This liquid has the density of ethanol.

Based on your last lesson, when the density of a substance is known, that provides us with unit factors that we can use in calculations. For example, the density of ethanol is 0.789 g/mL. That means that 0.789 g of ethanol occupies 1 mL of volume. In other words, 0.789 g of ethanol and 1 mL of ethanol are the same amount of ethanol! Since there is an equivalence here, then we have two available unit factors:

$$\frac{0.789 \text{ g ethanol}}{1 \text{ mL ethanol}} \quad \text{and} \quad \frac{1 \text{ mL ethanol}}{0.789 \text{ g ethanol}}$$

That can be useful.

Suppose you need 50.0 g of ethanol for an experiment, and you don't have a balance to weigh it on, but you do have a graduated cylinder to determine its volume. How many mL of ethanol would you need?

(1) What is given? A mass of 50.0 g of ethanol.
(2) What is asked for? That mass expressed in mL.
(3) Do we have a unit factor that will effect this transformation? Yes—the density.

$$? \text{ mL ethanol} = \frac{50.0 \text{ g ethanol}}{} \left| \frac{1 \text{ mL ethanol}}{0.789 \text{ g ethanol}} \right. = 63.4 \text{ mL ethanol}$$

You can also do the opposite transformation. Suppose you needed 95.0 mL of glycerin for an experiment. How many grams would you need? From the Table, we find that glycerin has a density of 1.26 g/mL. Therefore,

**Table.** Densities of various substances.

| Solids | | Liquids | | Gases | |
|---|---|---|---|---|---|
| Substance | Density (g/mL at 20°) | Substance | Density (g/mL at 20°) | Substance | Density (g/L at 0°) |
| aluminum | 2.70 | chloroform | 1.49 | air | 1.29 |
| copper | 8.92 | ethanol | 0.789 | ammonia | 0.771 |
| gold | 19.3 | glycerin | 1.26 | argon | 1.78 |
| iron | 7.86 | Karo syrup | 1.37 | carbon dioxide | 1.96 |
| lead | 11.34 | sulfuric acid | 1.84 | carbon monoxide | 1.25 |
| magnesium | 1.74 | turpentine | 0.87 | chlorine | 3.17 |
| mercury | 13.95 | vegetable oil | 0.91 | helium | 0.178 |
| sand (silica) | 2.32 | water | 1.000 | hydrogen | 0.090 |
| silver | 10.5 | | | methane | 0.714 |
| sodium chloride | 2.16 | | | neon | 0.90 |
| sucrose | 1.59 | | | nitrogen | 1.25 |
| sulfur | 2.07 | | | oxygen | 1.43 |

$$? \text{ g glycerin} = \frac{95.0 \text{ mL glycerin} \mid 1.26 \text{ g glycerin}}{\mid \text{mL glycerin}} = 120. \text{ g glycerin}$$

### Density and Boniness

The cells that make up living things have a density that is about the same as that of water. Most cells tend to be ever so slightly more dense than water. Bone has dense mineral deposits in it that account for its greater density and hardness. Fat, on the other hand, is less dense than water. Just as vegetable oil and animal fat float on water, fatty tissues in animals and people also float.

Although it sounds cruel to say, the result is that bony people are more dense than water. There is, nevertheless, a way for bony people to float. If they fill their lungs with air, the low density of the air tends to reduce the overall density of the person so that the person/air mixture will float. Now, if that same person blows all the air out of his lungs, he will sink. There are, in fact, some people that will not float, even with their lungs full of air. These people are generally bony and have little lung capacity. A bony man who has been smoking for many years often fits this category.

The reason you don't see many fat snorkelers is because, to go snorkeling, you have to be able to swim deep and stay down long. The low density of fat makes it very difficult for a fat person to dive and stay down, even with no air in his lungs.

It is largely fat that accounts for the difference between the shape of a woman and the shape of a man. Because women have more fat, they are notoriously better floaters. Men are better sinkers. Are you a floater or a sinker? Can you remain sitting on the bottom of a pool if you blow the air out of your lungs?

**Exercises**

1) One liter of homogenized whole milk has a mass of 1,032 grams. What is the density of the milk?

2) What volume of ethanol has the same mass as 54.0 mL of water? (Hint: Use the Table.)

3) When a piece of copper is placed in a graduated cylinder that already contains 43 mL of water, the total volume increases by 17.43 mL. What is the mass of the copper?

4) Please calculate the volume of 50.0 pounds of gold.

5) Concentrated hydrochloric acid has a density of 1.19 g/mL. Please calculate the volume of 300.0 g of this acid.

6) The density of osmium is 22.57 g/cm$^3$, the highest value for any element. What is the volume and mass of a cube of osmium if it measures 2.54 cm on each edge?

7) An unknown piece of metal had a mass of 4.827 g. When the piece of metal was added to a graduated cylinder containing 52.70 mL of water, the total volume of the metal and water was 54.19 mL. Please calculate the experimental density of the metal.

Your next lesson is a quiz. Read ahead on the following page to see what will be required of you. Be sure you are prepared.

## 6: Quiz 1

The following is a list of review items that you should know well before taking the first quiz. If you are not prepared to be tested on these items, don't take the quiz until you are. The quizzes are located in an envelope in your lab set. Each quiz is contained on a separate page. You will be allowed to use your calculator on all quizzes.

1) Know the list of common elements and symbols tabulated in Lesson 2.
2) Know how to express measurements and calculated values in the correct number of significant digits and in scientific notation. Know how to tell whether or not zeros are significant. From this day forward, all of your answers to exercises and quiz answers should be reported in the correct precision (i.e., with the correct number of significant digits).
3) Be able to convert among the three temperature scales—Fahrenheit, Celsius and Kelvin. The formulas will not be provided for you. You should know them by heart (or at least by brain).
4) Be able to carry out calculations using the unit factor system.
5) Know how to compute densities and how to use density as a unit factor in calculations.
6) Always learn anything in bold or italic typeface and anything in purple type. Don't just memorize the words, rules and definitions, but understand the meanings as well.

The future of chemistry is depending on you. Don't disappoint the entire world! (No pressure.)

Gratuitous cat photo to make you think that chemistry is a cute, warm and fuzzy subject.

## 7: What's the Matter?

We said in the introduction to this book that chemistry is the study of the composition and structure of matter, but we didn't tell you just what matter is. Well, allow me to formally introduce you. **Matter** is anything that has mass and occupies space. We speak of matter as existing in three **states: solid, liquid and gas**. When matter moves from one state to another (e.g., a block of solid ice melts to become liquid water), that matter is said to have undergone a "change of state." (What they won't think of next!)

**Figure 1.** This is a hydraulic jack. Liquids are useful in hydraulics because they are unique in two respects. First, unlike solids, they are fluid. For this reason we can place pressure on a small volume of fluid and it can redistribute that pressure over a larger area. This works much like a lever or pulley to magnify our force. (This jack is capable of lifting 6 tons using the force applied by a human hand.) Second, fluids are not compressible. If they were, some of that hand force would be wasted on compression and decompression rather than being applied directly to lifting.

**Solid** matter has a *definite shape* and a *definite volume* because the particles that make up solid matter, in general, cling rigidly to one another. For example, a block of wood, a nail, the hammer that might be used to force the nail into the block of wood—these are all solid matter. Solids may be **crystalline**—that is, their molecules may exist in regular, repeating, geometric patterns. Salt or sugar crystals are examples of crystalline solids. Many minerals (substances from the Earth) are also crystalline in structure. However, when a solid does not exist in a regular geometric pattern, it is considered to be **amorphous** (no + shape). Most plastics and glasses are amorphous.

But matter can also be in a **liquid** state, where it has a *definite volume*—but an *indefinite shape*. In fact, liquids take on the shape of the container in which they are placed. You see this every day when you pour yourself a glass of milk, cola or orange juice. Liquids possess this property because the particles that make them up are more energetic than the particles in the corresponding solid. Therefore, they are held tightly, but not as rigidly as in solids. This is what permits the liquid to flow. But while the shape of liquids is easily changed, the volume is not. Liquids are not compressible to a great degree. That's why they are useful in hydraulic equipment. (See Figure 1.)

Finally, matter can also exist as a **gas**. The particles that make up gases are highly energetic (as compared to solid and liquid particles) and move mostly independently of each other. This means that gases also have an *indefinite shape*. They also take on the shape of the container in which they are placed (and it had better be a *closed* container, or soon you won't have any gas at all). Also, the actual volume of gas particles is small compared to the amount of space among the particles. So, in addition to having no fixed shape, gases also have an *indefinite volume*—they can be expanded and compressed. By the way, when a gas is enclosed, it is the collective force of many individual gas particles bumping on the sides of the container that produces pressure on the container wall. Oxygen, nitrogen and natural gas (methane) are among the most common examples of gases.

Figure 2. This is a true solution. It's usually pretty easy to tell by looking. A true solution is crystal clear. That's because there are no large particles floating around to scatter light. Light passes through a true solution because it's completely homogeneous, right down to its molecules.

## Combinations of Substances

There are other terms used to describe matter when different types combine. When two or more substances are mixed, the resulting condition is called (appropriately enough) a **mixture**. A mixture may be **homogeneous** (from the Greek words *homo*, which means *same*, and *gene*, which means *beginning*) when it is uniform in appearance and possesses the same physical and chemical properties throughout. If that mixture is homogeneous on the atomic or molecular scale—that is, if its molecules or atoms are randomly dispersed among one another—such a mixture is called a **solution**. For example, the air we breathe is actually a mixture of randomly dispersed gas molecules (nitrogen, oxygen and small amounts of some others). Likewise, when sugar or salt is dissolved in water, a mixture is formed wherein the solid **solute** is dissolved in the liquid **solvent**. So random and fine is the mixture of the particles making up these substances that every drop of solution has the same physical and chemical properties as every other drop. (See Figure 2.)

Metal **alloys** are homogeneous mixtures of atoms of metallic elements. They are prepared by melting metals to their liquid states then mixing them completely. As they cool, they keep their homogeneity. (See Figure 3.) Homogeneous mixtures in any state may be referred to as solutions. However, we usually use the word *solution* to refer to homogeneous mixtures in either liquid or gaseous states. It is used less commonly to refer to well-dispersed mixtures of solids, such as metal alloys.

Figure 3. Most useful metals are made of two or more elements. For example, this 14 karat gold ring is 14/24 gold and 10/24 nickel, zinc, palladium or other white metal. The metals are melted, combined, thoroughly mixed and cooled. The result is a permanent, solid solution.

When a mixture lacks homogeneity, it is referred to as **heterogeneous** (from the Greek words *heteros*, which means *different*, and *gene*, which still means *beginning*). Heterogeneous mixtures consist of two or more distinct **phases**. A phase is that portion of a mixture having separate physical/chemical properties from other portions. One phase is separated from other phases by a physical boundary called an **interface**. (See Figure 4.) When water and vegetable oil are combined, they form a two-layered or two-*phase* system. Each layer of the mixture is homogeneous within itself, but there is an interface between the two phases.

Homogeneity and heterogeneity are relative terms. We refer to separations involving large, distinct types of matter as **gross heterogeneity**. Examples of grossly heterogeneous mixtures might include lumps of coal mixed with scraps of iron or a mixture of wood, hay and water. Another example of a heterogeneous system would be a mixture of table salt and fine sand. While this mixture is heterogeneous (the particles are not all uniform, and there is air in the spaces in between particles), it's clearly not as grossly heterogeneous as the previous example of iron and coal. In other words, this well-dispersed mixture is heterogeneous on a *fine* scale. This heterogeneity might be called **fine heterogeneity**. Such a mixture clearly cannot be called a *solution* because the molecules of the different substances are not evenly dispersed (Figure 5.)

Finally, on the subject of combining substances, although both ice and liquid water have the same molecular composition, they form a mixture that's heterogeneous. That's because they are in different phases. It's appropriate to refer to a mixture of different phases of the same substance as a heterogeneous mixture. The molecules of solid ice and liquid water cannot disperse in one another until the ice melts.

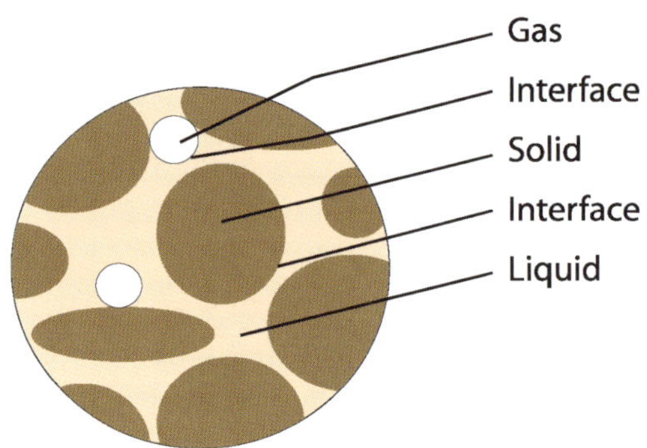


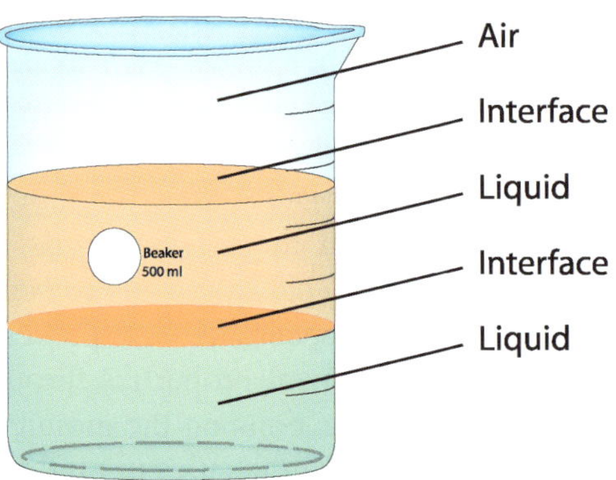


**Figure 4.** The upper drawing shows what microscopic heterogeneity might look like. There are three phases present: solid, liquid and gas. Interfaces are the places where any two of those phases meet. The lower drawing shows large-scale or gross heterogeneity. This beaker contains three phases: two liquid phases and one gas phase. Once again, the interface is a place where any one phase meets another.

## Physical and Chemical Properties

Every solid, liquid and gas has a set of unique properties that can be measured. Basically, there are two kinds of properties. **Physical properties** are those which describe physical attributes or physical changes. They are the characteristics of a substance that can be determined without altering its chemical makeup. Physical properties include things like: color, taste, odor, density, melting point, boiling point, hardness, consistency, etc. For example, if we wanted to determine the density of a solid, we could use a balance to obtain its mass, and we could measure its volume by submerging it in liquid in a graduated container. Then we could calculate its density, and none of our work on this object would have changed its chemical composition (makeup) in any way.

In addition to physical properties, substances also have **chemical properties**—those that describe its ability to react with other chemicals. These properties can be determined only by experimentation, and they always alter the chemical makeup of the material. For example, a chemical property of gasoline is that it can burn in the presence of oxygen to form carbon dioxide and water. Obviously, when it does so, you don't have gasoline anymore. Another example of a chemical property is the ability of hydrochloric acid to react with sodium hydroxide to form water and table salt.

**Figure 5.** On the left are two different heterogeneous systems. The first consists of air and a heterogeneous mixture of solids. The second consists of two liquids--corn syrup and water. This illustrates that heterogeneity may involve any combination of solids, liquids or gases. On the right are those same systems after they are homogenized. Note that neither is entirely clear. There must be some fine heterogeneity in each. In the real world, there are few true solutions that have no fine particles in them.

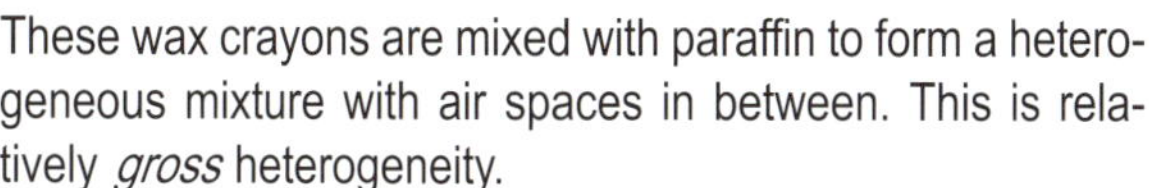

These wax crayons are mixed with paraffin to form a heterogeneous mixture with air spaces in between. This is relatively *gross* heterogeneity.

When the mixture is melted and stirred the result is a more homogeneous mixture. However, there must be some fine particles that are not dispersed or the mixture would be clear.

This is a two-phase, heterogeneous system of corn syrup (mostly soluble sugar) and water. Corn syrup is more dense than water, so it lies on the bottom. However, because it is soluble in water, it is slowly becoming homogenized.

Although the homogenized solution of syrup and water is somewhat clear, there are impurities that prevent it from being transparent. The mixture is not *completely* homogeneous although it's close. Compare this with the clear solution in Figure 2.

You need to cultivate the ability to distinguish between physical changes and chemical changes. A **physical change** is one in which a change in a physical property or in a state of matter occurs, but no change in chemical composition takes place. A **chemical change** is one in which one or more substances is either consumed or formed. Note that either physical or chemical changes may be accompanied by the absorption or release of energy. Take our example of burning gasoline, for example. The gasoline was used up, carbon dioxide and water were formed, and energy was released. Therefore, this was clearly a chemical change. What about the melting of ice? The ice changes its physical state from solid to liquid, and energy is absorbed. However, you start with $H_2O$ and you end up with $H_2O$. Therefore, this "change of state" is simply a *physical* change.

Let's say a sample of platinum wire, which has a silvery color, is heated over a Bunsen burner. As the wire gets hot it begins to give off a reddish glow. When the heat is removed, the wire resumes its former appearance. What kind of change took place—physical or chemical? Well, color is a *physical* property that clearly changed during heating, and changed again during cooling. But no platinum was consumed and no new products formed. We began with platinum and we ended with platinum; therefore, the change described was only a physical one.

The key question to ask yourself in determining whether a change is chemical or simply physical is whether the resulting substance(s) was different from the beginning substance(s). That is, were the **products** of the reaction different from the **reactants**? If so, the observed change was *chemical*. Note that any change in chemical composition will include a physical change as well. This is because new substances always have new physical properties. Even if you are unable to detect those changes with your eyes, physical changes will always have occurred.

As a final example, is the rusting of iron a physical or chemical change? It represents a reaction of the iron with oxygen to produce iron oxide. The products are different from the reactants; therefore, a chemical change takes place.

**Exercises**

1) We said in the lesson that particles of a substance in the solid state are less energetic than the corresponding particles in the liquid state. Likewise, particles in the liquid state are less energetic than the corresponding particles in the gaseous state. Therefore, would a given substance be solid at a lower or higher temperature than the temperature at which it is a liquid?

2) Would the following materials be considered crystalline or amorphous?

   a) A plastic bag
   b) A polished diamond
   c) A rubber ball
   d) A piece of paper

3) Classify each of the following either as a physical property or a chemical property:

   a) the melting point of iron
   b) ability to neutralize an acid
   c) hardness of an emerald
   d) color of an aluminum bar

4) Classify each of the following processes as either a physical change or a chemical change:

   a) tarnishing silver
   b) burning of natural gas
   c) cutting a copper tube in two
   d) freezing water

5) Classify each of the following mixtures as either homogeneous or heterogeneous:

   a) chocolate chip ice cream
   b) a glass of iced tea
   c) oil-and-vinegar salad dressing
   d) brewed coffee

## 8: Separation of Mixtures

In the last lesson we considered that when more than one kind of matter is mixed with another, we call it a mixture (what else would we call it?!). I should also point out to you that, in nature, most matter exists as mixtures. Finding a pure substance in nature is the exception, not the rule. For example, sea water contains dissolved salts and other minerals. Another example is the air we breathe; it is a mixture of gases. Most metals in the earth's crust are found there in mixtures with other ores and minerals. This means that, if we want to isolate pure substances, we have to separate them from the mixtures in which they are found. This lesson will provide an introduction to some elementary methods for separating mixtures.

One very important separation technique is **filtration**: *a process involving separation of solids suspended in liquids by passing the liquid phase through a filter that the solids cannot pass.* Suppose we scooped up a sample of sea water from the beach and accidentally got some sand with our sample. If we wanted to separate the solids from the liquid we could pour the mixture onto a filter. The filter might be paper, plastic or some other material – but it would be something through which the liquid could pass. The sand particles, on the other hand, would be too large to pass through, and so would be left behind. Filtration is often used in the laboratory to separate solid reaction products from liquids. Notice that filtration takes advantage of the physical properties of substances to effect separation: state of matter and particle size.

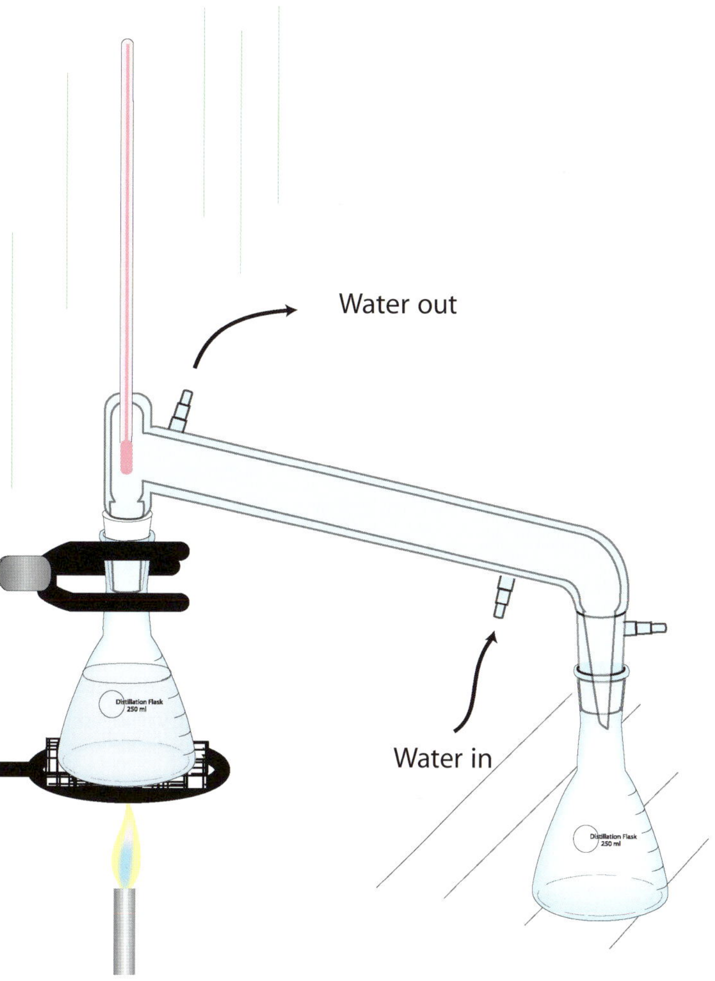


**Figure 1.** Distillation apparatus. The flask on the left is heated so that the liquid boils and evaporates. Cold water flowing through the condenser jacket cools the vapor and condenses it. The condensed liquid is collected in the flask on the right. Only the desired liquids having the correct boiling point (as measured on the thermometer) are collected. Others are allowed to pass through uncollected.

Finer filters are used to remove finer particles. These days the precise size of the pores in the filter (the openings through which liquid passes) can be controlled by the manufacturing process. Filters with pore sizes down to molecule-sized openings can selectively allow one small molecule to pass while retaining another. The process of filtering molecules is called **ultrafiltration**. One step up from ultrafilters is the thin, film-like **membrane filter**. These filters typically have pore sizes in a range from several micrometers (also called a micron, and abbreviated μm) down to a tenth of a micrometer. They can filter out microscopic particles. For example, they are often used to remove bacteria from a liquid suspension. **Paper filters** are usually made of cellulose (wood) fibers. The fibers are wetted, pressed together and cemented. When the filter paper dries it forms a network of crisscrossing fibers. The pore spaces among the fibers allow liquid to pass through, but the fibrous net catches any particles that are too large to pass through.

Another separation technique is **distillation**: *a process involving separation of a liquid from a mixture by boiling away the liquid, then condensing it back to the liquid state in a separate container.* Again, think about our sample of sea water. If we filter off the sand, the resulting liquid is still a mixture—water, and dissolved salt and other minerals. But we can separate the water from the dissolved salts through distillation. If we heat the sea water in a flask, like that shown in Figure 1, we can get the water to evaporate. That removes it from the mixture, leaving behind the undissolved salt and other mineral components. If, when we evaporate the water, we channel the vapor through a cold tube, then the steam will condense and liquid water will reform. Distillation is often used in the laboratory to separate liquids from mixtures. Distillation also takes advantage of a physical property—the unique boiling points of different substances—to achieve separation.

Separations of mixtures can be effected by taking advantage of other properties as well. For example, a mixture of iron filings and yellow sulfur powder may contain particles so small that we'd never be able to separate them by hand. However, a magnet can be used to remove the magnetic iron particles while leaving the non-magnetic sulfur behind. This technique, called **magnetic separation**, is used on a large scale in many industries to remove magnetic impurities. Alternately, the separation of this mixture might be achieved by adding carbon disulfide to the mixture to dissolve the sulfur, followed by filtration to separate the non-soluble iron filings, and then distillation to separate the sulfur from the carbon disulfide.

Here's another example. Suppose you took a plastic straw, cut it into very small pieces, and put the pieces onto a metal pan. Then you took a piece of notebook paper or a paper towel, cut it into pieces of similar size, put them on the same metal pan, and mixed the plastic pieces and the paper pieces together. If you then decided you wanted to separate the materials, you could probably pick them out one-by-one, but that would be a lot of trouble. There might be an easier way.... What if you took an inflated balloon, rubbed it on your hair a few times, then held it near the mixture? What do you think would happen? Chances are, unless the humidity is very high, the small pieces of paper would be electrostatically attracted to the balloon, separating them from the non-attracted plastic pieces. This technique, also used in industry, is called (you guessed it!) **electrostatic separation**.

**Figure 2.** Separatory funnel. This device is used to separate two liquids having different densities. The denser liquid is allowed to drain out of the "sep funnel" while the less dense fluid is retained. Separatory funnels come in volumes on the order of 1 L and less, down to a few tens of milliliters.

Separations can even be accomplished by taking advantage of density differences. Remember our mixture of water, oil and Karo syrup back in Lesson 5? Individual liquids from mixtures which separate into phases can be recovered by using a **separatory funnel** (See Figure 2). The liquid having the highest density will naturally gravitate to the bottom of the funnel and by carefully opening the funnel at the bottom, we can recover the liquids, one phase at a time. Density differences can be used even to separate mixtures of solids. In fact, that's exactly why gold panning works. Prospectors scooped up samples from river beds and gently swirled the mixtures around in a pan. Since gold has a pretty high density (19.3 g/mL), if any was present, it would settle to the bottom of the pan. As the light sand and dirt were washed away, the nuggets of gold would remain. (Now don't get too excited – the only large pieces of gold left in this country are found in jewelry stores, or in the homes of rich people.)

**Exercises**

1) The amount of silver in a solution can be determined by adding hydrochloric acid. This forms insoluble silver chloride. Being insoluble the silver chloride falls out of solution ("precipitates"). What technique would you recommend for separating the $AgCl$ from the mixture?

2) How would you go about separating a mixture of charcoal and table sugar?

3) A junkyard dealer buys a great deal of metal scrap. How could he recover the scrap iron from the scrap he obtains?

## 9: Specific Heat

In previous lessons we've emphasized that every measurement results in both a number and a unit. We've discussed the standard units for length, volume, mass and temperature. In this lesson we introduce a new set of units—those used in measuring heat energy. Of course, concerns about heat energy are as much about physics as they are about chemistry, since energy is a physical property. But it's a property that all substances possess and one that changes when both physical and chemical changes take place. Don't be surprised if you see these units again in a physics course.

A common unit for measuring heat energy is the **calorie** (cal). Now I'm sure you've heard of calories before in reference to food and dieting. These are called **dietary calories** or **large calories** (which strikes me as a poor choice of words). Pick up any packaged food and look on the back and you'll see that the number of calories per serving are listed there. (It may not say dietary calories, but you can be sure that the calories listed will be the dietary kind.) Chemists are not at all concerned with dietary calories (unless they're overweight). We often deal with smaller energy changes than a dietician would be concerned with. That's why a true, **chemical calorie** or **small calorie)** is smaller—by a factor of 1,000. There are 1,000 calories (1 kilocalorie) per dietary calorie.

What is a calorie? It is the amount of heat energy required to heat 1 g of water by 1°C. Think about that—that's not very much heat. A second common unit for heat energy, the **joule** (J), is even smaller. The joule (pronounced like "jewel") is the SI unit for heat, and it takes 4.184 J to equal 1 cal. Since these units describe very small amounts of heat, kJ and kcal are often used in heat change calculations.

So, it takes 1 calorie of heat to raise the temperature of 1 gram of water by 1°C. What if you were heating something else? What if you had 1 g of aluminum, or lead, or iron, and you wanted to raise its temperature by 1°C. How many calories of heat would it take? In fact, different substances require different amounts of heat to raise their temperatures. The quantity of heat required to raise 1 g of any substance by that same degree is called its **specific heat**. Clearly, we could have come up with a standard of comparison based on raising two grams by a half degree, but this one is a little tidier. After a hundred years of everyone using it, you'll look a little weird if you use a different one.

**Table.** Specific heats of some selected substances.

| Substance | Specific Heat | |
|---|---|---|
| | cal/g°C | J/g°C |
| liquid water | 1.000 | 4.184 |
| ethyl alcohol | 0.511 | 2.138 |
| ice | 0.492 | 2.059 |
| aluminum | 0.215 | 0.900 |
| carbon | 0.170 | 0.710 |
| iron | 0.113 | 0.473 |
| copper | 0.092 | 0.385 |
| mercury | 0.033 | 0.033 |
| gold | 0.031 | 0.031 |
| lead | 0.030 | 0.128 |

Well, given this definition, what is the specific heat of water? How much heat does it take to change the temperature of 1 g of water by 1°C? It takes one calorie, right? Therefore, the specific heat of $H_2O$ is 1 cal/g °C or 4.184 J/g°C. The specific heats of some other substances are given in Table 1 in order of decreasing specific heat. Notice that the specific heat of water is high compared to all these substances. Water gains and loses heat stingily.

This is all very nice, but of what use is it? Don't you realize what we have done? We have come up with a way to figure out how much

**specific heat** - the amount of heat required to raise the temperature of 1 g of a substance by 1°C.

### What do calories have to do with food?

In living things food does two things: It provides a source of material for structure and it provides a source of energy. Calories are a measure of the amount of energy that can be gained from a particular food by a person. We would measure that energy in calories.

But we've always been taught that too many calories are bad for us! Now we learn that calories are a measure of energy. How can too much energy be bad for us?! The answer has to do with how bodies use and store energy.

When living beings take in food, there are a limited number of things that can happen to it. It can be broken into molecules to be used for growing or maintaining the body, it can be broken down into carbon dioxide and water to extract the energy from the molecules, it can pass through undigested and be wasted, or it can be stored for use at a time when food is scarce. Unless a body is diseased, it always takes what it needs from the food and stores a little extra for a rainy day. There are mainly two ways it stores food—in the form of glycogen (a polymer of sugar) and in the form of fat. Glycogen is a short-term, get-me-to-my-next-meal form of stored food; fat is a long-term, what-if-I-don't-see-another-meal-for-months form of stored food.

In circumstances where food is plenteous, some people eat continually. If the body is in a habit of putting food into storage, but we continue to eat the plenteous food, the storage barns get bigger. There are only three ways to get rid of that stored food: (1) eat food having a lower energy value (fewer calories), (2) eat less food, or (3) use more of that energy store by burning it off (be more active—exercise).

Dietary calories are a means of measuring the energy value of food. The greater the energy value, the greater the chance it will end up stored as fat. There are, however, people who are concerned with getting too little energy value from their food: elderly people, people with high rates of metabolism (that burn energy more quickly than others), people who can't eat often or digest food well, and people who live in circumstances where food is in short supply. To these people, high-calorie food is better than low-calorie food.

For the moment, most of us here in the USA live in a country where food is plenteous. Be thankful, treat it wisely, and help the less fortunate. Don't overuse it to your own detriment.

The two photos show two entirely different types of food--four full heads of lettuce against one candy bar. What do they have in common? The same calorie content. You pick.

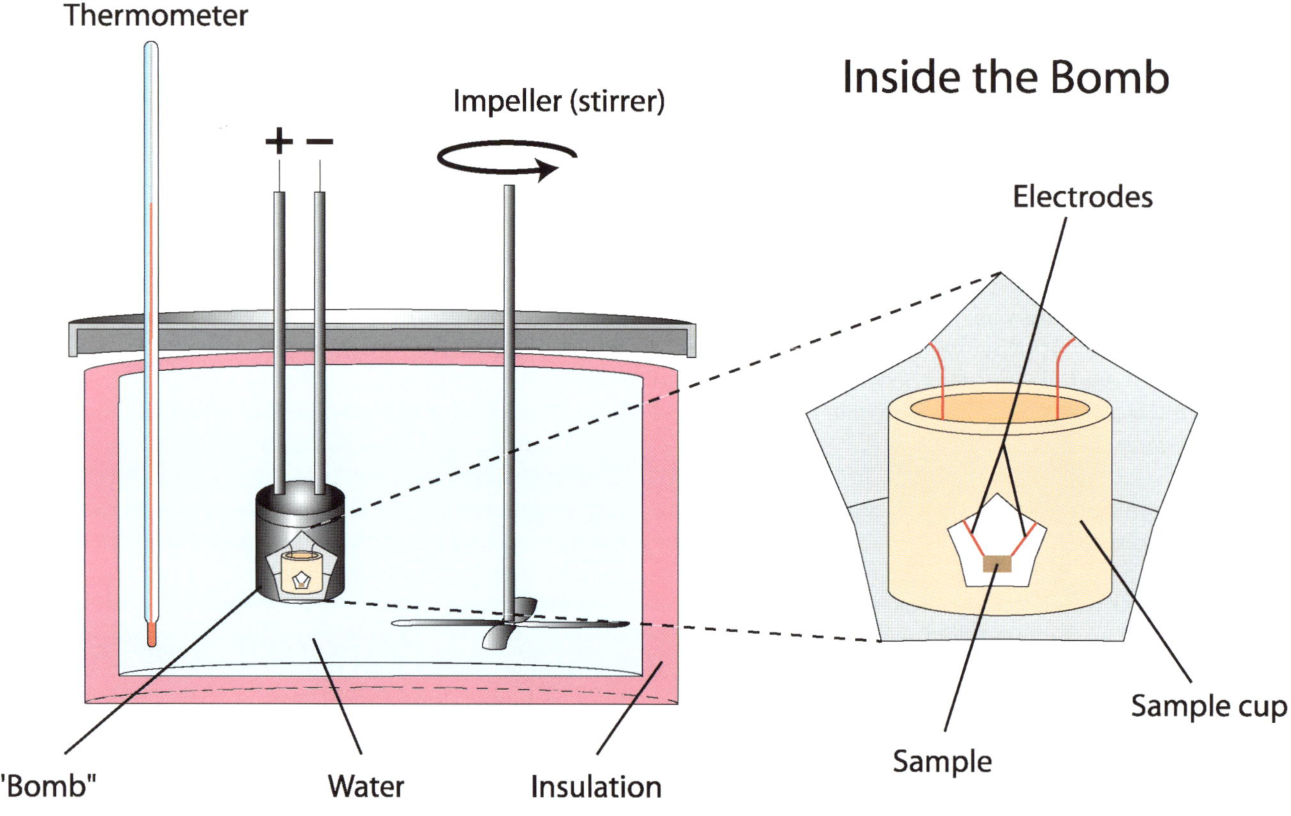


**Figure.** This is a diagram of an oxygen bomb calorimeter (*heat+ measure*). This is a laboratory device that measures the amount of heat given off during the complete combustion of a substance. The sample is placed in the bomb in a high-oxygen atmosphere and hit with high voltage causing it to be completely oxidized. Because the sample is completely contained all the energy given off is in the form of heat. The heat is absorbed by the water and the change in the water temperature is recorded. Knowing the specific heat of the water, the total amount of heat given off by the oxidation of the sample can be calculated.

energy will be required to change any mass of any substance by any temperature. To do this, we will need to know three things: the mass of the substance, its specific heat and the change in temperature we want to see (we'll call this $\Delta t$, "delta t"). Those factors relate to a change in the heat of a substance as follows:

$$\text{heat energy} = \text{mass of substance} \times \text{specific heat} \times \text{change in temperature}$$

or:

$$\Delta\varepsilon = mh_s\Delta t$$

This equation has never been named, so, as we go on, we'll call it "George."

For a given substance, then, if you have any three of the four variables above, you should be able to calculate the fourth. For example, how much heat would be needed to raise the temperature of 200 g of water by

10.0°C? Well, we are given the mass of the water (200 g) and the temperature change (10.0°C), and we know the specific heat of water to be 1 cal/g°C. (Don't forget the units!) This should be a piece of cake:

$$\Delta\varepsilon = \frac{200.0\text{ g} \mid 1\text{ cal} \mid 10.0°\text{C}}{\mid \text{g}°\text{C} \mid} = 2.00 \times 10^3 \text{ cal}$$

Yep…cake.

Since we were not asked for any specific units of heat energy, the problem could also have been solved using the other value for the specific heat of water:

$$\Delta\varepsilon = \frac{200.0\text{ g} \mid 4.184\text{ J} \mid 10.0°\text{C}}{\mid \text{g}°\text{C} \mid} = 8.37 \times 10^3 \text{ J}$$

We might observe that if our calculations are correct, then 2,000 calories must be equivalent to $8.37 \times 10^3$ J. And, in fact, they are. Multiplying 2,000 cal by a conversion factor of 4.184 J/cal gives us $8.37 \times 10^3$ J.

Congratulations! You can now estimate how much heat would be required to heat an entire lake to any temperature of your choosing if you can provide an estimate of the amount of water in the lake. Let's see. Suppose you were having a pool party with your friends, but the water was too cold…

As we said, if we have any three of the factors in George, we can calculate the fourth. So then, we could determine experimentally the specific heat of any substance by measuring the temperature change after putting in a known amount of heat. Let's calculate the specific heat of a solid, in J/g°C, if 1,638 J were required to raise the temperature of 125 g of the solid from 25.0°C to 52.6°C.

First, let's rearrange George to calculate the specific heat:

$$\Delta\varepsilon = mh_s\Delta t$$

so,

$$h_s = \frac{\Delta\varepsilon}{m\Delta t}$$

We can now use the given information to calculate $h_s$:

$$h_s = \frac{1{,}638\text{ J}}{(125\text{ g})(27.6°\text{C})} = 0.475\frac{\text{J}}{\text{g}°\text{C}}$$

Oh, surprise of surprises. That just happens to be about the same specific heat of pure iron. (See Table.) All you need to do is plug the numbers into the equation (unless of course you have to manipulate the equation).

**Exercises**

1) Calculate the energy required (in calories) to heat 145 g of water from 22.3°C to 75.0°C.

2) Calculate the heat necessary to raise the temperature of 40.0 g of Al from 20.0°C to 32.3°C.

3) A gold bar weighing 33.5 g and having an initial temperature of 25.0°C, absorbs 5,00$\underline{0}$ J of heat. What would be its new temperature? (Hint: Replace $\Delta t$ with $t_2$ - $t_1$ then solve for $t_2$.)

4) A metal bar having a mass of 212 g is heated to 125.0°C and then dropped into 375 g of water at 24.0°C. After a while the temperature of the water and the metal is 34.2°C. If the specific heat of the metal is 0.831 J/g°C, how much heat did the metal lose? How much heat did the water absorb?

## 10: Who Thought Up "Atoms"?

In this lesson you will embark upon a study of atoms.

Way back in Lesson 2 we told you that all matter is made up of substances called "elements." We said that elements are the fundamental building blocks of matter. We also indicated that the smallest part of an element that would retain the properties of that element would be one **atom** of it. You might wonder who figured that out. Well, it's a long story, and we won't give it all to you at once, but in this lesson we want to introduce you to **atomic theory**.

Notice that the word *theory* is displayed prominently. It's easy when you work as a chemist to forget that your model for thinking is only a well-supported series of thoughts. As time goes on, information is added that modifies the theory and changes your thinking slightly. There is much work to do before the properties and makeup of atoms are fully understood. That's why a historical development telling who added what to our present thinking on the subject is the best approach to understanding current thought.

A Greek philosopher named Democritus (who lived about 400 years before the coming of the Christ) is credited with actually coining the term "atom." He claimed that all matter was made up of tiny, indivisible particles—so he called them atoms, from the Greek word *atomos*, which means "indivisible." Of course, he had no way of knowing what these particles looked like or how they functioned.

It was over two thousand years later that the first substantial evidence was provided. That evidence is owed to a man named John Dalton. Here is a summary of **Dalton's Atomic Theory**, which he proposed in 1808:

**Elements are composed of minute, indivisible particles called atoms.** Well, there's nothing new here—this is the same thing that Democritus said.

**Atoms of the same element have the same mass and size.** In other words, Dalton suggested that every copper atom has the same size and mass as every other copper atom. He said the same would be true for iron atoms, or oxygen atoms, or any other atoms of a given element.

**Atoms of one element have a different mass and size from those of another element.** For example, Fe atoms have a different mass and size than Pb atoms.

**Atoms are immutable in chemical reactions.** That is, they do not disappear or change into other atoms. This postulate in Dalton's theory gave support to the Law of Conservation of Mass, which was widely recognized to be true in Dalton's time. This law states that during an ordinary chemical reaction, matter is neither created nor destroyed. In other words, the mass of all substances present at the beginning of a reaction is always equal to the mass of all substances present at the end of a reaction.

**Compounds are formed by the combination of atoms in simple numeric ratios.** Dalton theorized that the mass proportions of elements in compounds (which were known in his time) would translate into simple numeric ratios of atoms. For example, water has a mass ratio of 11.2% H to 88.8% O. As it turns out, this is the mass ratio of two atoms of H to one atom of O. Therefore, water is $H_2O$. So Dalton offered an atom-based explanation for the Law of Definite Composition which was becoming widely accepted at this same time (early 1800's).

**Atoms may combine in different ratios to form different compounds.** A theory which incorporates the notion of atoms is also helpful in explaining the Law of Multiple Proportions which would later be tested and proven by Dalton and others in his time. Recall that we gave a comparison between the different element compositions of water ($H_2O$) and hydrogen peroxide ($H_2O_2$) as an example of this law back in Lesson 2.

The fact that Dalton's theory provided such a convincing explanation for the composition of compounds was largely responsible for its general acceptance. Of course, Dalton's theory wasn't perfect. You see, atoms themselves are composed of *smaller* particles and *are* divisible—although it takes some effort to divide them. Also, all atoms of a given element do not necessarily have the same size and mass. When atoms of a given element have different masses, they are called *isotopes;* we'll talk more about them later. Dalton's theory did not provide explanations for charged particles (atoms are theoretically neutral) nor did it consider the "decay" of unstable isotopes that would give chemists and physicists grand discoveries in the centuries to come. In spite of these exceptions, though, Dalton's theory did a remarkable job of modeling chemical structure and reactions—not a small achievement for anyone. This "anyone" happened to be a high school science teacher!

### The Structure of Atoms

So what did atoms look like? Dalton didn't know. The question was a matter of much speculation among the scientists who adopted Dalton's model. It wasn't until around 1900 that experimental evidence provided some clues. A physicist named Joseph Thomson discovered that elements contain tiny negative particles. These subatomic particles came to be known as **electrons**. Thomson was also the one who later described the nature of **protons**—the tiny, positively charged particles. Since Thomson had shown that elements contained both negatively- and positively-charged particles, it was clear that atoms were not indivisible, as Dalton had suggested, and as the name "atom" implies. Later, another physicist named Ernest Rutherford showed that the positively-charged particles must be concentrated in a small area within the atom—in other words, within a **nucleus**. Still later, Rutherford and James Chadwick showed that the nuclei of most atoms also contain neutrally-charged particles—**neutrons**.

Although there have been many refinements to atomic theory since its inception, and we'll talk about some of them later, just about all the chemistry of matter can be understood in terms of this concept: atoms are composed of electrons, protons and neutrons. Here are some of the particulars about electrons, protons and neutrons that you need to know.

**electrons** – negatively-charged particles, carrying a relative electrical charge of –1. These particles are not located in the nucleus but are constantly moving in "orbitals" (more on this later) around the nucleus. Electrons are much smaller than protons or neutrons. They have a mass equal to $9.110 \times 10^{-28}$ g. Since electrons are about 1/1,800 times the size of protons or neutrons, their mass is negligible in calculating the total atomic mass of an atom. Therefore, the actual mass of an atom is essentially the mass of the protons and neutrons present in the nucleus.

**protons** – positively-charged particles, carrying a relative electrical charge of 1+. They are located in the nucleus of atoms and have a mass of $1.673 \times 10^{-24}$ g, or 1 atomic mass unit (amu).

**neutrons** – neutral particles. Like protons, they are located in the nucleus of an atom and have a mass of 1 amu ($1.675 \times 10^{-24}$ g). Notice that this mass is nearly, but not exactly, that of a proton.

### Atomic Number

Simply put, the atomic number of an element represents the number of protons in a given atom of that element. Inside the front cover of your book you'll find the **periodic table of the elements**. Each of the known elements is listed there in the sequence of their atomic numbers, beginning with hydrogen, which has an atomic number of 1. The atomic number corresponds with the number of protons in one atom of that element. For example, tin has an atomic number of 50; therefore, there are 50 protons in the nucleus of a tin atom. Pretty easy, huh?

And there is more that we can conclude. Since elements are electrically neutral, if an atom contains a certain number of protons, it must also contain that same number of electrons. For example, a neutral boron atom, having an atomic number of 5, contains 5 protons. Each of those protons possesses a 1+ charge, so the total of positive charges equals 5+. That means that there must also be negative charges that total 5– in order to have electrical neutrality. Therefore, there must also be 5 electrons.

### The Discovery of Electrons

Near the turn of the century after Dalton, J.J. Thomson, the British physicist mentioned earlier, offered some insight into the nature of an atom. The "cathode ray tube" (shown below) had been invented, but nobody was exactly sure how it worked. This was under investigation. By applying an electrical charge across a gas-filled tube, a beam could be detected jumping from one electrode to another. The beam could be detected only if a luminous substance, such as zinc sulfide, was placed in the tube. A "screen" for viewing the cathode ray was made by coating a sheet of glass with the zinc sulfide and inserting it in the tube before it was sealed. By applying an electrical or magnetic field across the tube, perpendicular to the flow of electricity, Thompson showed that the ray of light could be directed. The direction assumed by the beam under these influences demonstrated that it consisted of negatively-charged particles, which he dubbed *electrons*.

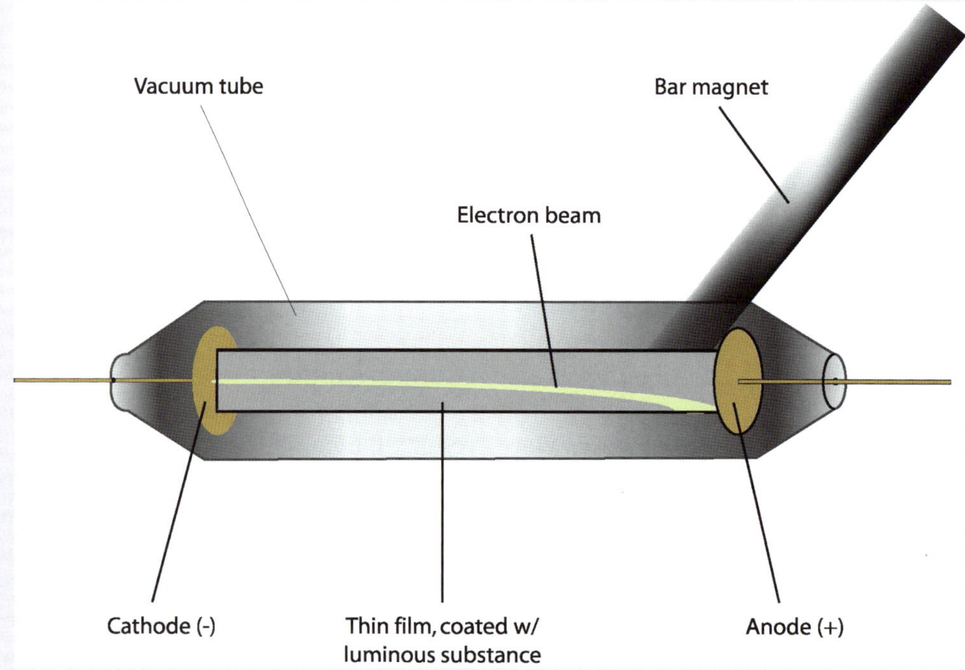


This diagram illustrates a cathode ray tube like the one used by Joseph Thomson to demonstrate the nature of electrons. The cathode ray generally projects a linear image on the "screen" (the coated film). However, when a bar magnet approaches the ray, it bends in a manner that is predicted if the beam consists of a stream of negative-charged particles.

The elements are arranged in the periodic table in a series of **periods** (rows) and **groups** (columns). The periods are merely known by their number in sequence (period 1, 2, 3, etc.). The groups are known by group numbers, which are shown at the top of each group, and also by certain group names which are shown in the simplified periodic table that is provided in the Figure on the last page of this lesson. We'll use roman numerals to refer to the groups throughout this text.

## Ions

Even though an atom of an element is electrically neutral, it is possible to produce a charged species, called an **ion**, by either adding or removing electrons from it. If electrons are added, then that means the species now has extra negative charges. This produces a negatively charged species called an **anion**. If electrons are removed from an atom, then there are fewer negative charges. This

## The Discovery of the Atomic Nucleus

Fourteen years after Thomson's discovery, Ernest Rutherford, a New Zealander working in England demonstrated the internal structure of atoms. He bombarded a thin foil of gold with a stream of positively charged, alpha particles (which would later come to be known as helium nuclei). He encased the foil with a zinc sulfide screen (see below). If the gold was solid throughout, you might expect all those alpha particles to have bounced off of it but, instead, most of them went straight through it. A few particles did bounce back at angles that would be predicted if it were assumed that there were tiny, hard centers arranged throughout this thin film of gold atoms. By the proportion of passing particles to rebounding particles, he was also able to estimate the size of the nucleus at 1/10,000 of the size of the entire gold atom. Thus, atoms consisted of mostly empty space, but had a tiny core that repelled positively-charged particles. This provided a foundation for our understanding of the nature of the atomic nucleus.

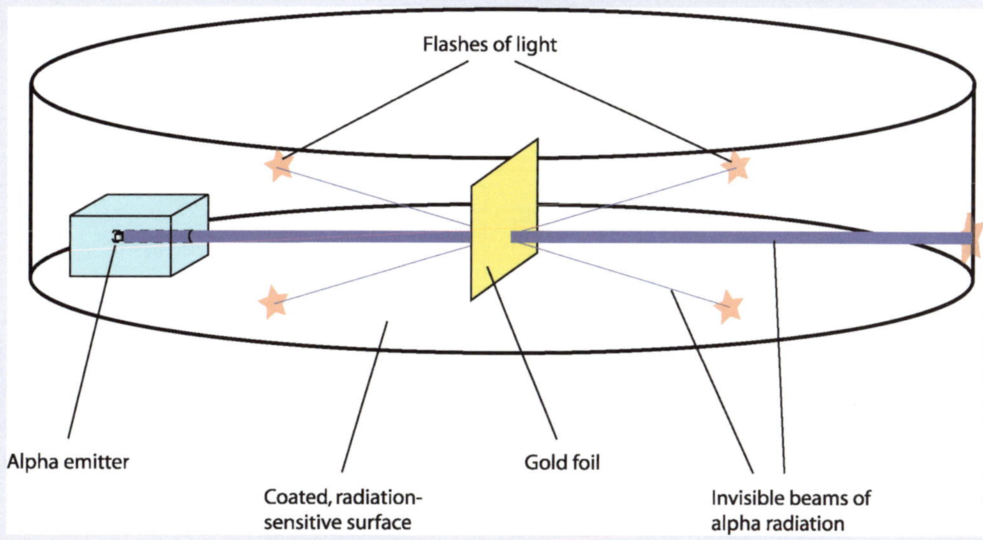


This figure illustrates the apparatus used by Rutherford to demonstrate the nature of the atomic nucleus. Most of the positive-charged alpha particles pass directly through the thin foil of gold. However, a few of them are deflected at large angles indicating they have struck something that causes them to bounce in a distinct pattern. That pattern is predicted if it is assumed that atoms are mostly empty space, but contain a dense nucleus on the order of 1/10,000 the size of the entire atom. That nucleus repels positive charges, so it might be presumed to carry a positive charge itself.

means that the positive charges of the nucleus are no longer neutralized. The result is a positively charged species called a **cation**. Remember, anions are negative and cations are positive. (You can keep it straight if you remember that "anion" has an extra "n," which can stand for "negative."). Ions are designated by writing the symbol of the element and putting the appropriate + or – charge at the upper right-hand corner. The magnitude (size) of the charge on an ion depends on how many electrons were added or removed.

If you'll look at your periodic table, you'll see that chlorine has an atomic number of 17. That means there are 17 protons in the nucleus of a chlorine atom, and 17 electrons orbiting the nucleus. However, chlorine has a pretty strong tendency to accept an extra elec-

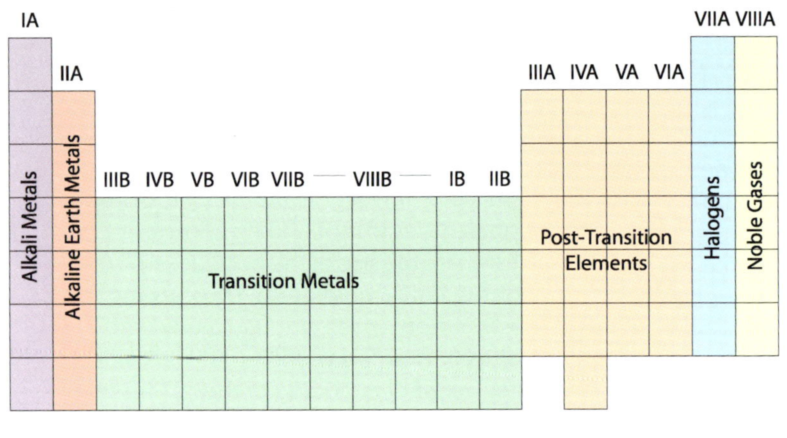


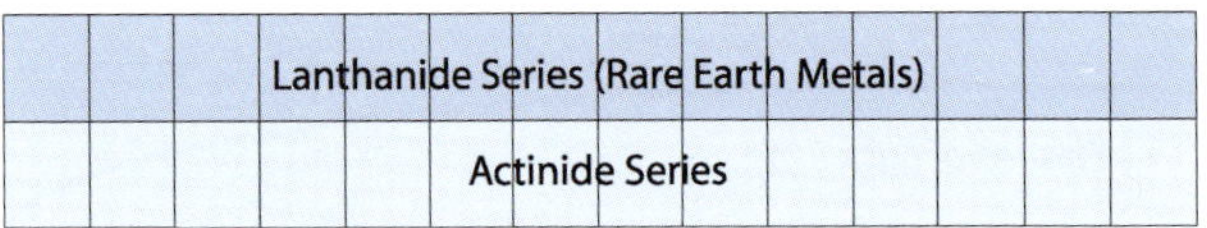


**Figure.** This table shows the names by which the groups are commonly called. Each group is headed by its group designation (1A through 8A, or in this book, IA through VIIIA). Beneath the main body of the table are the lanthanide and actinide series elements (sometimes "lanthanoid" and "actinoid"). These are named after the elements lanthanum and actinium which they follow in atomic number. You'll do well to memorize these group and series names. You'll be hearing several of them throughout the course.

tron. When it does so, what results is the chloride ion, $Cl^-$. The process may be written out like this:

$$Cl + e^- \rightarrow Cl^-$$

Magnesium has an atomic number of 12 and a pretty strong tendency to give up electrons. In fact, it can readily give up two. If it does so, then the nucleus still has 12 "1+" charges, but there are only 10 "1–" charges orbiting the nucleus. So the resulting magnesium ion would have a 2+ charge—it is designated as $Mg^{2+}$. The magnesium ionization process would be written thus:

$$Mg \rightarrow Mg^{2+} + 2e^-$$

Alternatively it could be written this way:

$$Mg - 2e^- \rightarrow Mg^{2+}$$

It's time to exercise your new knowledge.

**Exercises**

1) Since a hydrogen atom consists of just one proton and one electron, it should have a mass of $1.673 \times 10^{-24}$ g. And so it does. Using your unit-factor technique, calculate how many hydrogen atoms you'd have if you had 5.6 g of hydrogen.

2) The mass of a helium atom is $6.65 \times 10^{-24}$ g. What would be the mass of 5 million He atoms?

3) Complete the following, putting the appropriate charge on the resulting species.

a) $K - e^-$ c) $O^- + e^-$
b) $Cl + e^-$ d) $Ca - e^-$

4) For the following elements and ions, give the number of protons and the number of electrons.

a) U c) $Cu^{2+}$
b) $Br^-$ d) $S^{2-}$

## 11: What's New?

For its time, Dalton's theory did remarkably well at explaining so many things in chemistry. You can understand why most scientists adopted it. But although Dalton's theory does do a good job of modeling chemical behavior, atomic theory has come a long way since the 1800's. In this lesson, I want us to consider the current thinking with regard to atomic theory, especially in terms of how electrons are arranged in atoms.

### Bohr's Model

As a starting point for our discussion, let's consider the fact that substances have the ability to absorb energy. You know that from your previous studies in science, and because we told you in Lesson 1. For example, anyone who tries to pick up a pan from the stove realizes very quickly that the pan has absorbed energy—it's hot. What you may not have considered is that the absorption of energy takes place in the atoms themselves. Then, as the pan cools down, that energy is given off as **radiated heat** (electromagnetic radiation in the infrared region of the spectrum). It is even possible to heat some substances to such an extent that they glow. In those cases the absorbed heat is given off as both visible light and infrared light.

Similarly, if you take gaseous atoms of an element and subject them to high voltages or high temperatures; they absorb energy. And just as a piece of heated metal releases electromagnetic energy, energized or "excited" atoms release the energy that they have absorbed.

In the early 1900's a man named Niels Bohr (who was far more interesting than his name would suggest) found that he could excite hydrogen atoms. When he did so they would release the energy they had absorbed, by emitting electromagnetic radiation. Now, you know that you can direct white light through a prism and observe all the colors of the rainbow. The prism separates the light into the various wavelengths making it up. But when Bohr put the radiation emitted from excited hydrogen gas through a prism, it was separated into several very *distinct* and *separate* wavelengths. In essence, he obtained what is called a **line spectrum**, where the lines represented different colors that showed up at specific distances from one another (See Figure 1). Bohr concluded that each of the lines on the *line spectrum of hydrogen corresponded to the energy possessed by the hydrogen electron*. This means that *the energy levels that could be assumed by hydrogen were also separate and distinct.*

**Figure 1.** When energy is given to an atom its electrons absorb that energy and jump to a higher energy level. Eventually they will give up that excess energy in the form of electromagnetic radiation (visible and invisible "light") as they return to their ground states. The light energy given up by those atoms is unique. It represents the difference between their ground state energy and their excited state energy. When the light is passed through a prism, it is scattered. Instead of a continuous spectrum of wavelengths, however, individual, distinct lines of light are produced. Because that pattern of lines is unique to the element that produced it, this so-called "emission spectrum" can be used to positively identify an element.

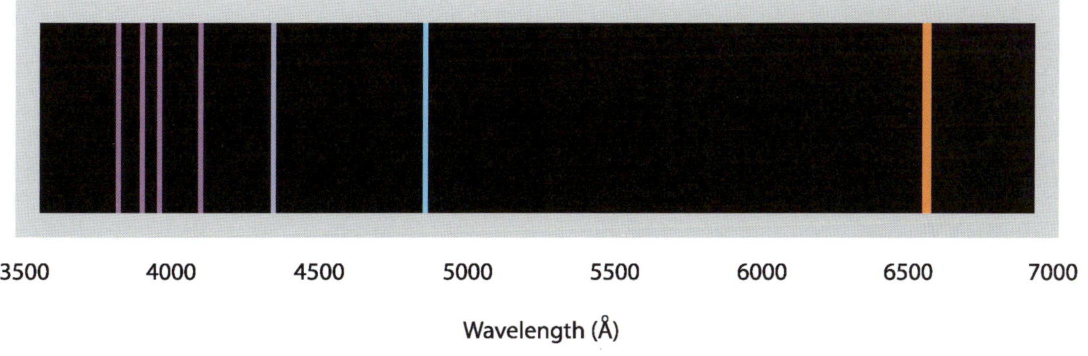

This suggested to Bohr that maybe the best model for atoms was a planetary model, where electrons occupy circular orbits around the nucleus like planets occupy orbits around the sun. In Bohr's model, electrons that occupied orbits close to the nucleus possessed less energy than electrons that occupied orbits farther away from the nucleus. The orbit that had the lowest energy was called $n = 1$. The orbit that had the next highest energy was $n = 2$, etc. When the electrons occupied only the lowest energy level ($n = 1$), they were said to be in the **ground state**. When energy was applied, ground state electrons could jump to a higher energy level—one farther away from the nucleus. Then they were said to be in an **excited state**. (See Figure 2.)

The Bohr model is often called the "planetary model" because it suggests that electrons orbit nuclei in the same way that planets orbit the sun (Figure 3). While this is not entirely correct, it did introduce the idea that the energy possessed by electrons is "quantized," or graduated—divided into distinct, fixed levels. It also taught us that electrons could, in fact, move from one energy level to another.

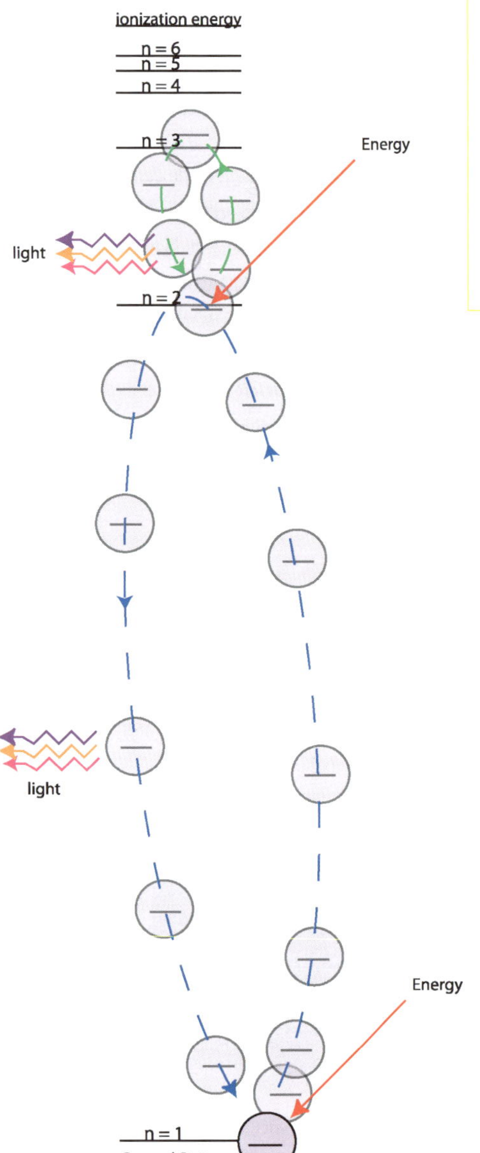


**Figure 2.** When an electron in its ground state absorbs energy it is elevated to a higher level of kinetic energy. At this point, it may absorb additional energy, elevating it to an even greater state of energy, or it may fall back to its ground state. As it returns to its ground state, it must give off energy. That energy leaves in the form of light (visible or invisible). Light of different wavelengths will be given off for each energy level the electron has been elevated to. If the absorbed energy exceeds the ionization energy for the electron, the electron will be lost by the atom.

**Figure 3.** The Bohr "planetary model" portrayed electrons as moving about the nucleus in distinct orbits. Orbits would account for the different energy states of electrons observed by Bohr in his experiments with hydrogen. Here we see four different orbits, each representing a different energy level that can be assumed by the electron in hydrogen.

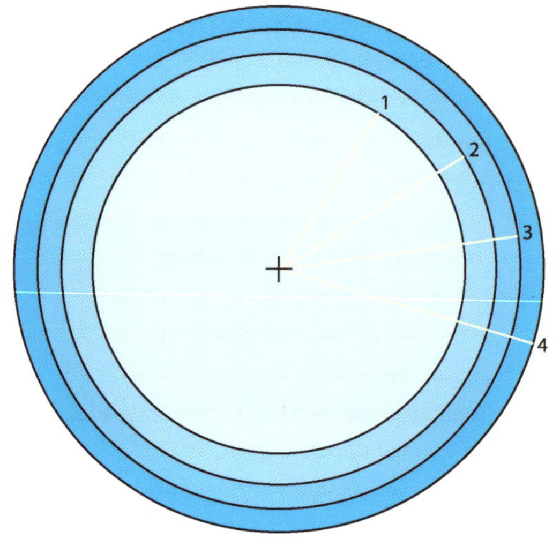

## Orbitals, Not Orbits

That bit of work by Bohr, as enlightening as it was, was something of an oversimplification because it was based only on hydrogen. Electrons do indeed occupy principal energy levels as Bohr observed hydrogen doing. However, it is now understood that Bohr's principal energy levels (remember, $n = 1, 2, 3, ...$) are divided into sublevels. In fact, $n = 1$ has only one sublevel, but $n = 2$ has two sublevels, and so on.

To further complicate the matter, these sublevels are not orbits, as Bohr suggested. This is because electrons are not entirely like planets—they do not act entirely like solid matter. Thanks to the work of Louis Victor de Broglie, we now understand that electrons really have dual properties: in some ways they act like particles of matter, but in other ways they behave more like waves of energy. If you know the position of a planet on a given day and time, and if you understand gravity, you can calculate exactly where that planet will be at another moment in time. The path of that planet is the same, year after year after year.
This is not the case with a wave.

Think of a situation where a wave of light is traveling around the nucleus of an atom. Its path is not the same on every passage. You can't draw a circle to represent the path of that wave—it oscillates (weaves) while it circles. So then, the space occupied by the electron is better thought of as a realm rather than an orbit. The realm of an electron is called an **orbital**. A couple of years after de Broglie postulated the wave theory of electrons, Erwin Schrödinger developed a series of equations by which the movement of electrons could be modeled. The result was the branch of physics that we now call **quantum mechanics** or **wave mechanics**.

Perhaps, if you could detect the exact location of an electron in a given moment and do that several times in sequence, you could, by developing a pattern, come to know exactly where the electron would show up next. However, there is no way to detect the position of an electron without altering its energy. When you alter its energy, you affect the course of its travel. Therefore, it is impossible to simultaneously determine both the position and the energy of an electron. This dilemma was first stated by a German physicist named Werner Heisenberg and is called the **Heisenberg Uncertainty Principle**.

Since we can't tell exactly what the pathway of an electron will be, nor precisely locate it at a given moment, all we can really do is determine the *probability* of finding it in a given place at a given time using Schrödinger's equations. The orbital is the region associated with a certain probability of the presence of an electron. What does an orbital look like? Well, the boundaries are somewhat indistinct. (See Figure 4.) Sometimes chemists and physicists refer to these areas where electrons are located as "**electron clouds**", because they are best represented as fuzzy areas made

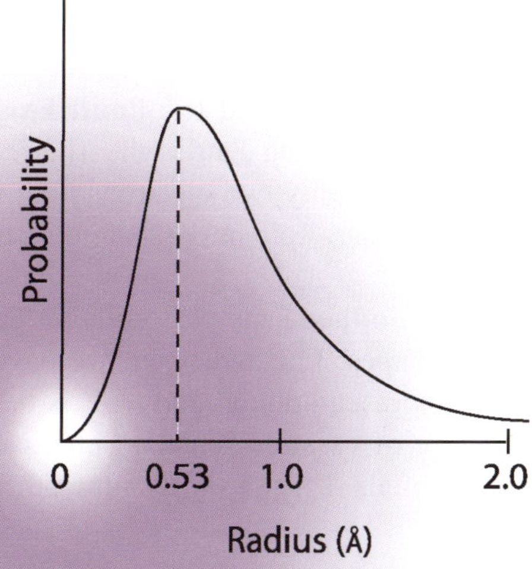


**Figure 4.** In this representation of an electron cloud of hydrogen, the density of the cloud at a given location represents the probability of finding an electron in that location. The graph shows that the density of the cloud peaks at 0.53 Å from the nucleus.

up of dots. The dots are most highly concentrated in those areas where the electrons are most likely to be found—at a small fixed distance from the nucleus. As the distance from the nucleus increases or decreases from that fixed distance, the probability decreases.

How big are orbitals? Well, we know that the higher energy levels are located at a greater distance from the nucleus. Therefore, atoms that have electrons in those levels are "larger." But how can we even determine the radius of a hydrogen atom since the electrons do not stay at a constant distance from the nucleus? Well, it became necessary to agree on some kind of standard, so it was decided that orbital size would be defined as the region where the electron will be located 90% of the time. The other 10% of the time the electron will be outside of that area. So remember, when you see drawings of orbitals, they are simply representations of the shape and size of the area where electrons can be found the majority of the time.

How many electrons can occupy an orbital? Two. This is because electrons "spin." They spin about their axes in one of two directions. We refer to these using up arrows (↑) and down arrows (↓). Two electrons can fill a given orbital (↑↓), provided they have opposite spins. This rule is called the **Pauli Exclusion Principle**, after Wolfgang Pauli who figured this out in 1925—two and only two electrons may fill an orbital if and only if they spin in opposite directions.

What are the shapes of orbitals? They can have different shapes. For hydrogen, we have shown you our best representation already—it is best thought of as a sphere of variable density. Spherical orbitals are called *s* orbitals. I'll show you some others in the next lesson.

### Electron Configurations

Believe it or not, you now have all the information you need to begin constructing the electron structures or "electron configurations" of atoms. We'll begin with the simplest atoms, hydrogen and helium. We know that hydrogen has a single electron, and that, in its ground state, that electron occupies the lowest energy level, $n = 1$. This first, principal energy level has only one sublevel—the *s* orbital. So the electron configuration for hydrogen can be written as:

$$1s^1$$

That notation says that it is the $n = 1$ energy level, that it is the *s* orbital which is being occupied, and that there is 1 electron in that orbital.

Now what about helium? How many electrons? Two. Since there is room in the *1s* orbital for another electron (remember, each can hold two), both electrons for helium are found there. The electron configuration is:

$$1s^2$$

In the first energy level, the *s* orbital has two electrons.

I want you to notice something now about the periodic table. Have you ever wondered why hydrogen and helium are the only two elements in the first period? The reason is that they are the only two elements in which electrons reside in only the principal energy level, $n = 1$. The next element in the table, lithium, has three electrons. But the *s* orbital in the first energy level can hold a maximum of two. Therefore, the third electron must occupy an orbital in the $n = 2$ energy level.

As you can see, we are beginning to develop an understanding of the behavior of element groups and periods in the periodic table based on our theory of electrons and their energy levels. In the next lesson we'll look at the electron configurations of several more elements and develop the periodic table more completely. This is what's new with atomic theory over the last 50 to 100 years, so what's new with you?

**Exercises**

Enough, already!

## 12: The Periodic Table—To Be This Ugly, You Need a Reason!

Now that you know a little bit about the concepts of principal energy levels and sublevels, we can examine the electron configurations of elements that are more complex than hydrogen and helium. Remember, hydrogen and helium are the only elements in the first period of the periodic table. That's because they are the only two elements in which electrons reside *only* in the lowest energy level, $n = 1$. But there are other principal energy levels. For example, let's now consider $n = 2$. Electrons filling this energy level are farther away from the nucleus than those in the $n = 1$ level; therefore, they possess more energy. Also, just as the $n = 1$ level had one sublevel, the $n = 2$ level has 2 sublevels—an *s* sublevel and a *p* sublevel. The *s* sublevel has only one orbital, so it is called *2s*. The *2s* orbital is very much like the *1s* orbital; it is spherical in shape (Figure 1), but has a larger radius. Like every other orbital, it can hold only two electrons having opposite spins.

If you look at the periodic table you will see that the first element listed in the second period is lithium. The atomic number of lithium is three. Therefore, lithium has a total of three protons, and an electrically neutral lithium atom must also contain three electrons. Since these electrons will naturally seek the lowest energy level, two of them will occupy the *1s* orbital. Since that orbital can contain only two electrons, the third electron must occupy the *2s* orbital. So we would write the electron configuration of lithium as:

$$1s^2\ 2s^1$$

Now, what would you expect the electron configuration of beryllium to be? It's the next element on the periodic table after lithium. It has four electrons—two will occupy the *1s* orbital, and the other two will occupy the *2s* orbital. The electron configuration of beryllium is written as $1s^2\ 2s^2$.

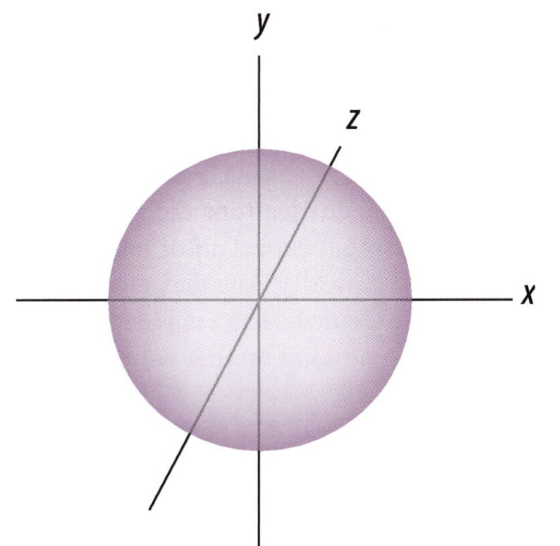


**Figure 1.** An *s* orbital is a spherical and holds only two electrons. The *2s* orbital has a greater radius than the *1s* orbital to accommodate the greater energy of its electrons.

### The *p* Sublevel

Now things begin to get fun. If we continue to the next element in the periodic table, we come to boron, which has five electrons. Again, the first four will occupy the *1s* and *2s* orbitals. That means one electron must occupy the *p* sublevel. While each *s* sublevel (*1s* and *2s*) contains one orbital, each *p* sublevel contains *three* orbitals. So, while each *s* sublevel holds only two electrons, each *p* sublevel holds six. The three orbitals making up the *p* sublevel are designated $p_x$, $p_y$ and $p_z$.

Unlike the orbitals in the *s* sublevel, *p* orbitals are not spherical. They look like a three-dimensional "figure eight." See (Figure 2.) When these orbitals are aligned on three axes having their origin at the nucleus, as shown in the Figure, you can readily see why they are designated as $p_x$, $p_y$, and $p_z$— each orbital is on the axis which bears its name.

What then would be the electron configuration of boron? It is given as $1s^2\ 2s^2\ 2p^1$. If you think about it, you should now be able to write the electron configurations for the next 5 elements: carbon, nitrogen, oxygen, fluorine and neon.

| | |
|---|---|
| carbon (6 electrons): | $1s^2\ 2s^2\ 2p^2$ |
| nitrogen (7 electrons): | $1s^2\ 2s^2\ 2p^3$ |
| oxygen (8 electrons): | $1s^2\ 2s^2\ 2p^4$ |
| fluorine (9 electrons): | $1s^2\ 2s^2\ 2p^5$ |
| neon (10 electrons): | $1s^2\ 2s^2\ 2p^6$ |

**Figure 2.** There are three *p* orbitals in every principal energy level starting at $n = 2$. The shape is roughly that of a "figure eight" in three dimensions. One orbital lies on each of three perpendicular axes. For this reason they are referred to as the $p_x$, $p_y$ and $p_z$ orbitals. Because each orbital holds two electrons, the entire *p* sublevel holds six electrons.

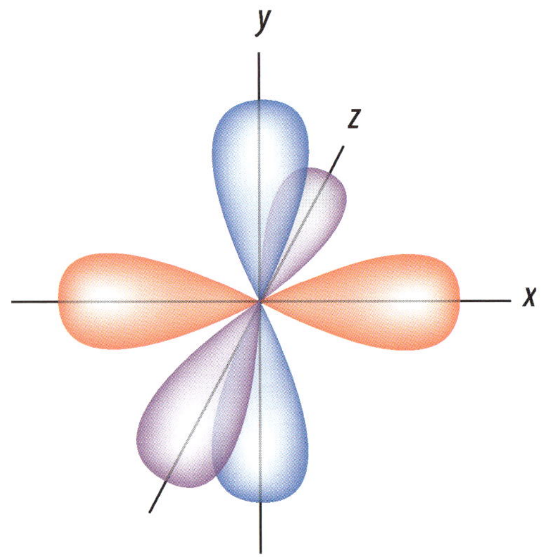


There's something else you should know about filling the *p* sublevel. Each of the orbitals receives an electron before any receives a second one. So nitrogen, which has three electrons in the *p* sublevel, actually has one electron in $p_x$, one electron in $p_y$, and one electron in $p_z$. The result is that nitrogen has three "unpaired" electrons. (In a few days this will help you to understand that nitrogen likes to form three bonds by sharing its unpaired electrons with another element.)

We observed that, since the *p* sublevel contains three orbitals ($p_x$, $p_y$ and $p_z$) and each orbital can contain two electrons having opposite spins, then the *p* sublevel can contain a total of six electrons. We also noticed that neon, at the end of the second period, has its *p* sublevel full. This gives it the familiar *noble gas configuration*. This atom is so self-satisfied it will not bond with any other atom. At this point there is no more room in the *p* sublevel, and therefore, no more room in the second principal energy level. So if we move on to the next element—sodium—which has 11 electrons, where must the 11th electron go? Into the next principal energy level, $n = 3$. In fact, moving from left to right along the third period of the periodic table, that's what we find—electrons filling up the third energy level, which also contains *s* and *p* sublevels.

| | |
|---|---|
| sodium (11 electrons): | $1s^2\ 2s^2\ 2p^6\ 3s^1$ |
| magnesium (12 electrons): | $1s^2\ 2s^2\ 2p^6\ 3s^2$ |
| aluminum (13 electrons): | $1s^2\ 2s^2\ 2p^6\ 3s^2\ 3p^1$ |
| silicon (14 electrons): | $1s^2\ 2s^2\ 2p^6\ 3s^2\ 3p^2$ |
| phosphorus (15 electrons): | $1s^2\ 2s^2\ 2p^6\ 3s^2\ 3p^3$ |
| sulfur (16 electrons): | $1s^2\ 2s^2\ 2p^6\ 3s^2\ 3p^4$ |
| chlorine (17 electrons): | $1s^2\ 2s^2\ 2p^6\ 3s^2\ 3p^5$ |
| argon (18 electrons): | $1s^2\ 2s^2\ 2p^6\ 3s^2\ 3p^6$ |

Now let's make a few observations:

1) Look at the electron configurations of hydrogen, lithium and sodium:

| | |
|---|---|
| hydrogen | $1s^1$ |
| lithium | $1s^2\ 2s^1$ |
| sodium | $1s^2\ 2s^2\ 2p^6\ 3s^1$ |

Do you see any similarity? Each of these group IA elements has only one electron in the highest energy level that contains any electrons at all. This is true for each of the other elements in group IA—each of them has one electron in its highest occupied energy level. Electrons in the outermost energy level of an atom are called its **valence electrons**. These are the electrons that are largely responsible for their chemical reactivities . Since each group IA element has one valence electron, and since valence electrons dictate chemical behavior—we might expect all of these elements to be similar, chemically. And they are. For example, each of these elements has a tendency to form only a 1+ cation by giving up that valence electron.

2) I already mentioned that neon has 8 electrons in the highest occupied energy level. In fact, each of the group VIII elements (except He) contains 8 electrons in its highest occupied energy level. We said back in Lesson 10 that helium, neon, argon, krypton, xenon, and radon are called the noble gases. They just don't form compounds with other elements very easily. When atoms have an elec-

tron configuration that ends in $ns^2\ np^6$ (where $n$ is the number of the highest principal energy level filled) then they are pretty unreactive. Helium is slightly different in that it has only two electrons, but its outer ($1s$) shell is full, just as the outer (valence) shells of the other noble gases are full.

3) What we have observed for group IA elements and group VIIIA elements is true for the elements in the other "A" groups as well. Their electron configurations are such that they contain the same number of valence electrons, so there are significant similarities in their chemistry.

## *d* Orbitals

Now things get even *more* fun. (How much fun can a chemistry student take?) You see, the $n = 1$ principal energy level has one sublevel ($s$); the $n = 2$ principal energy level has two sublevels ($s$ and $p$). Now you find out that the $n = 3$ principal energy level has three sublevels ($s$, $p$ and $d$). The $d$ sublevel contains five orbitals, $d_{xy}$, $d_{xz}$, $d_{yz}$, $d_{z^2}$, and $d_{x^2-y^2}$. Notice that, as the average distance from the nucleus increases, the number of orbitals that can fit into the allotted, increased space also increases. Don't let the funky names of these orbitals throw you. Once again, they correspond to a mathematical description of the space occupied by these orbitals. They are calculated values and, by the time we get to the $d$ orbitals, should not be taken too literally. (After all, these are theoretical shapes that do not account for the interactions among the large numbers of electrons that they possess.) While $s$ orbitals are roughly spherical, and $p$ orbitals are two-lobed, $d$ orbitals are thought to look like four-lobed versions of the $p$ orbitals. (See Figure 3 on the following page.)

When do electrons start filling the $d$ orbitals in principal energy level $n = 3$? Well, it turns out that once we get this far from the nucleus, there is some overlap of the orbitals of different principal energy levels. The $4s$ orbital actually represents a lower energy than the $3d$ orbitals. What that means is that the electron configuration for potassium (atomic no. 19) is actually $1s^2\ 2s^2\ 2p^6\ 3s^2\ 3p^6\ 4s^1$. The 19th electron goes into the $4s$ orbital rather than the first $3d$ orbital. In the same way, the electron configuration for calcium (atomic no. 20) is $1s^2\ 2s^2\ 2p^6\ 3s^2\ 3p^6\ 4s^2$, so that $4s$ orbital fills before a $3d$ orbital sees its first electron.

It is not until we get to scandium (atomic no. 21) that we find electrons occupying the $3d$ orbitals. Its electron configuration is $1s^2\ 2s^2\ 2p^6\ 3s^2\ 3p^6\ 4s^2\ 3d^1$. Now remember, there are five $d$ orbitals and each of them can hold two electrons. Therefore, starting with scandium and ending with zinc, there are a total of ten elements that have electrons filling the $d$ orbitals. Elements in which the $d$ orbitals are being filled are often called **transition elements**. And I suppose one of the main things I want you to see with regard to the $d$ orbitals is that they are the reason why the periodic table has that block, 10 elements long, between the group IA and group IIIA elements.

## The *f* Sublevel

We won't talk about $f$ orbitals in much detail because too much fun isn't good for us. Let's be satisfied to learn that, as we might expect, the $n = 4$ principal energy level has four sublevels- $s$, $p$, $d$ and $f$. The $f$ sublevel has seven orbitals and can therefore hold 14 electrons. The elements in which the $f$ orbitals are being filled are called the lanthanides (where the $4f$ orbitals are being filled) and the actinides (where the $5f$ orbitals are being filled). These elements are usually placed in two separate rows at the bottom of the periodic table. Well, that's a lot of orbitals. Are there $g$ and $h$ orbitals? Presumably, but we have not yet discovered elements with atomic numbers high enough to enter any of these orbitals. (If you insist on learning about f orbitals, go to the library and pull out an upper college-level book on quantum chemistry or quantum physics. That'll fix ya!

Perhaps the most crucial information that you will have gained from this lesson for immediate application is that the most stable atoms are those having their outer $s$ and $p$ sublevels filled. This is the so-called "noble gas configuration." Atoms that do not exist in this state will seek it by either removing an electron from another atom or by sharing valence electrons with another.

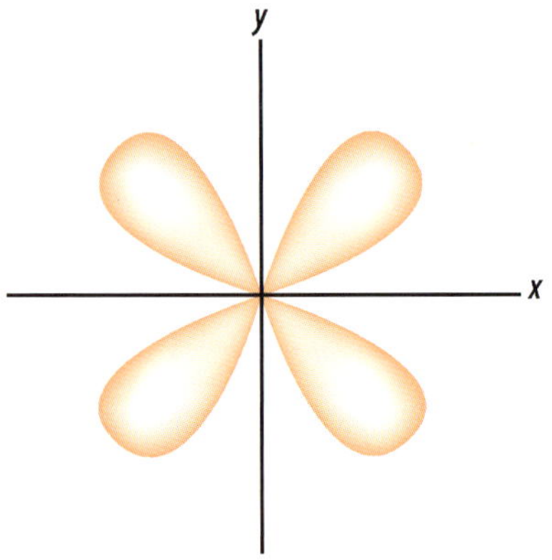

$d_{xy}$ 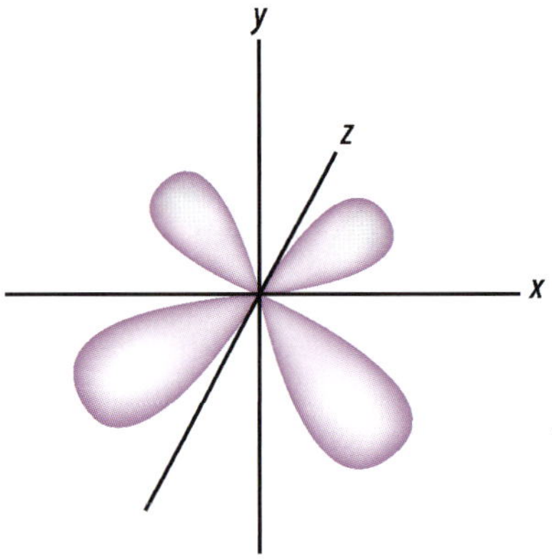 $d_{xz}$ 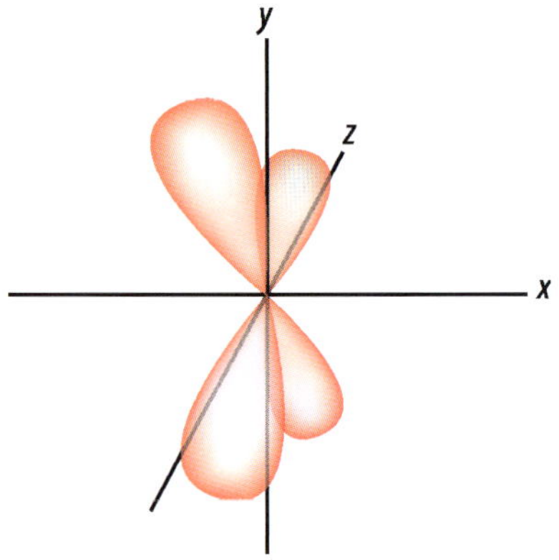 $d_{yz}$

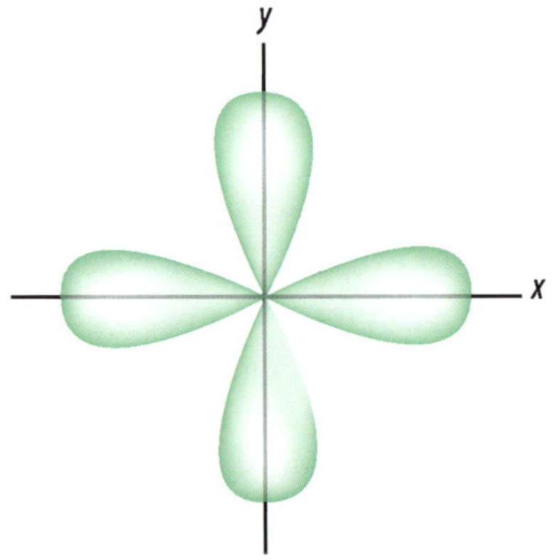

$d_{x^2-y^2}$ 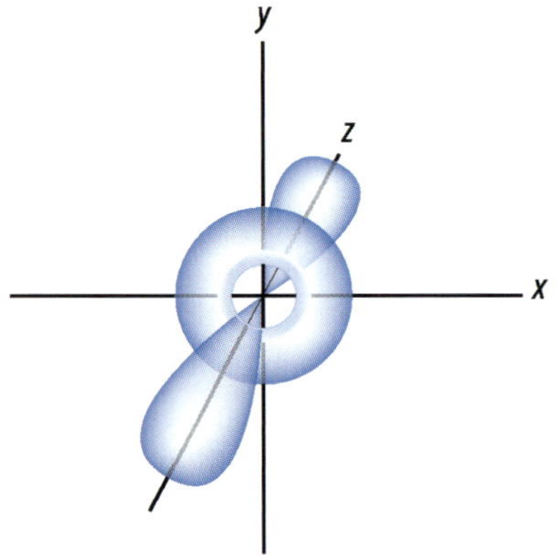 $d_{z^2}$

**Figure 3.** Here are the d orbitals. There are five of them. Because each holds two electrons, any d sublevel can hold up to twelve electrons. When they are all combined around a single nucleus, they might appear as shown on the bottom left.

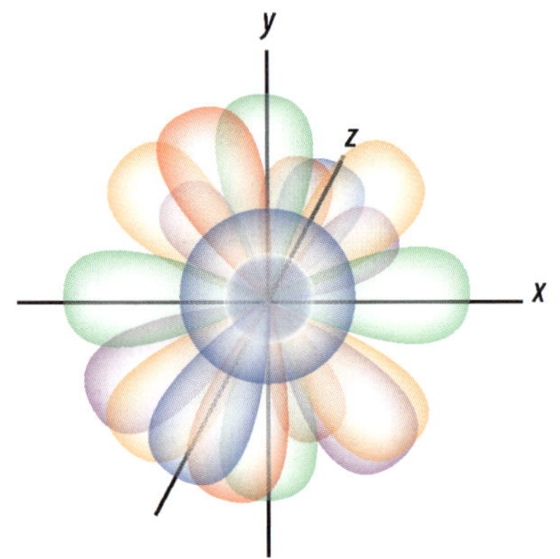


All d orbitals

An atom that has found this state will have eight electrons in its valence shell. This is known as the "octet rule." Atoms "react" with other atoms in such a way as to fill the outer *s* and *p* sublevels of both. To understand this basic fact is to understand the most important premise of reaction chemistry. Group IA elements react with group VIIA elements because together they satisfy the octet rule. Two group IA atoms react with one group VIA atom for precisely the same reason.

**Exercises**

Don't worry. In your lab you'll get plenty of exercise.

## 13: With the Exception of Hydrogen

Hopefully, you can now understand the organization of the periodic table. Elements are placed where they are because of the numbers of principal energy levels (which make up the periods) and because of the distribution of those electrons among orbitals within each energy level. The groups result from the filling of various orbitals in the highest principal energy level. (See Figure 1.) The electrons in this highest energy level are the "valence electrons" of that element. This is a good way to organize the table because electrons with similar valence electron configurations fall into the same groups and behave in a similar fashion.

The objective of this lesson is to examine some of the atomic properties of the elements that change in predictable ways within a group or period. During this lesson we will be talking a lot about the sizes of atoms and how easily they give up their electrons. Why should you care about any of this? Because knowing these trends can be very helpful in predicting the behavior of atoms. And what is chemistry if it's not understanding the behavior of atoms? Away we go…

**Figure 1.** This diagram shows the meaning of the periodic table in terms of electron orbitals as they fill. The orbitals fill in the order given as you read the periodic table from left to right and from top to bottom. The 1st principal energy level is represented in the periodic table as the 1st period (the first row). It includes only hydrogen and helium. The first principal energy level has only one sublevel (the *s* sublevel) which has only one orbital. This orbital is full when it contains two electrons--the condition that describes helium. The 2nd and 3rd periods represent the 2nd and 3rd principal energy levels. However, notice that the *3d* sublevel doesn't fill until after the *4s*. To date, new elements have been found that have all of their *6d* orbitals filled. Physicists are presently finding elements with their outer electrons in the *7p* sublevel. Can you anticipate what sublevel will fill after *7p*?

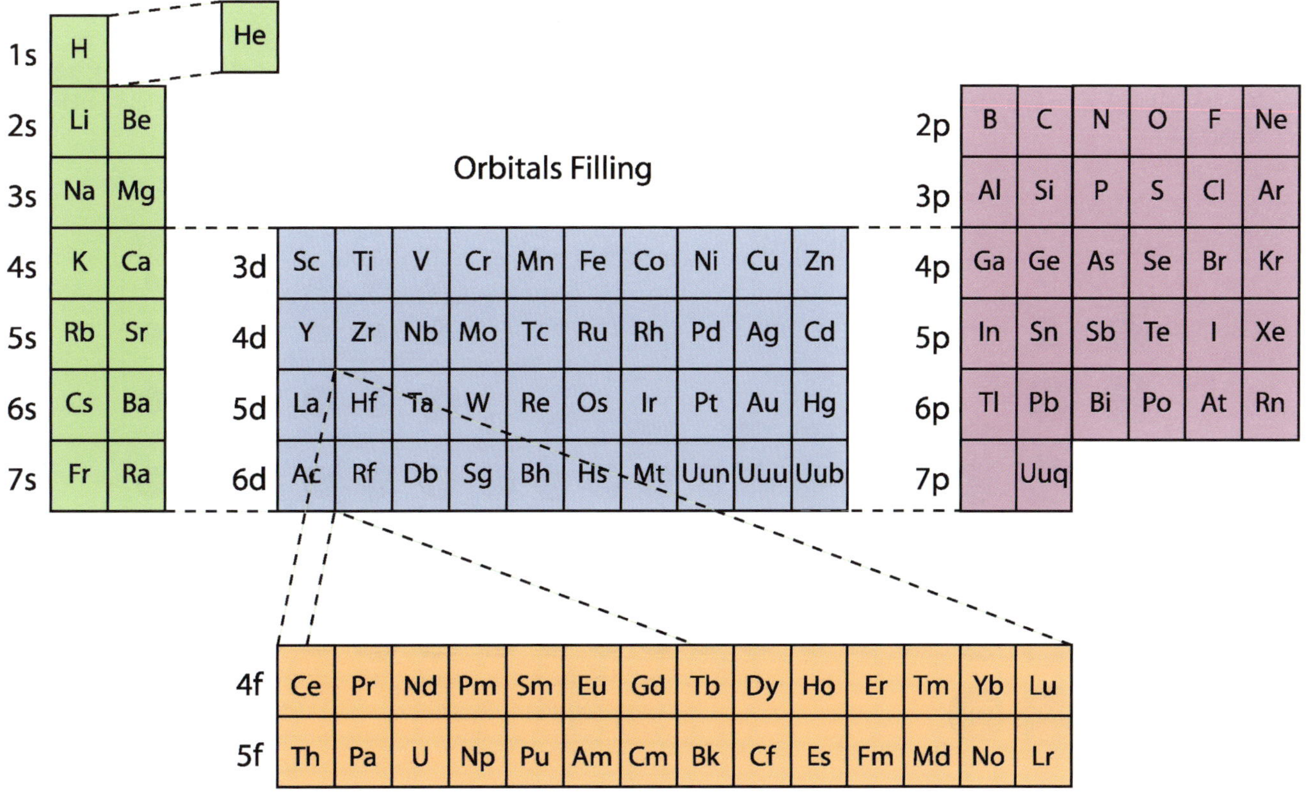

## Atomic Radii

Look at Figure 2. It shows the relative atomic radii for the first 89 elements (except for the actinide and lanthanide series elements). The radii are not all the same, are they? It probably shouldn't surprise you, knowing all you do about atoms, that they come in a range of sizes. It's not a huge range, however. The smallest is hydrogen, which has a radius of about 1/3 Ångström (Å), and the largest (among known atoms) is Francium, which has a radius that is estimated at 2.8 Å. So the radius of the largest neutral atom is less than seven times larger than that of the smallest. All but 13 of the atoms illustrated have radii between 1 and 2 Å. Those having radii smaller than 1 Å have a green fence drawn around them, while those having radii greater than 2 Å are enclosed in a purple fence.

Before we go on, there is something you should be reminded about. Although the radius is a good tool for describing the size of something that is approximately spherical, it can be a little misleading. That's because a twofold increase in radius has an eightfold effect on volume. So an atom that has a radius of 2 Å takes up *eight times* the volume of an atom having a radius of 1 Å.

Figure 3 shows the A group elements separately, because the trends among atoms are more obvious if we take out the transition elements. (Adding electrons to the *d* sublevel doesn't have as much impact on the dimensions of an atom as adding electrons to *s* and *p* sublevels. Adding electrons to the *f* sublevel—in the lanthanide and actinide series—will have even less impact.) Looking at the Figure, then, notice that potassium (group IA, period 4) is clearly larger that lithium (group IA, period 2). This is precisely what you would expect, since period 2 atoms have electrons in only two principal energy levels, while period 4 atoms have electrons in four. The $n = 4$ energy level is higher in energy and farther away from the nucleus than $n = 2$. Since the size of an atom is determined by its outermost electrons, we would expect the potassium atom to have a larger radius than the lithium atom. In fact, *as we move down any given group*, we are filling higher and higher energy levels with electrons, so *the atomic radius increases*.

**Figure 2.** These are the relative atomic radii (in Å) of the neutral atoms. Except for a few large atoms and a few small ones, most of them fall within a pretty limited size range. As a point of reference, for comparison, chlorine (group VIIA, period 3) has a radius of about 1 Å.

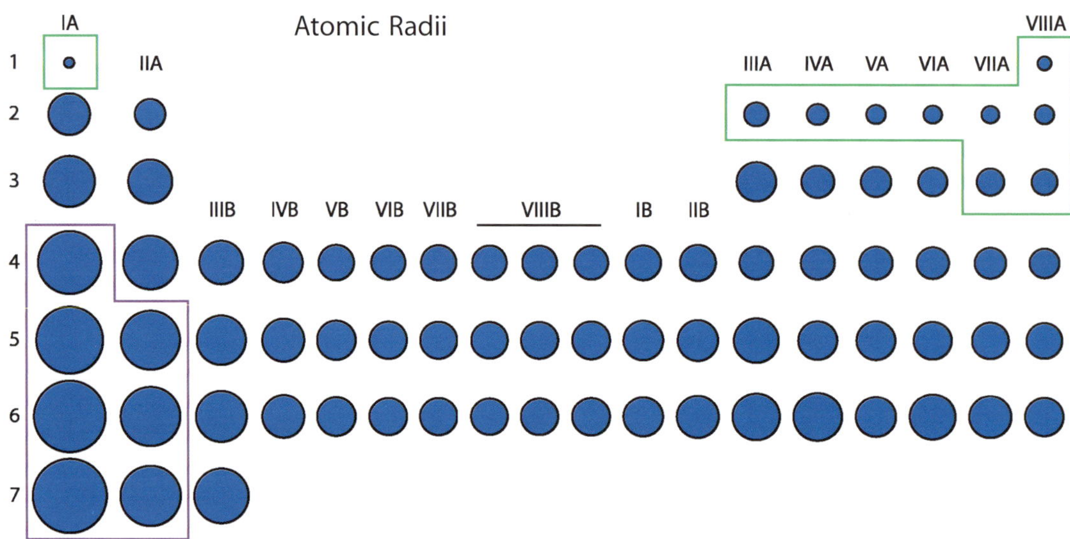

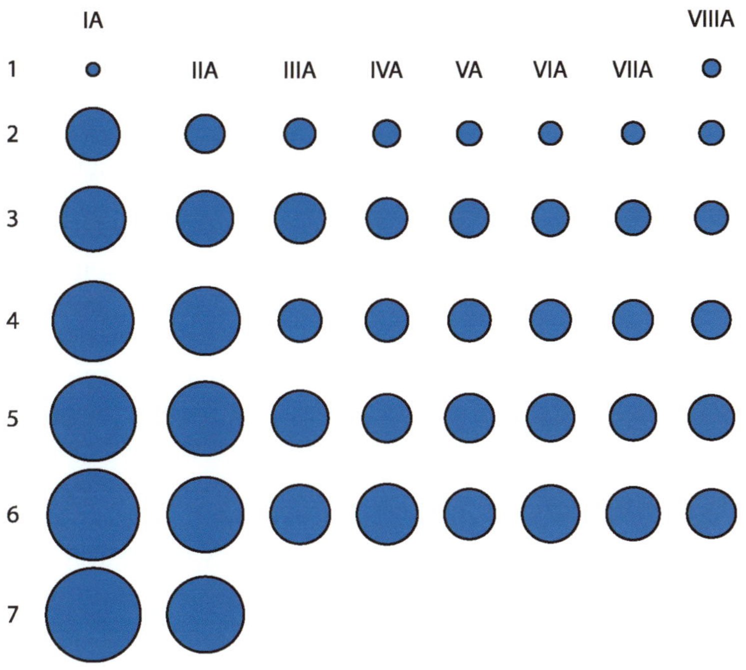


**Figure 3.** You will notice two trends in the radii of A-group atoms. First, within a group, the radii increase as the principal energy level of the outer electrons increases. Second, within a period, the radii decrease as the numbers of protons increase.

Atomic radius also changes as you move along a period. *As you move from left to right within a given period the atomic radii decrease.* That's because, as you move from left to right, the number of positively-charged protons in the nucleus increases. Therefore, the electrons, attracted by this increasing charge concentration, will be held more and more tightly.

Overall, notice how the atomic radii increase as you near the lower left corner of the table (francium) and how they decrease as you move upward and to the right (toward helium). That is, with the most notable exception of hydrogen, which has the smallest of all atomic radii.

## Ionic Radii

Would you expect an ion, having either one more or one less electron than its neutral atom, to occupy exactly the same amount of space as the neutral atom itself? Of course not! Direct your attention to Figure 4, which compares the radii of the A-group elements. The noble gases were left off because they form ions only rarely and with difficulty. First, notice that, as you go down a group, the trend is the same for both positive ions and negative ions as it was for the neutral elements—the radius increases. It does so for the same reason—there are more and more electrons filling higher and higher energy levels.

However, notice some differences. First of all, notice that when an element forms a cation (shown in red), its radius decreases. In fact, *the cation of an element will always be smaller than the element itself.* Why? You could answer this question yourself—to form a cation, one or more electrons must be removed, while the number of protons remains constant. Therefore, the same positive charge that was attracting a certain number of electrons in the neutral element is now pulling on fewer negative charges in the ion. The fewer electrons that remain can be held more tightly. Thus the radii decrease.

The same logic applies to anions (shown in yellow), but the effect is reversed. They have *gained* one or more *extra* electrons. What effect should this have on the radius? Well, again the number of protons remains the same, so the same number of positively-charged particles is now pulling on more negatively-charged ones. This means that the electrons cannot be held quite as tightly. That extra electron will also assume a high energy level. Both of these factors tend to increase the radius of the ion. *The anion of an element will always have a larger radius than the corresponding neutral element.*

If we consider only those elements that form cations (shown in red), then the trend as we move from left to right within a period is the same as for the neutral elements—the radius decreases. Remember, though, as you move from left to right within a period that you are forming cations by removing more and more electrons. Sodium loses one electron to form $Na^{+}$;

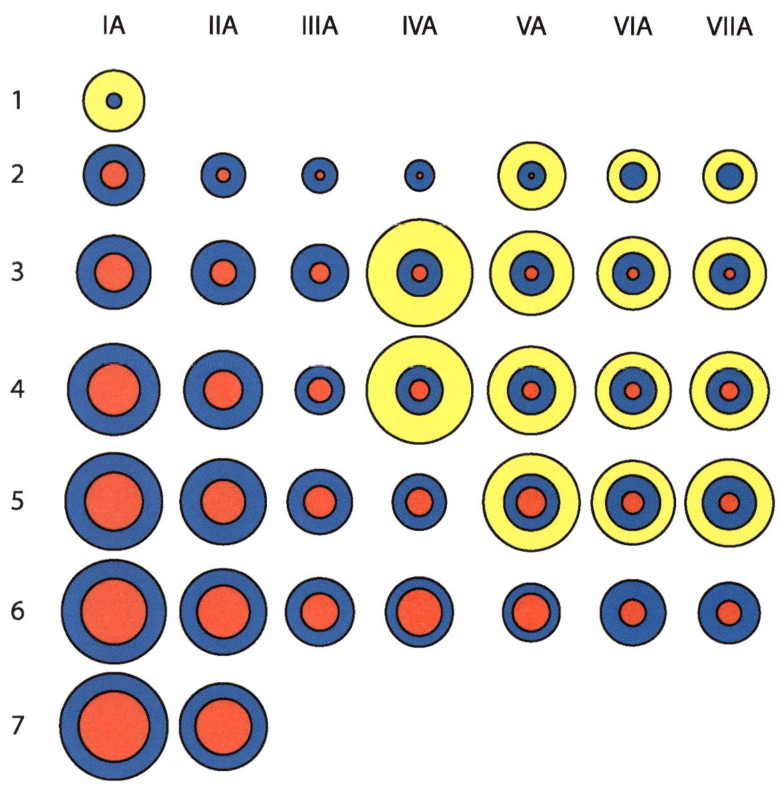


**Figure 4.** Positive ions (red) are smaller than the neutral atoms that form them (blue), because the excess positive charge holds the remaining electrons more tightly. Negative ions (yellow) are larger than their neutral atoms, because they have an extra electron relative to their numbers of protons. The trends in the radii as you move down a group or across a period are the same for the ions as for the neutral atoms--smaller from left to right, larger from top to bottom.

magnesium loses two electrons to form $Mg^{2+}$; and aluminum loses three electrons to form $Al^{3+}$. So the cations should get tighter and tighter as you proceed to the right. This is precisely what we see in the Figure. From IA to IIA to IIIA, there are more and more protons holding onto fewer and fewer electrons—the radii decrease rapidly. Because of this trend, it gets harder and harder to form positive ions as you move to the right. You'll notice that nitrogen and fluorine have no positive ions reported in the Figure. They won't give up those electrons.

So then, the cations will be seen to decrease in size, as you move from the lower left of the table to the upper right. That is, of course, with the notable exception of hydrogen. It has only one electron. When that electron is removed, its radius becomes vanishingly small—the radius of a single proton.

Enough about *cations*. Is there a trend among those elements that form *anions*? Yes. First of all, notice that the elements on the left of the table will not gain electrons. You must be in group IVA or above to have any desire to add an electron to your quiver. That is, with the notable exception of hydrogen, which has only one electron in its *1s* orbital. It will form an anion.

As was the case with cations, we have to remember that some of the ions shown on the table represent elements that are gaining more than one electron. So fluorine gains only one electron to form $F^-$ oxygen gains two electrons to form $O^{2-}$; and nitrogen gains three electrons to form $N^{3-}$. So the anions should get looser as you move from right to left—fewer protons—looser, more energetic electrons. And this is generally what we see (except for hydrogen—it has no anions in its period with which to compare it.)

Just for the fun of it, we'll now put the transition elements back into the Figure to see what effect they have on the trends. This is done in Figure 5. Now you can see what I meant when I said that adding *d*- sublevel electrons doesn't have quite the same amount of impact. There are some interesting atoms and ions that stand out here that we could talk about individually (although we won't), but generally the transition metals have a similar look about them: (1) their radii are all in the range from 1.24 Å (Ni) to 2.0 Å (Ac); (2) none of them form anions; and (3) the ions they form are a little more than half the size (radius) of the corresponding neutral atoms.

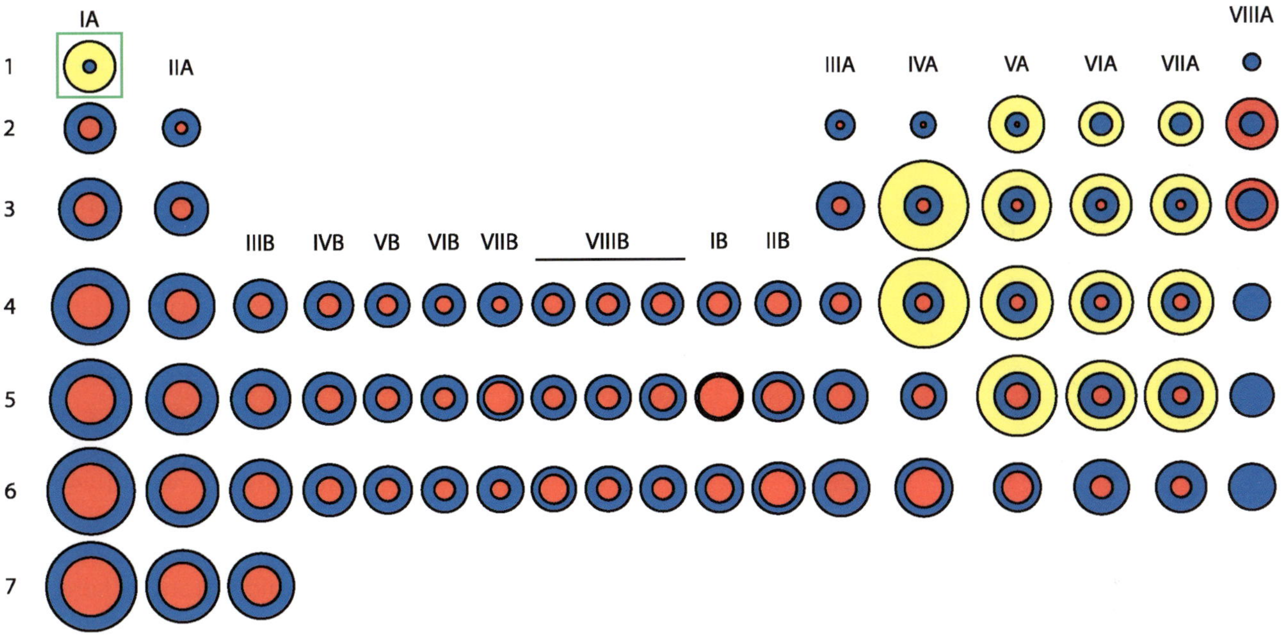


**Figure 5.** This is the master figure of the atomic radii of all the atoms and their most common ions. Once again, positive ions are shown in red and negative ions are shown in yellow. The radii of the neutral metals are all the same, but there is more variation in the radii of their positive cations. Hmmm. Do you suppose this could have any relationship to the way they behave? By the way, hydrogen has a positive ion--the famous $H^+$ ion. Why do you suppose its radius doesn't show up on the figure? (Hint: How many electrons and protons does a positive hydrogen ion have?)

## Ionization Energy

Understanding the trends in atomic radius makes it easier to understand other atomic properties of the elements. For example, consider the **ionization energy** of an element. Ionization energy is defined as the amount of energy required to remove an electron from an atom to form a cation. Why do ionization energies matter? They are the measure of how easily an atom will give up electrons. This will tell us much about its reactivity as we move ahead.

As an example, consider the formation of the lithium cation:

$$Li + \text{energy} \rightarrow Li^+ + e^-$$

The energy required to remove an electron from an atom is related to the energy the electron already has. If an electron has a great deal of energy to start with, relatively little is required to remove it from its nucleus.

As you would expect, an electron having relatively high energy and having a location farther from the positive charge of the nucleus would require less energy to remove it. An atom having such an electron would have a lower ionization energy than one having a smaller radius. Look at Figure 6, which shows the first ionization energies of the group IA elements. (The first ionization energy is the energy required to remove the outermost electron.) In group IA, you can see that the *ionization energies decrease as you move down the table*. This is true for all the group A elements.

Hydrogen also follows this trend except that, because it has only one electron, and it is in the *1s* sub-

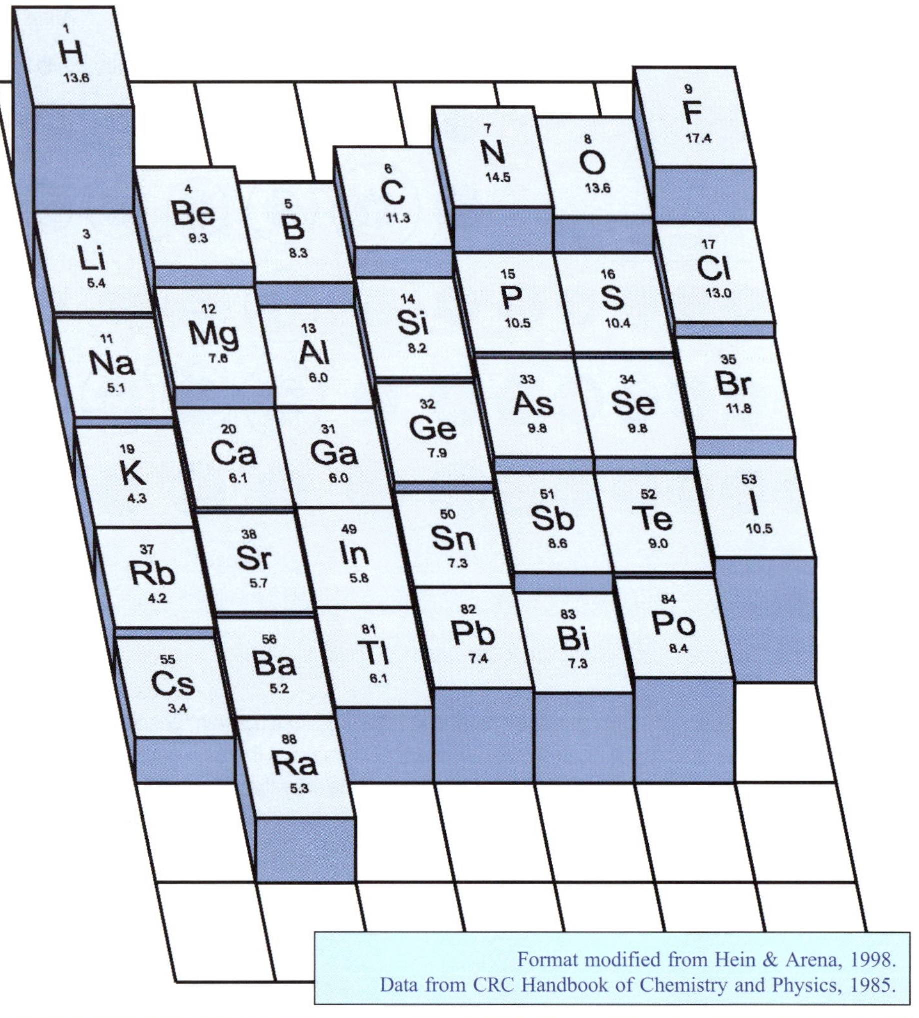


**Figure 6.** These are the ionization energies (in "electron volts") of the group-A elements Fluorine tends to hold onto its electrons tightly. Given a chance it'll take yours too. Cesium is at the opposite end of the spectrum. It takes little energy to steal away an electron from a cesium atom. The trends in this table look remarkably like the trends in atomic radii. It seems that the greater the atomic radius is, the easier it is to remove an electron. Does this seem right to you? In this figure, does any atom stand out as being out of place in either its group or its period?

level, it is the hardest one to remove. Its ionization energy is far greater than any other element in its period—it's way out of line. Our figure shows only *first* ionization energies. Every element *except hydrogen* has a series of ionization energies because they all have more than one electron that may be removed and each of those electrons has its own characteristic ionization energy.

What about any trend across the periods? Well, what was the trend among the atomic radii? As you moved from left to right, the electrons were held more tightly because of a more concentrated positive charge in the nucleus. If electrons are held more tightly, would you expect to need more energy or less to remove one? I hope you said, "More." Therefore, *as you move from left to right within a period, ionization energy tends to increase*. You need to be aware, though, that this trend is not as consistent as it is within groups. As you can see in the Figure, beryllium and magnesium are the major trend breakers among the group A elements. As usual, hydrogen is in a class by itself.

Now, just for the fun of it, let's put the transition elements back in and see their effects on these trends. (See Figure 7.) The ionization energies of the transition elements don't change much as you go down the table, and they increase very gradually as you go across. When the *d* sublevels are full (Zn, Cd, Hg), there is a noticeable decrease in ionization energy as you jump into group IIIA (Ga, In, Tl). Of course, this makes sense. You might expect it to be a little harder to remove electrons from a full sublevel than from one that contains only a single electron.

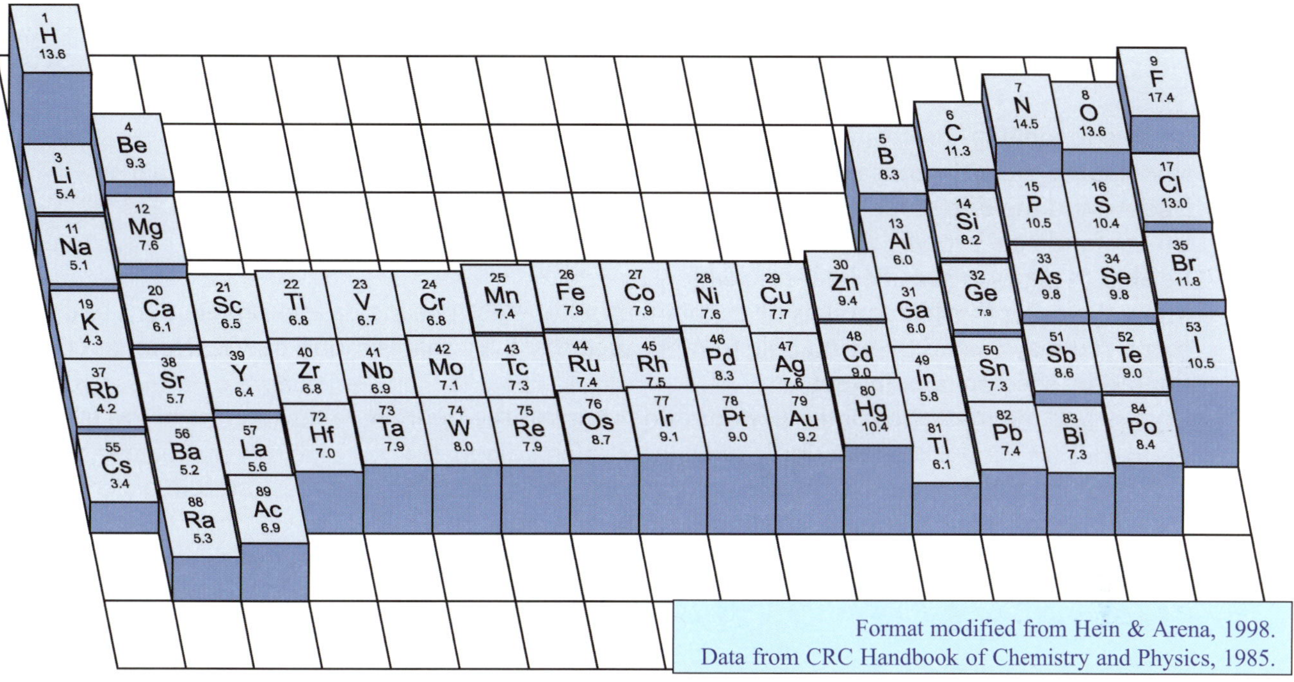


**Figure 6.** These are the ionization energies (in "electron volts") of all the elements. There is a much neater trend among the metals in this table than we have seen before. As a rule, electrons get more and more difficult to remove as you move across the table from left to right. However, as you jump from the transition metals to the post-transition elements ionization energies take a dip--that d sublevel is full at the end of the transition metals, and happy to stay that way, thank you very much! You see the same effect (to a lesser extent) in moving from the alkaline earth metals to the transition metals. Once the *s* sublevel is filled, the next added electron will be slightly easier to remove.

## Exercises

1) Which element has the highest ionization energy? Why?

2) Which has the lowest ionization energy? Why?

3) Which neutral element has the greatest atomic radius? Why?

4) Which has the smallest atomic radius? Why?

5) Which cation has the greatest radius? Why?

6) Which cation has the smallest radius? Why?

7) Which anion has the largest radius? Why?

8) Which anion has the smallest radius? Why?

9) Which has the smaller radius, Cl or $Cl^-$? Why?

10) Which has the greater ionization energy, Mg or $Mg^{2-}$? Why?

## 14: Quiz 2

1) Be able to identify the crystalline and amorphous substances mentioned.
2) Understand the difference between physical and chemical properties and processes.
3) Understand the difference between homogeneous and heterogeneous substances.
4) Understand how properties of substances are used in separating mixtures.
5) Be able to perform specific heat calculations.
6) Know the order in which orbitals fill as the number of electrons increases from element to element. You may do this by being able to reproduce the chart that is found on the "electron configuration" sheet in your lab set.
7) Understand the meanings of the periods and groups in the periodic table in terms of the filling of orbitals (as shown in the periodic table inserted in Lesson 13).
8) Know the informal names of the element groups and series. They are illustrated in a periodic table in Lesson 8 (alkali metals, alkaline earth metals, etc.).
9) As in every other lesson (whether or not we say) know and understand anything in bold or italic typeface.

# 15: Isotopes and Atomic Mass

## Isotopes

In previous lessons we have pointed out that the atomic number of an element always tells us the number of protons in the nucleus of an atom of that element. However, all atoms of a given element may not have the same mass. Why? Remember we said that the mass of an atom is essentially the mass of its nucleus. So if the mass of one atom of an element is different from the mass of another atom of the same element, and yet they both contain the same number of protons, then it must necessarily follow that the number of neutrons is different in the two atoms. Atoms which contain the same number of protons but have different masses are called **isotopes**. For example, there are three known isotopes of hydrogen:

protium (1 proton, 0 neutrons) - "normal" hydrogen
deuterium (1 proton, 1 neutron)
tritium (1 proton, 2 neutrons)

Notice that each of these isotopes has the same atomic number—the number *1*—which corresponds to the number of protons. Since there is only one proton, there is only one positive charge in the nucleus. In order to be charge balanced, there must be only one electron present in each of the isotopes. But because of the added neutrons, the mass of deuterium is very close to twice that of protium, and the mass of tritium is very close to 3 times that of protium.

Because we occasionally need to communicate the different isotopes of the elements, we need a method to distinguish them. **Isotope notation** was invented for that purpose. The general notation is given by:

$${}^{A}_{Z}X$$

where:

X = the element's symbol
A = mass number (number of protons + neutrons)
Z = atomic number

Here is an example of isotope notation:

$${}^{25}_{11}Na$$

To help you remember which letter corresponds to the mass number and which corresponds to the atomic number, just remember that there may be many mass numbers (isotopes) for a single element, so the one you are writing is just "*A* mass number." But there is only one atomic number for any element, so the one you are writing is "*Z only* atomic number." Always put the big number on top. It'll be the mass number. Here are isotope notations for the three isotopes of carbon:

${}^{12}_{6}C$ ${}^{13}_{6}C$ ${}^{14}_{6}C$

Sometimes, since the atomic number is often known or so easily obtained from the periodic table, the Z is omitted. So the notation for carbon-14 would simply look like this:

$${}^{14}C$$

Also, if you are given the notation for an isotope, you should be able to figure out how many protons, electrons and neutrons are present. For example, if you were asked how many protons, electrons and neutrons are in the isotope of fluorine represented by ${}^{19}F$, you would first look up F in the periodic table to see that the atomic number of fluorine is 9. Therefore, there are 9 protons. Since it is electrically neutral, there must also be 9 electrons. Finally, because the mass number is 19, the number of neutrons can be determined by subtracting the number of protons from it. There are 19 – 9 = 10 neutrons.

All the common elements have known isotopes; several elements have as many as 30 to 50 isotopes each. However, only a few of these will account for more than one percent of the element found in nature. As in the case of zinc, there are even some elements of which less than half of the naturally occurring element is in the form of its most prominent isotope. The rest is made up from small amounts of each of a multitude of isotopes.

Figure 1. The substance shown is practically 100% carbon. Natural carbon consists of mostly $^{12}C$ with a little $^{13}C$ mixed in. To use it in the lab, we'll need to know its *average* atomic weight.

## Atomic Mass

As we mentioned before, individual atoms have very small masses (remember the mass of a hydrogen atom was $1.673 \times 10^{-24}$ g). Therefore, it is simply not practical to compare the actual masses of various atoms when the numbers are so small. For this reason a *relative scale* was devised so that the masses of different elements can be more conveniently compared. Carbon-12 ($^{12}C$) was chosen as the standard against which everything else would be measured, and it was assigned a mass of 12 **atomic mass units** (**amu** or **AMU**). Therefore, one amu equals 1/12 the mass of a $^{12}C$ atom. Just be aware that when you read atomic mass values for a given element on the periodic table, these values were calculated relative to $^{12}C$.

Even though the scale of atomic mass is built around $^{12}C$, natural carbon is not *exclusively* made up of $^{12}C$. It also has a little bit of $^{13}C$ in it. If we want to do chemistry in the real world we won't be doing chemistry with pure $^{12}C$, but with a mixture of $^{12}C$ and $^{13}C$! (See Figure 1.) We are not so much concerned with the atomic mass of $^{12}C$, but rather the *average* atomic mass of *all* carbon. There are two ways we could calculate this average atomic mass.

Look now at Figure 2. The fishing sinkers shown here are of two different masses. What is the average mass of all the sinkers? Well, you could figure it out the way you have always calculated averages. That is, by simply adding the

Figure 2. The pile of ten fishing sinkers shown is made up of two different kinds of sinkers. There are seven 4-g sinkers and three 10-g sinkers. What is their average mass? There are two different ways of finding out!

masses of the individual sinkers and dividing by the total number of sinkers like this:

| Sinker | Mass |
|---|---|
| 1 | 4 g |
| 2 | 4 g |
| 3 | 4 g |
| 4 | 4 g |
| 5 | 4 g |
| 6 | 4 g |
| 7 | 4 g |
| 8 | 10 g |
| 9 | 10 g |
| 10 | 10 g |
| **Total** | **58 g** |

$$\text{Average mass (g)} = \frac{58\text{ g}}{10} = 5.8\text{ g}$$

This is a little tougher to do with atoms than it is with sinkers (unless you are asked to calculate the average mass of a trillion sinkers.)

The other way we could do it is by multiplying each of those masses by the portion of all sinkers it represents. In this case 7/10 of the sinkers were 4-g sinkers and 3/10 were 10-g sinkers. So:

$$\text{Average mass (g)} = 0.7\,(4\text{ g}) + 0.3\,(10\text{ g}) = 5.8\text{ g}$$

Both methods give the same answer. When working with atoms, the first method is not an option. There are few circumstances under which we would ever count and weigh individual atoms. (We'll tell you how this can be done later.) Using the second method you needn't count and weigh individual atoms. You just need to know the masses of the individual atoms that make up the two different kinds, and the proportion of the entire mass made up by each.

To weigh out one trillion sinkers would require 10,000 workers, working 8 hours per day, each weighing one sinker per second, over 13.35 years to complete the task, not including restroom breaks.

This second method of calculating the average is called the **weighted average** method. The individual masses are "weighted" by their corresponding proportions to calculate an average. Look at your periodic table and find the atomic mass of carbon. It is given as 12.01, not simply as 12. Why isn't the atomic mass of elemental carbon recorded as 12? Because carbon doesn't exist in nature as only $^{12}C$, but as a mixture of 98.89% $^{12}C$, and 1.11% $^{13}C$. So the mass given in the periodic table is a weighted average of these two isotopes.

You would calculate the weighted average mass of carbon like this. $^{12}C$ has a mass of 12 amu by definition. But how much importance or "weight" should we give to it in calculating the average atomic mass? Well, it accounts for 98.89% of all the mass, so it should get 0.9889 times the total importance in determining the average atomic mass. $^{13}C$ has a known mass of 13.0033 amu (to six significant digits) but it accounts for only 1.110% of the total mass of carbon. It should get only 0.01110 times the total importance in our calculation. So we should *weight* each atomic mass by its relative abundance to get its "weighted mass." The sum of the weighted masses is the *average* mass.

Here it is, plain and simple:

Average mass (amu) =

$$0.9889\,(\underline{12}\text{ amu}) + 0.01110\,(13.0033\text{ amu})$$

$$= 12.01\text{ g}$$

(Remember, the 12 is a number that is known by definition, so its precision is infinite. If you need help remembering not to let it limit the precision of your calculations, underline it when you write it down, like I did.)

All of the atomic masses of the elements that appear in the periodic table are the weighted averages of the isotopes that make them up. If you know the relative abundances of all the naturally-occurring isotopes of an element and the atomic mass of each isotope, you can calculate the average atomic mass for that element.

Now it's your turn!

**Exercises**

1) For the following isotopes, give the number of protons, electrons and neutrons.

   a) $^{238}_{92}U$

   b) $^{81}_{35}Br$

   c) $^{63}_{29}Cu$

2) Complete the following table:

| isotope notation | A | Z | no. of protons | no. of electrons | no. of neutrons |
|---|---|---|---|---|---|
| $^{79}Br$ | | | | | |
| | | 11 | | | 12 |
| | 90 | | 38 | | |
| | 3 | | | | 2 |
| | 39 | 19 | | | |
| $_{7}N$ | | | | | 8 |

3) Magnesium occurs in nature as a mixture of three isotopes, $^{24}Mg$ (23.98504 amu), $^{25}Mg$ (24.98584 amu), and $^{26}Mg$ (25.98259). Their relative abundances are 78.70%, 10.13%, and 11.17%, respectively. Calculate the average atomic mass of Mg.

4) There are only two naturally-occurring isotopes of iridium, $^{191}Ir$ and $^{193}Ir$. $^{191}Ir$ has a mass of 190.9609 amu and $^{193}Ir$ has a mass of 192.9633 amu. Their relative abundances are 37.30% and 62.70%, respectively. Calculate the average atomic mass of Ir.

# Bonds and Molecules

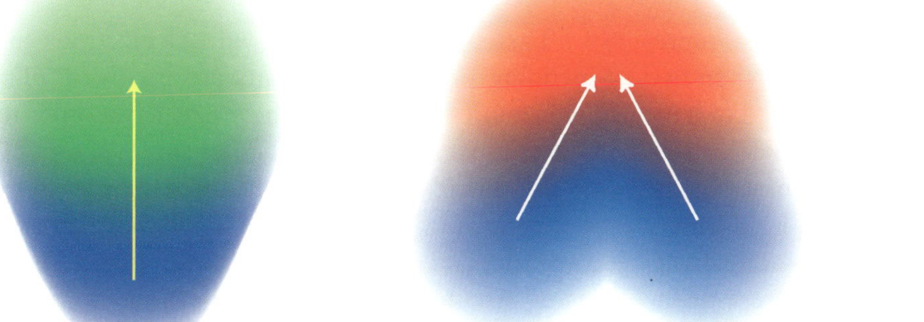

## 16: Chemical Bonding and Electron Pork

Although we think of atoms as separate entities, most do not exist that way in nature. In fact, there are only a few (primarily the noble gases) that do. For the sake of simplicity, let's think about what might take place between two individual hydrogen atoms that encounter one another. (Remember, though, that individual hydrogen atoms are actually a rarity in nature. A single hydrogen atom is a very reactive beast—not content to live alone.)

Two hydrogen atoms will come together to form a molecule when the positively-charged proton of one atom is attracted to the negatively-charged electron of another, and vice versa. (See Figure 1.) The two atoms pull together until the two protons exert a push on each other that prevents the nuclei from coming closer. They will find a fixed distance from one another that represents the balance of these opposing forces. When two atoms come together in this way, they are said to have **bonded**. The distance between the two nuclei of the bonded atoms—the **bond length**—is determined by the strengths of the opposing forces, and will be different for any specific pair (or group) of elements. The nuclei assume a distance from one another that represents the *lowest possible potential energy* among the forces affecting them.

Once two atoms come in close contact, their valence electrons will assume a new pattern of behavior. That pattern will be the result of the new charge environment of the bonded pair. (Figure 2.) What were once two atoms have now become one molecule. The molecule is held together by the attractions among the positive nuclei and the negative electrons. The nuclei are held apart by their opposite charges. Just as importantly, the outer (valence) electrons will redistribute in an attempt to provide a "**noble gas configuration**" for each atom. In other words, electrons will be shared so that they "fill" the outer valence shell of both atoms. In the case of hydrogen gas ($H_2$), this means that both electrons (one from each atom) are shared between the two atoms, filling the *1s* shell of each.

There are two ways that valence electrons can act, depending on which elements are involved: (1) they may be stolen away by the new partner in order to satisfy the hunger of its empty, outer sublevel, or (2) they may continually encircle both atoms in order to satisfy the demands of both. Of course, which they actually do will depend on just how demanding each of the atoms is. An atom that has a high affinity for electrons will steal them from an atom with a much lesser affinity. In cases where atoms are unmatched, but the difference between their affinities is not as great, they will share. However, in those cases the sharing is unequal—

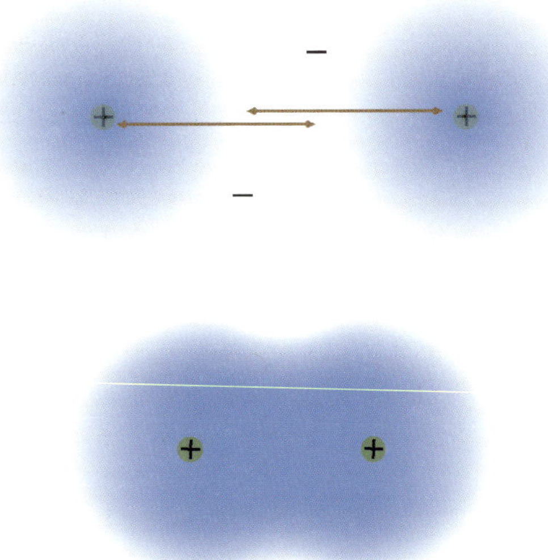


**Figure 1.** The attraction between the electron of one atom and the proton of another draws hydrogen atoms near to one another.

**Figure 2.** Both electrons are shared between the two atoms causing them to be bonded together. The bond length (the distance between the two nuclei) is the balance point of (1) the attraction between the protons and electrons, and (2) the repulsion of the two nuclei. Because the shared electrons spend a high portion of their time between the two nuclei, the positive charges of the two nuclei are shielded from one another, so they can draw closer than they otherwise could.

one atom will hold the electrons for a greater portion of the time.

### Equal Sharing

Which do you think will happen in cases where the two atoms are identical, as in $Cl_2$ or $H_2$? Can the electrical charge of one of those atoms have more influence on electrons than the other? Impossible. They have identical charge environments; they can't influence electrons differently. They will share equally to satisfy the demands of both atoms. Although the electrons are constantly in rapid motion, the place in the molecule where the electrons are most likely to be found at a given time is where the positive charges are the strongest—right between the two nuclei. Of course, they are less likely to occupy that space at the same time because of their strong repulsion for one another.

Take a hard look at that hydrogen molecule again (Figure 2). It consists of two protons held at a stable distance from one another. There are two electrons forming a cloud around both nuclei. This is the simplest example of **covalent bonding**—two shared electrons in a single molecular electron cloud. Two protons are charge balanced by two electrons, and the two electrons "fill" the valence electron shell of both.

Nevertheless, these are still two atoms, not one. It's not as if the two protons had combined into one nucleus having two electrons to balance them (as in the case of the helium atom). The $H_2$ *molecule* will be far more stable than a lone H *atom*, but not as stable as a helium atom which has a full valence shell. Helium has two electrons of its very own and doesn't have to share with anyone. Furthermore, the helium nucleus can't come apart because of the immensely strong nuclear forces holding it together. The hydrogen molecule, on the other hand, consists of two separate nuclei that really don't like each other. This is the nature of molecules—less stable than noble gases, but more stable than the individual atoms making them up.

Can you think of some molecules that would likely have an equal distribution of electrons around their nuclei? Any diatomic molecule, consisting of two identical atoms, would. For atoms that were different from one another you would have to consider the similarities between the two nuclei and compare their electron structures. The more alike they were in these respects, the more equal their sharing would be. Since no two elements are exactly alike, some differences in electron sharing are likely to be seen among any two dissimilar bonded atoms. This brings us to our next point.

### Unequal Sharing and Polarity

Atoms that are covalently bound needn't necessarily share alike. Some atoms (oxygen, for example) are notorious for unfair sharing. (Oxygen seems to be a pig.) Any compound that includes one of these atoms is likely to suffer from the effects of unequal electron distribution. In our next lesson, we'll discuss what makes an atom hold on to electrons longer than others do. But first, let's consider the effect of this unequal sharing.

When a compound has an uneven distribution of electrons, it will have some unique chemical properties. The side of the molecule that holds the electrons the greater portion of the time will have a negative charge (referred to by the lower case Greek letter "delta" with a minus sign: $\delta-$). How negative that charge is will depend on the number of electrons involved and how unbalanced the attractions are. Of course, anytime an atom's electrons are withheld from it, it takes on a net positive charge ($\delta+$). Molecules that have such positive and negative charge distributions are referred to as **polar**. Any such bond that results from unequal sharing of electrons is called a ***polar* covalent bond**. Figure 3 shows several examples of such molecules. Water is a notable example. We'll have much more to say about its polarity in later lessons. Molecules characterized by *equal* sharing of electrons are **nonpolar**.

There are two cases where dissimilar atoms bonded together give rise to nonpolar behavior. First, if two different atoms are alike in terms of their attraction for electrons, their molecules will be nonpolar for practical purposes. For example, hydrogen and iodine are dissimilar molecules, but their attractions for electrons are similar. For this reason, HI is practically nonpolar and acts like other nonpolar molecules. (See Figure 4.)

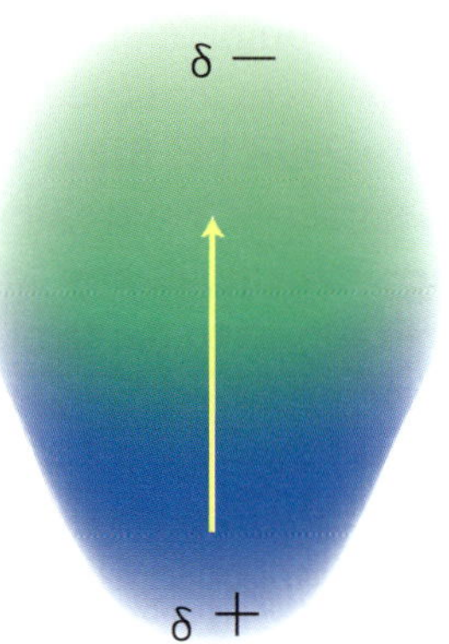


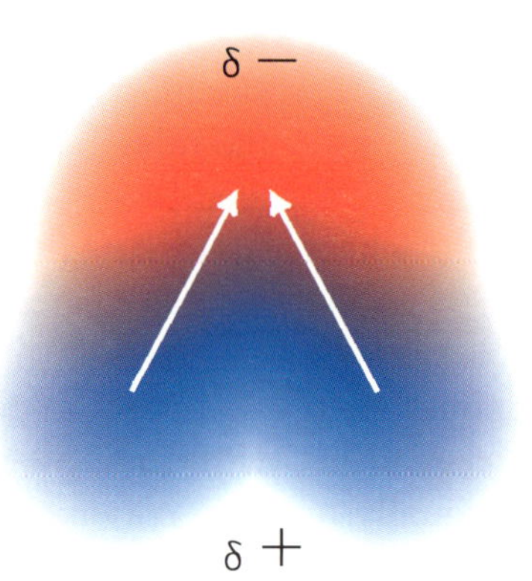

**Figure 3.** All of the molecules shown here are polar because electrons are pulled in the direction of the arrows by elements that tend to hold them longer. This places a slight negative charge on that end of the molecule and a positive charge on the opposite end. These molecules are, from left to right, hydrogen chloride, water and barium sulfide.

Second, even where strong polar attractions exist, if the overall molecule has an even distribution of charges over its surface, the molecule as a whole will not show any signs of polarity. (Figure 4 again.) For example, in carbon tetrafluoride ($CF_4$), the fluorine atoms have a strong tendency to pull electrons away from carbon. However, there are four fluorine atoms that are distributed evenly about the carbon atom, giving it a surface that is uniform in regard to its charge. A nearby particle approaching this molecule would experience it as nonpolar, even though the bonds holding the molecule together are polar.

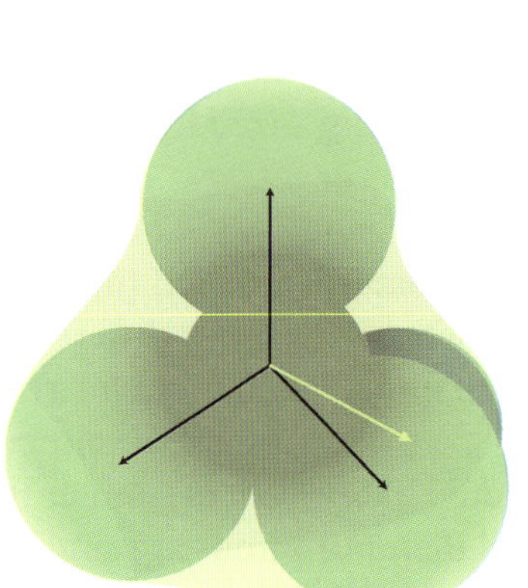

**Figure 4.** The HI molecule (top), although it is made up of two different atoms, is not polar. That's because neither of the atoms draws on the electrons with greater strength than the other. It's a tug-of-war between two well-matched teams. The carbon tetrafluoride molecule (bottom) is made up of different atoms having different electron withdrawing tendencies. All four of its bonds are polar. However, there is no positive and negative quality to the surface of the molecule because all the surface atoms are identical. The electrons tend to be drawn outward toward the four fluorine atoms (in the directions of the arrows). The only atom that's different—the carbon atom—is surrounded by fluorine. Even with four polar bonds the overall molecule is nonpolar.

**Exercises**

Do fifty push-ups.

# 17: Unequal Sharing and Unequal *Not* Sharing

## Electronegativity

As we said in our last lesson, some elements tend to share electrons equally and thus form nonpolar covalent bonds. Other elements share them unequally, so their bonds are polar. Still others tear electrons away from their bonding partners. These bonds are not "covalent" since the term implies that electrons are being shared, and stealing isn't *usually* thought of as sharing. Instead, two ions form: (1) an anion that has an extra electron, and (2) a cation that has one too few. These two ions will then be attracted to one another because of their charge difference. What we did not explain in the last lesson is *why*. Why do some elements hold onto electrons more tenaciously than others?

## Smaller Atomic Radius

Return in your mind to our discussion about ionization energies—the energies required to remove electrons from atoms. We learned that the energy required to remove an electron from its nucleus decreases as the distance from the nucleus increases. And, it stands to reason that electrons that are farther away from the protons will be easier to remove. Can we say then, that the greater the atomic radius the easier it will be to pull away its electrons? Yes, we can. While this is generally true, it is not, as you'll recall, the end of the story.

## Less Electron Shielding

We also learned in that lesson that the removal of an electron from a positive ion is more difficult than removal of an electron from a neutral atom. So the balance of charges between a nucleus and its electrons will have an influence over how tightly the electrons are held. Now, although every neutral atom has the same number of protons as it has electrons, different electrons are controlled by different influences. There are tightly-held inner electrons and loosely-held outer ones. The negative charges of those inner electrons shield the outer ones from the positive charge of the nucleus. This makes them hang on a little less tightly.

## Full Valence Shells

Finally, recall that removal of an atom from a valence shell that is full is much more difficult than removal of an electron from a valence shell that has only a single electron in it. The full outer shell is the most stable electron configuration for an atom or ion. That's why a halogen tends to gain an electron to form a negative ion, and an alkaline earth metal tends to lose an electron and form a positive ion. The noble gases, with their full outer shells, are exceedingly difficult to ionize.

These influences all tend to make atoms hold electrons more tightly. A high affinity for electrons created by these effects is called **electronegativity**. Why is the tendency to hold onto electrons called electronegativity? If you view a molecule that has an electronegative atom in it—one that draws electrons—that element will tend to have a greater negative "charge density" surrounding it (See Figure 1.) It's that local negative charge

**Figure 1.** The chlorine - carbon bond in chlorobenzene is polar because chlorine has a greater attraction for electrons than does carbon. This places a slightly more negative charge on chlorine and a slightly more positive charge on carbon. Chlorine is the more "electronegative" atom.

in molecules that earns these atoms the term *electronegative*.

## Electronegativity and the Periodic Table

Since the two properties of atoms—ionizability and electronegativity—are opposites, a value placed on one would practically be the inverse of the other. Linus Pauling, a two-time Nobel laureate who died as recently as 1994, assigned electronegativity values to the elements based on fluorine (the *most* electronegative element) having an assigned value of 4.

Compare Figure 2 with Figure 6 on page 59. You will see that the electronegativities of the elements behave as you would predict based on their ionization energies. If you'll recall, the ionization energy is the amount of energy required to ionize an atom. Therefore, atoms that give up electrons easily will have low ionization energies. While the most readily ionizable atoms (those with the lowest energies) are those in the lower left corner of the periodic table, the *most* electronegative ones are in the upper right corner.

This electron-gathering tendency that we call electronegativity is an important concept. You will hear of it over and over during your journey through chemistry. In the first place, as we said, it gives rise to polarity. Polarity determines which molecules will come into close contact with which others. A polar substance has little interaction with a nonpolar one, whereas two polar molecules will be all over each other. (See Figure 3.)

In the second place, the local charges on molecules help in deciding their reactivities. An electronegative area on a molecule has a strong attraction to, and is a prime target for, attack by an electropositive area on another molecule.

## Electron Thievery

At the extreme of unequal sharing, we mentioned in the last lesson and earlier in our introduction to this lesson that one atom may actually take away an electron from another. When this happens, two ions form that have opposite charges. Because of their opposite

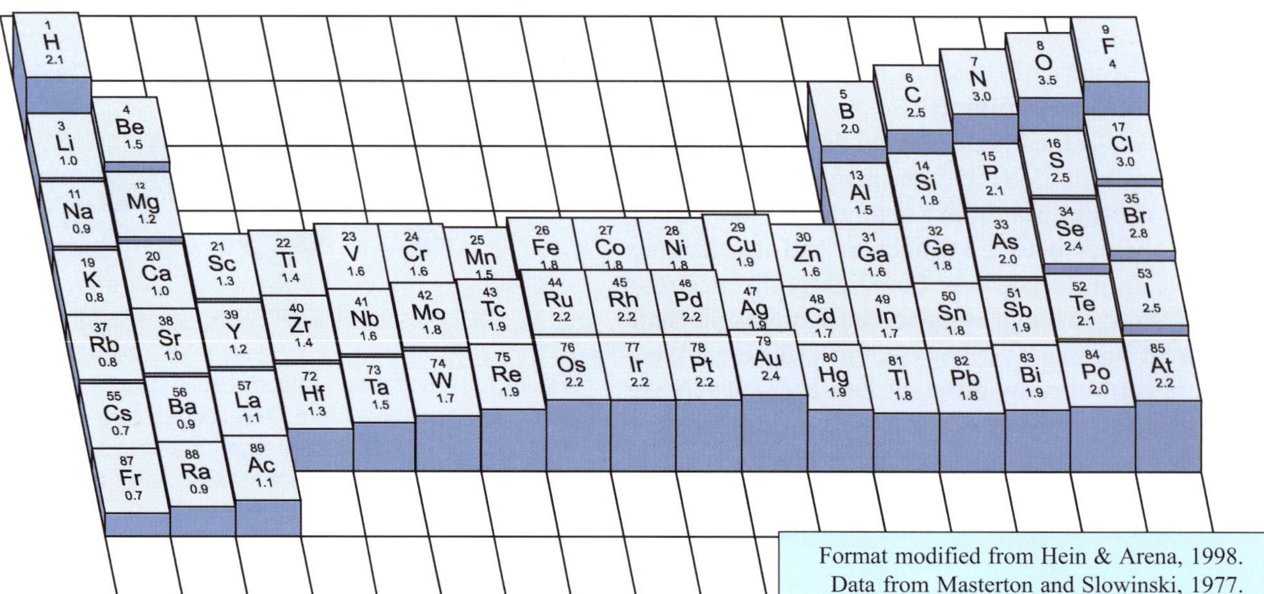


**Figure 2.** The electronegativities (the strength with which elements hold their electrons) are given based on fluorine having an electronegativity of 4. Notice once again that the elements that let go of their electrons the easiest are in the lower left corner. Those that hold onto their electrons most tightly are in the upper right. This corresponds closely to the trend in ionization energy.

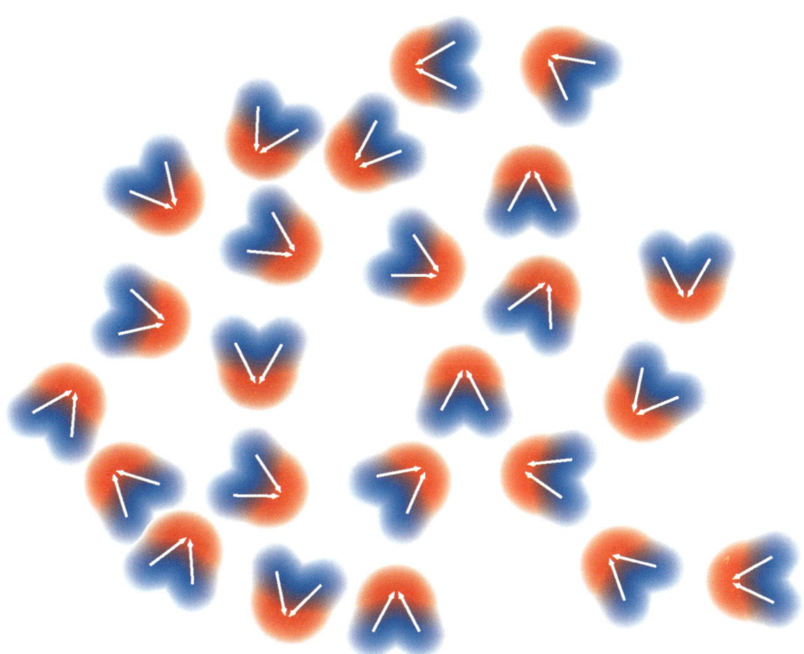

**Figure 3.** While there is little strength of attraction among nonpolar molecules, polar ones have surface charges that cause them to attract one another end-to-end. The molecules tend to orient themselves in response to the other molecules around them so that positive charges align with negative ones. This tendency, being strong in water, gives it unique properties that will be the topic of much discussion in later lessons.

charges, they have a strong attraction to one another. This attraction is responsible for a tight bond between the ions called an **ionic bond**. While chlorine can remove an electron from sodium, it can't completely remove the electron from hydrogen. The result is that sodium chloride is an ionic compound while hydrogen chloride is held together by a covalent bond (based on the sharing of a pair of electrons). See Figure 4.

## How Can I Predict?

Just how may we know if the bond between two atoms will be ionic or covalent? Furthermore, how will I know if a covalent bond will be polar? Of course, we go back to our explanation of the strength of attraction of atoms for electrons—electronegativity. Using the information from the periodic table in Figure 2, a bond that is mainly covalent will form between atoms having a difference in electronegativity values of less than 1.5 units. A bond that is mainly ionic will form between atoms having a difference in electronegativity greater than 2.0. The highest possible difference in electronegativities is that between fluorine (4) and francium (0.7), or 3.3. Any bond involving atoms that have an intermediate difference in their electronegativities will be polar covalent—that is, there will be some sharing of electrons, but that sharing will be unequal. (See the Table.)

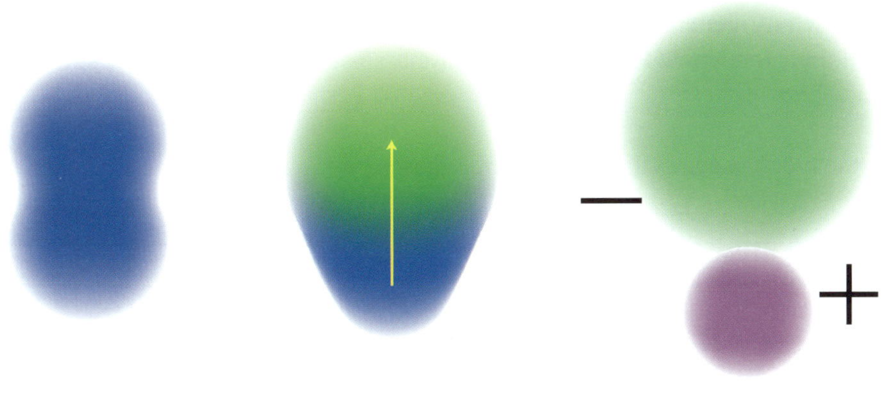


**Figure 4.** Three bond types are illustrated here: (1) the simple covalent bond of $H_2$, (2) the polar covalent bond of HCl and (3) the ionic bond of NaCl. Notice that the chlorine anion with its extra electron has a larger electron cloud than does the neutral chlorine atom. That extra, rather loose electron gives the ion almost twice the radius of neutral chlorine.

**Table.** Classifications of bonds based on electronegativities of the participating atoms.

| Difference in Electroneg. Values | Bond Class |
|---|---|
| 0 to < 1.5 units | Covalent |
| 1.5 to 2.0 units | Polar covalent |
| > 2.0 to 4 units | Ionic |

**Exercises**

1) Tell which of the following are ionic, which are polar covalent and which are non-polar covalent:

   a) MgS
   b) CaS
   c) MgO
   d) The C-H bonds in $CH_4$
   e) The N-H bonds in $NH_3$
   f) The C-N bonds in HCN

2) From your answers can you draw any generalizations about the different groups (e.g., halogens, alkaline earth metals, etc.) and their tendencies to form different classes of bonds with other specific groups?

3) What can you surmise about the electronegativities of the noble gases?

## 18: Striving for Nobility

Lewis structures (also called *electron dot structures* or *flyspeck structures*) are named after the American physical chemist, G.N. Lewis, who was very productive in the understanding of bonding and electron behavior in the early to mid 1900's. His insight into the behavior of substances came in part from a method he gave us for diagramming the outer, valence shell of an atom. He surrounded the atomic symbol of an atom with dots which represented its valence electrons.

For example, consider hydrogen which has a single electron in its valence shell:

$$H\cdot$$

Now recall what we said in our previous lesson about the desire of all atoms to have valence shells that are full like those of the noble gases. The nearest noble gas to hydrogen is helium. Having two valence electrons it has a full *1s* shell (*$1s^2$*). So hydrogen needs only one electron in order to satisfy its hunger.

You can now anticipate the importance of this simple set of observations. The question is "What will hydrogen combine with?" The answer is "Any atom that satisfies that hunger." We need another atom that will provide a second electron for hydrogen.

### What's Better Than One Electron?

Obviously, hydrogen should combine with any other atom that would give up a single electron. However, atoms like to hold onto their electrons to balance their positive protons. Only under the extreme condition where an a highly electronegative atom is bound with another of low electronegativity will there be the complete transfer of an electron from one element to another. Hydrogen is not electronegative enough to take away an electron from any other atom.

If hydrogen is not strong enough to steal an electron, it will have to settle for sharing. An atom that has partial ownership of an electron is almost as well satisfied as an atom that has complete control of one. Once again, we have described the conditions of covalent bonding. Two atoms in need of filling out their valence shells share electrons to the satisfaction of both. The easiest way to demonstrate covalent bonding is in the form of simple hydrogen gas:

$$H : H$$

Okay, so maybe $H_2$ is not as happy as helium, which has its very own pair of electrons and a full valence shell. But, compared to a lone hydrogen atom, $H_2$ is extremely stable.

Remember that every other atom in group IA has exactly the same electron configuration in its outer shell as the lone hydrogen atom. The only difference is, those electrons get easier and easier to remove as you go down the group. So then, not only will hydrogen combine with another hydrogen atom to fill its *1s* shell, it will combine with any other group IA element to form a metal hydride: LiH, NaH, KH, etc.

The rest of the atoms on the left side of the periodic table are of relatively low electronegativity. Oh, sure, these atoms would be happier if they had their valence shells filled, but they don't have the same drive for electrons that hydrogen has and that the atoms on the right-hand side of the periodic table have. For example, sodium atoms don't have any great need to combine with other sodium atoms to fill their valence requirements. But place sodium in the presence of chlorine gas and just watch the fireworks. (The reaction gives off visible light and a tremendous amount of heat.)

Why do sodium and chlorine react so vigorously? It's because chlorine is not passive with respect to electrons. It demands to gain the use of any loose electrons nearby. Any group IA element will react vigorously with chlorine. The farther you descend down group IA, the more vigorous will be the reaction. We have now laid the groundwork for you to understand simple reactions involving two atoms to form binary molecules.

## Simple Binary Molecules

Let's begin with the simple compound HCl. If we are to understand why this covalent bond between hydrogen and chlorine is stable we should look at its valence electrons. Of course, you know hydrogen (a IA element) has only one electron.

H·

To fill its 1*s* orbital would require a second electron. Chlorine is complementary to hydrogen in that it has seven valence electrons. (It's in group VIIA):

·C̈l:

For its valence shell to be as complete as that of a noble gas, it would require one more. Hydrogen has the one it needs. These two atoms fit nicely together to satisfy the valence shell requirements of each atom, giving each a noble gas configuration. Notice that, with the pair of shared electrons, each of the two atoms has its full complement of valence electrons. H has a full *1s* shell and Cl has its complete octet, obeying the octet rule.

H:C̈l:

Of course, as you know, the noble gas configuration is the most stable. This explains the stability of HCl. It also explains the stability of most other stable structures. It might be better to replace the two shared electrons with a hyphen to show that they form a bond, but you must remember that the hyphen represents *two* shared electrons, not one:

H-C̈l:

Now let's take a look at another simple molecule. Carbon tetrachloride is made up of a single carbon atom having four valence electrons (because it's in IVA):

·Ċ·

and of four chlorine atoms, each having seven (VIIA):

4 x ·C̈l:

Can they be put together such that each atom has eight valence electrons about it? The only way is for each of the four carbon electrons to be shared by a different chlorine atom and for each of those chlorine atoms to share one of its electrons with carbon:

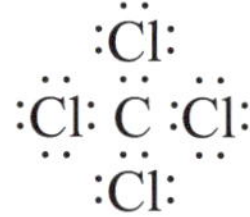

As you can see, each atom has its complete set of eight valence electrons. We'll now replace the shared electrons with bonds—one bond for each pair of shared electrons:

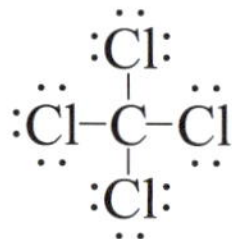

The most important thing to remember when making electron dot diagrams of molecules is that every nonmetal atom must have access to eight electrons, either shared or of its own. Every metal atom of group IA or IIA must share its valence electron(s) with a bonding partner. This gives the metal atoms full *s* sublevels. It also gives them stable relationships with a partners that desperately need those electrons. The group IIA elements will not be found sharing only one of their two electrons. It's either both or none.

Let's think for a moment about how many bonds an element can form in an attempt to fill its valence electron requirements. It would be easy to jump to the conclusion that, since elements are seeking a noble gas configuration, they will all form four bonds of two electrons each. That's not true. They will form as many bonds as their valence electrons and those of their partner will allow. Some elements, those in group IA, have only one valence electron, so they have only one electron to share. For example, each sodium atom can form only one covalent bond with sulfur (6 valence electrons) to make sodium sulfide ($Na_2S$):

Na:S̤̈:Na

Each sodium atom shares one of its electrons with sulfur; sulfur shares one of its six electrons with each atom of sodium. In this way, that single, awkward sodium electron is tied up while sulfur achieves its octet.

Oxygen is in the same group as sulfur, so it has the same six valence electrons. Lithium is in the same group as sodium. Once again, you see two lithium atoms, each sharing its electron and oxygen sharing one of its electron with each lithium atom to the satisfaction of both:

Li–Ö̤–Li

Here oxygen has achieved nobility while tying up the single electron in the valence shell of each lithium atom.

Although, in each of the previous examples, the group IVA atom forms two bonds this would not be true if it were to react with an element from a different group. For example, draw the Lewis structure of oxygen:

·Ö̤·

and the Lewis structure of magnesium:

·Mg·

Is it evident to you that these two atoms will combine to the satisfaction of both? It should be, because the number of valence electrons of the two atoms add up to a perfect octet. They will combine to form magnesium oxide:

Mg:Ö̤:

Unlike the two previous examples, oxygen forms not two bonds, but one. So you see how the number of bonds formed may depend on the bonding partner that the atom encounters. The results are that the two valence electrons of magnesium are tied up and oxygen has access to eight electrons. Both are satisfied.

## Where Does the Pickle Go?

Anytime you see two or more atoms of the same kind reacting with a single atom of a different kind, the single atom will nearly always insert itself in the middle of the foreign atoms—*the pickle goes in the middle*! That's mainly because any diatomic molecule such as $H_2$, $O_2$ or $N_2$ is already content to exist as it is. Any intruder must improve on that situation by pushing its way between them. You can't get:

Na-Na-S

because it would require the sodium atom in the middle to form two bonds. It can't. Bonds are formed from shared electrons, and that atom has only one electron to share. The unlikely structure shown above has that central sodium atom forming too many bonds to satisfy its valence shell and too few to satisfy the needs of sulfur. Neither will be satisfied. On the other hand:

Na-S-Na

makes plenty of sense. Each of the sodium atoms donates a single electron to sulfur which has a need for two electrons.

Based on this information, you should be able to tell me very quickly what the Lewis structure for $K_2S$ will look like. It won't be K-K-S, but K-S-K. The dot diagram will be:

K:S̤̈:K

Like oxygen, sulfur bonds with two group IA atoms or with one group IIA atom. The same is true of the other group VIA elements.

Will you ever see oxygen forming three bonds? No. It has only two unpaired electrons to share. You will no sooner see oxygen forming three bonds than you will see hydrogen or sodium forming two. You can't make three bonds with only two bonding electrons.

## It's All About Bonding

I want you to learn how to draw the Lewis diagrams of simple compounds for yourself to prove that you understand bonding before we start on chemical reactions. Suppose I were to ask you why chlorine forms a diatomic molecule. In order to explain it to me you would first draw the Lewis structures of a pair of chlorine atoms:

:C̈l· ·C̈l:

Then you would show that those two atoms fit together beautifully so that each gets a noble gas configuration:

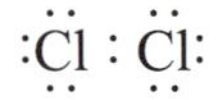

:C̈l : C̈l:

## Multiple Bonds

Next, consider that some compounds form stronger, double bonds by sharing two pairs of electrons. For example, carbon disulfide is composed of one carbon atom (four valence electrons) and two sulfur atoms (six valence electrons each). We put the pickle in the middle and we get:

:S̈:C:S̈:

Can you tell what is wrong with this structure? We have used up all 16 electrons and yet carbon is unsatisfied. It has access to only four electrons. If we give it four of the electrons belonging to sulfur, it will be happy, but sulfur will not. I have an idea. Watch:

S̈::C::S̈

This structure indicates that two pairs of electrons between sulfur and carbon are being shared. Now each sulfur atom has access to eight electrons and each carbon atom has access to eight. Each of those is a bonding pair, so we can write them in bond form:

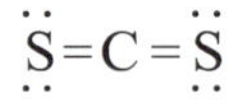

S̈=C=S̈

While a single bond is formed from the sharing of one pair of electrons, a double bond is formed from the sharing of two pairs. Can two atoms share three pairs of electrons? Yes. That's a triple bond. Can they share four? No. That's a dream. We're trying to cram too many electrons into too little space. Their repulsion is strong enough to prevent four bonds (eight electrons) from holding two atoms together. The bond will be unstable. Three bonds—okay; four bonds—no go.

Triple bonds are most likely to be found among C and N atoms. An example of a triple-bonded compound is acetylene, $C_2H_2$. I'm going to lead you through the steps you will go through to draw the Lewis structure of any covalent compound. As you go through these steps you will come to understand why a triple bond is needed to stabilize the acetylene molecule. More importantly you will learn a technique that will aid you in drawing the electron structures of most molecules.

**First, draw the atoms in a logical arrangement connected with single bonds.** You can't put a hydrogen atom between two other atoms because it forms only one bond. There is only one possible arrangement of $C_2H_2$:

H–C–C–H

**Second, add up all of the valence electrons available in the compound.** Each hydrogen atom has one; each carbon atom has four. So:

$$2\,(1) + 2\,(4) = 10$$

Each single bond you have already drawn represents two electrons. We must account for those, so, **third, take away two electrons for each single bond shown in your structure.**

$$10 - 3\,(2) = 4$$

This is the number of electrons we still have to dispose of.

Looking back at our structure, we see that hydrogen is satisfied, but carbon is not. **Fourth, distribute those electrons among the nonmetals that do not have a complete octet. If there are not enough electrons to give every nonmetal atom access to eight electrons, use double or triple bonds.** In our case, to distribute these four remaining electrons evenly would be to give two each to the carbon atoms:

$$H-\ddot{C}-\ddot{C}-H$$

But that isn't enough to make carbon obey the octet rule. Each of the carbon atoms has access to only six electrons as drawn. The only way to distribute our four electrons to provide eight for each carbon atom is to use those two unshared pairs of electrons to make two additional bonds:

$$H-C\equiv C-H$$

Now every atom has access to a full outer shell. Each of the hydrogen atoms has two electrons in its valence shell, while each of the carbon atoms has eight. Perfecto!

Before turning you loose on your own, I'm going to lead you through one more example. What you will learn from this example is that sometimes you have to play around a little bit to figure out the structure that works. Let's draw the Lewis structure of formaldehyde, given the molecular formula, $H_2OC$. There are two ways that this structure could be arranged before we begin distributing electrons. The next two structures you will see are actually the same structure:

$$H-C-O-H$$

and its equal,

$$H-O-C-H$$

(Do you understand why these two structures represent the exact same molecule? Molecules are not glued to paper; they can spin around in space. If you spun one of these molecules around on a vertical axis, it would be identical to the other.) The other way the structure could be arranged is:

$$\begin{array}{l} H \\ | \\ C-O \\ | \\ H \end{array}$$

Try as you might, you won't find any other possibilities. A nonmetal atom will usually form only as many bonds as it has unpaired valence electrons. Hydrogen can form one, carbon can form four and oxygen can form only two. (It has six electrons but four of them are paired.)

Let's start with the first of these skeletons. We have 2(1) + 6 + 4 = 12 valence electrons (Look at that first skeleton and be sure you understand how I got this.) and we have used six on single bonds. There are only 12 – 6 = 6 left. We now must begin by putting them where we know they will be needed. We might as well go ahead and put four of them on oxygen because oxygen only forms two bonds. It needs four more electrons to have an octet:

$$H-C-\ddot{\underset{\cdot\cdot}{O}}-H$$

Then we have only two electrons left over. Carbon is lacking electrons, so we'll put them there:

$$H-\ddot{C}-\ddot{\underset{\cdot\cdot}{O}}-H$$

Not only does carbon still lack two electrons, but it has also formed only two bonds. It needs two more bonds to satisfy the octet rule. We can't make any multiple bonds because oxygen and hydrogen already have as many as they can form. We're stuck.

Let's try the other structure:

$$\begin{array}{l} H \\ | \\ C-O \\ | \\ H \end{array}$$

The total number of valence electrons is the same (12). We have used six of them again, leaving six to distribute. However, this time we must give oxygen and carbon a double bond in order for carbon to get its four bonds and for oxygen to get its two:

```
H
|
C=O
|
H
```

Now we have only four electrons to distribute. Hydrogen is satisfied; carbon is satisfied. All four electrons go to oxygen to complete its valence shell:

```
H   ..
|
C=O
|   ..
H
```

Everybody is happy.

If you like puzzles, these can be a lot of fun. They're better than crosswords. What's more, if you sit drawing Lewis structures for a pastime, people will think you're totally off your rocker! Go ahead and try some. The best way to learn to do this is practice. Practice, practice, practice. Follow the steps in the procedure I gave you.

1) Write the symbols for the atoms, putting any odd atom in the middle. Draw as many different configurations as you can think of remembering that atoms can form only as many bonds as they have unpaired valence electrons:

- H atoms and the IA metals form only one bond because they have only one valence electron.
- The IIA metals form a maximum of two bonds because they have only two valence electrons.
- The IIIA metals form a maximum of three bonds because they have only three valence electrons. (These elements are bonding oddballs, so there are exceptions.)
- Carbon and other IVA atoms form four bonds (one for each valence electron).
- N, P and other VA atoms form three bonds and keep one unshared pair of electrons (5 total).
- Oxygen, sulfur and the other VIA elements form up to two bonds and keep two unshared pairs (6 total). (They form up to four bonds w/ each other.)
- The halogens usually form one bond and keep three unshared pairs (7 total).

2) Connect the symbols with single bonds.
3) Sum the valence electrons in all atoms and subtract two electrons for each single bond you drew.
4) Distribute the remaining electrons to give each atom a full valence shell.
5) If there are too few electrons to go around, create double bonds between atoms to make up for them. Remember to deduct two electrons for each additional bond you draw. (If there are too many electrons, you messed up somewhere.)
6) Check your structure atom by atom to be sure everybody's happy.

There is a tremendous amount of chemistry knowledge packed into a Lewis structure. When you draw such a structure, you should realize what an accomplishment it is. It means you know the elements, their valence shell configurations and their bonding habits. Then you can look at the structure you drew and have a sense of where its charges are concentrated and develop a sense of its overall shape. As a matter of fact, we'll talk about molecular shapes in our next lesson.

**Exercises**

Draw Lewis structures for the following compounds:

1) potassium iodide
2) hydrogen fluoride
3) methane ($CH_4$)
4) water
5) $N_2$
6) $CO_2$
7) HCN (hydrogen cyanide gas)
8) $C_2H_4$ (ethylene gas)
9) $CNH_5$

# 19: Molecules and Shape

## Molecules Have Shapes

The shape of a molecule depends (1) on the structure of its atomic skeleton and (2) on the charge influences of its electrons. This may not make you flip your wig, but it can be important to you if you happen to be a molecule. Your shape will ultimately decide what other structures you can interact with and how those interactions will go.

Having the Lewis structure in hand, we can have a clearer understanding of both factors: skeletal shape and electron influence. Take, for example, the structure of hydrogen cyanide gas (HCN). The Lewis structure is as follows:

$$H - C \equiv N:$$

The skeleton of this molecule is laid out in a straight line. Each of the atoms located along that line shares electrons with others located along the line, so, as you would imagine, the electron paths from atom to atom are laid out along that line as well.

How the atomic skeleton of a molecule affects its shape is pretty obvious. But how do the electrons affect the shape of a molecule? Of course, the answer has to do with the negative charge. You've known for a long time that two negative charges will repel one another. For this reason, two electron clouds will also repel one another. Electron clouds will push each other as far away as possible.

There is a single unshared pair of electrons at the nitrogen tip of the HCN molecule. The charges of the other electrons along the axis will force that electron pair outward. We call this charge influence among electrons "**valence shell electron pair repulsion**," or VSEPR, for short. (We do this only to frighten students away. Otherwise we would call it "**outer electrons pushing each other away**.")

The space occupied by the bonding electrons might look something like this:

$$H - C \equiv N:$$

(More detail on the specific shapes of the clouds of bonding electrons is available that is not presented here. If you are interested in knowing more, consult a college-level chemistry text on the subject of "molecular orbitals.") Note that the shape of this structure can best be drawn as falling on a straight line. We say that the molecule is **linear**. All of the **bond angles** in this molecule are equal to 180°.

180° 180°

$$H - C \equiv N:$$

That unshared pair of electrons on nitrogen will occupy the tip of that same axis, because that's the place where they will be farthest from the other electron clouds.

## Bent and Tetrahedral Molecules

Not all molecules have such a nice, linear distribution of electron charges. For example, take notice again of the Lewis structure of water:

$$H - \ddot{\underset{\cdot\cdot}{O}} - H$$

While this might be the easiest way to draw the Lewis diagram, it's not really the most accurate. Just looking at the skeleton you might think it would be linear. But pay close attention to those two unshared pairs of electrons on the oxygen atom. Because electron pairs repel each other, this molecule is not linear as it might appear. To draw the electron clouds of the hydrogen atoms and the unshared pairs of electrons as far from one another as possible is to draw them, not in a straight line, but staggered about the molecule in three dimensions as you'll see in Figure 1.

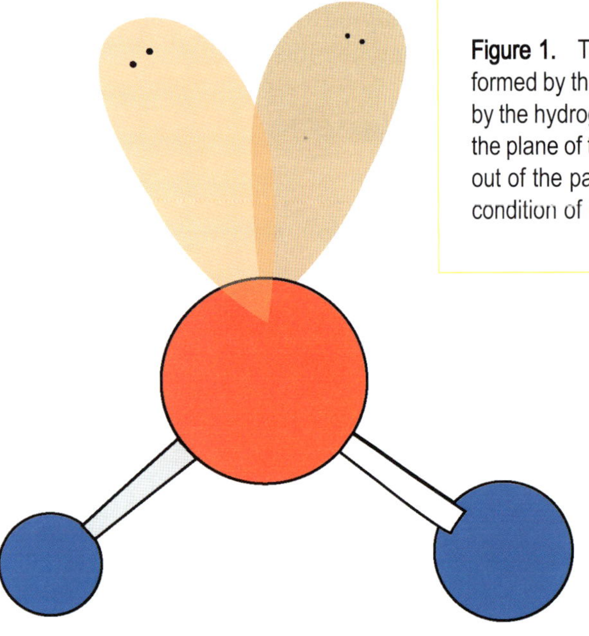

Figure 1. The water molecule is not straight and it's not flat. The electron clouds formed by those extra pairs of electrons push away from the electron clouds formed by the hydrogen electrons. This makes them rotate ninety degrees in relationship to the plane of the molecule. In this figure, one of the electron clouds would be coming out of the page at you and the other would be pointing away from you. This is the condition of least stress the molecule can take on.

A more accurate picture of the electron clouds would not be a two-dimensional, flat picture, but a three-dimensional one:

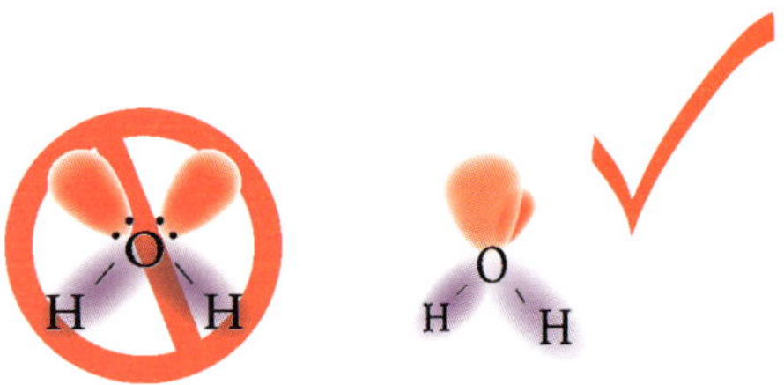


And the Lewis diagram, instead of being drawn with an unshared pair of electrons on each side of the molecule, should more accurately be drawn with both of these unshared pairs on the same side:

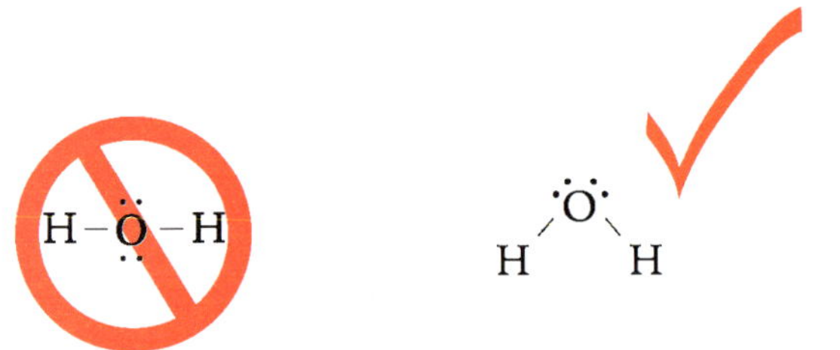


So what can we conclude about the shape or "molecular geometry" of water? There are two different ways we could consider it. One way is to consider only the shape formed by the atoms themselves. According to this view, water is a **bent** molecule.

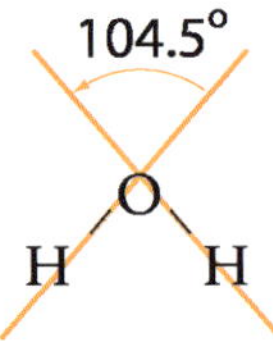


The molecule is not linear. It is bent at an angle of 104.5°. (This angle can be calculated using methods that find the minimum potential energy of electron charge stresses on the molecule.)

The second way of looking at the molecular geometry of water takes into account not just the atoms, but also the unshared pairs of electrons. The axes of the electron clouds of this molecule form the axes of a **tetrahedron** as shown in Figure 2. A tetrahedron is a three-dimensional shape with four faces. The axes of the bonds and of the electron clouds form the **axes** of the tetrahedron. The oxygen atom is in the interior of the structure.

So, in summary, we may think of water as a bent molecule when considering only the arrangement of its atoms, or we may think of it as a tetrahedron when we consider its electron clouds. In either case, water is a bipolar molecule having a strong electronegative charge on the side of those unshared pairs of electrons.

Although the atoms of water do not, themselves, form a tetrahedron (there are only three of them), many molecules have tetrahedral arrangements of atoms. Consider methane and carbon tetrachloride in Figures 3 and 4, respectively. The atoms themselves form the "corners" or **vertices** (singular, *vertex*) of the tetrahedron, while carbon is at its **centroid**.

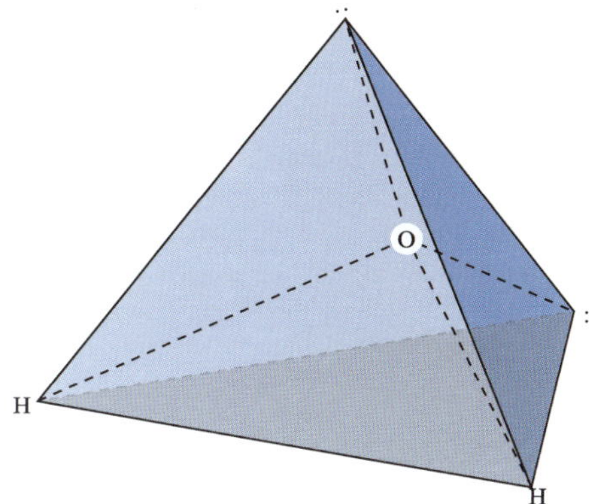


**Figure 2.** The dashed lines are the axes of the water molecule. Two of the axes are formed by the bonds between oxygen and hydrogen and the other two are formed by the electron clouds of the unshared pairs of electrons (represented by two dots at the corners). In the case of water, the tetrahedron is not perfect, because the bond angle of the hydrogen atoms is slightly smaller than the angle of the unshared pairs. That means oxygen is not right smack dab in the middle (at the "centroid") of the tetrahedron.

**Figure 3.** Unlike water, the methane molecule is a *regular* tetrahedron. Rather than unshared pairs of electrons at its vertices, it has four hydrogen atoms. Carbon, being equally influenced by the four hydrogen atoms, is precisely at its centroid.

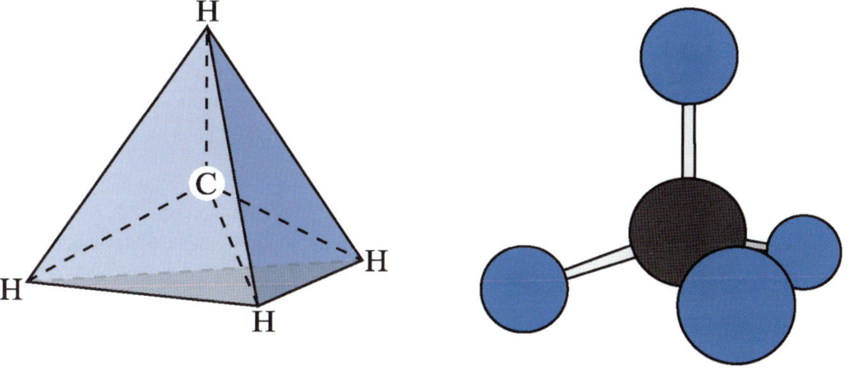


**Figure 4.** Notice any similarities between this carbon tetrachloride molecule and the previous methane molecule? Except for the fact that all four hydrogen atoms are replaced by chlorine atoms, they are identical in their geometries.

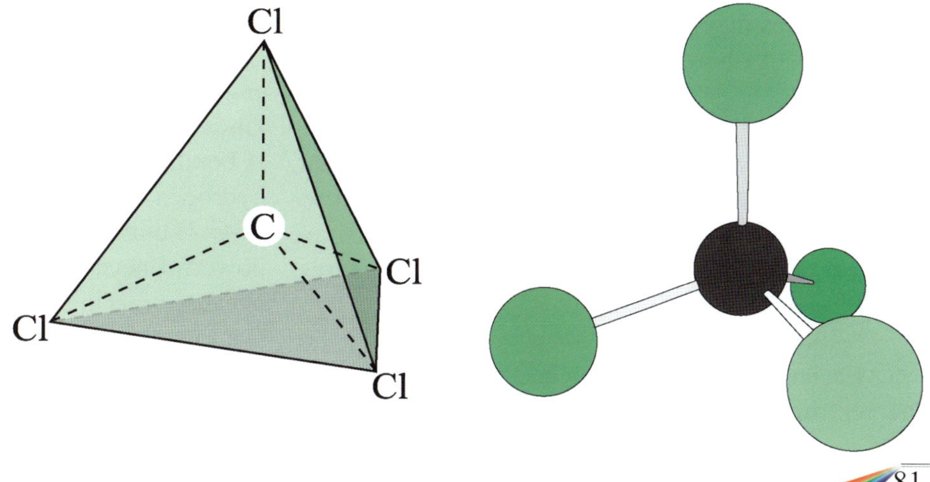

## Pyramidal Molecules

We have discussed linear, bent and tetrahedral shapes. There are several other less common shapes among molecules. One example is found in the shape of ammonia. The Lewis structure of this molecule is:

$$\begin{array}{c} H-\ddot{N}-H \\ | \\ H \end{array}$$

Once again, that single pair of unshared electrons will influence all the other clouds around the hydrogen atoms and will repel them. In order to put these electron clouds as far apart as possible, a tetrahedral arrangement will again be seen. (Figure 5.)

Similar to water, that unshared pair of electrons changes the overall geometry of the molecule. If we take away the unshared pair of electrons and consider the geometry of the atoms alone, we get a different situation which can be viewed in Figure 6.

Although this shape has come to be commonly called **pyramidal** (having the shape of a pyramid), you'll notice that it's still a tetrahedron. The nitrogen atom now occupies the apex (tip) of the tetrahedron and there is no atom at its centroid.

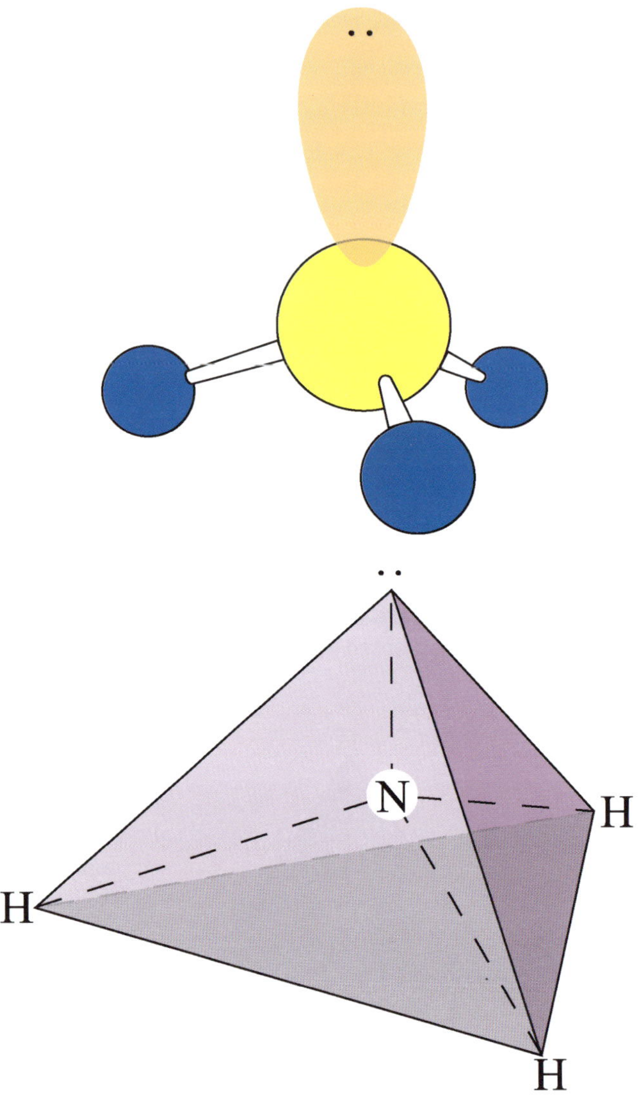


**Figure 5.** If you consider the electron cloud of the unshared pair of electrons, ammonia appears tetrahedral.

## Planar Molecules

There are also planar molecules—molecules whose atoms all lie on a single plane. Several have been drawn for your viewing pleasure at Figure 7. The bent structure of water already discussed is one planar molecule. Planar molecules are named according to how many angles meet the eye as you look at them. So planar molecules may be (1) biangular (bent) like water, (2) trigonal like boron trifluoride or (3) tetragonal like ethylene.

But aside from the number of angles that appear, the bond angles of molecules having these geometries tend to be rather specific. Since all of the bonds are identical in $BF_3$, they are simply 360° divided by three, or 120° each (Figure 7).

Each carbon atom in ethylene acts as if it were at the centroid of a trigonal planar molecule. The bond angles are very nearly 120°. Notice that there are two kinds of bond angles in ethylene. The first is an H–C–H bond angle (the angle formed by the connection of hydrogen to carbon to hydrogen), and the other is a C=C–H bond angle. The C=C bond has a slightly more powerful negative charge than the H–H single bonds because of its higher concentration of shared electrons. For this reason, one of these two types of bond angles is smaller than the other by a couple of degrees. Can you decide which will be smaller?

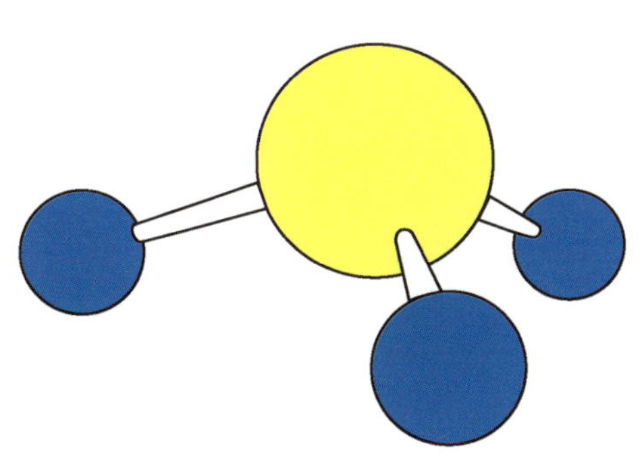

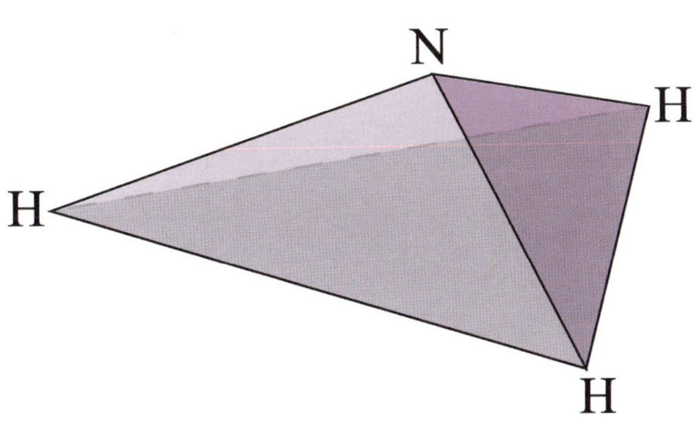


Figure 6. If you pay no attention to the unshared pair of electrons, it looks like a shorter tetrahedron. In this case, however, nitrogen is at the apex rather than the centroid of the tetrahedron. This kind of arrangement has come to be called "pyramidal."

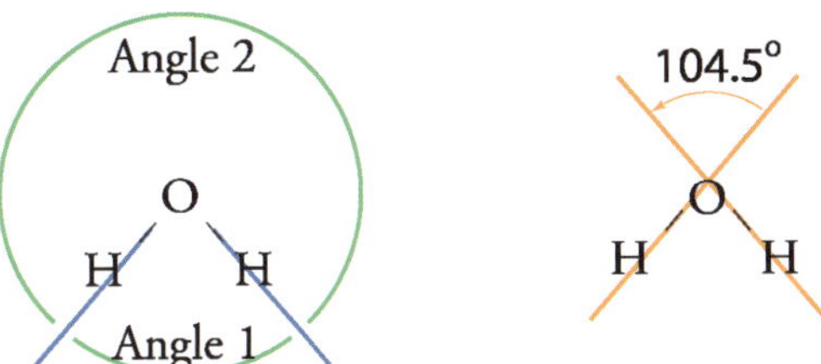


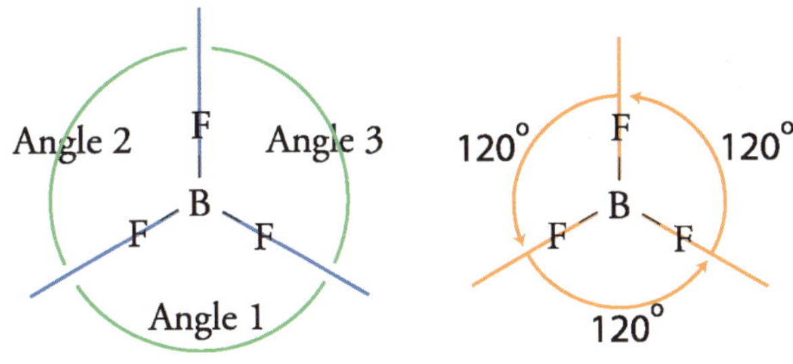


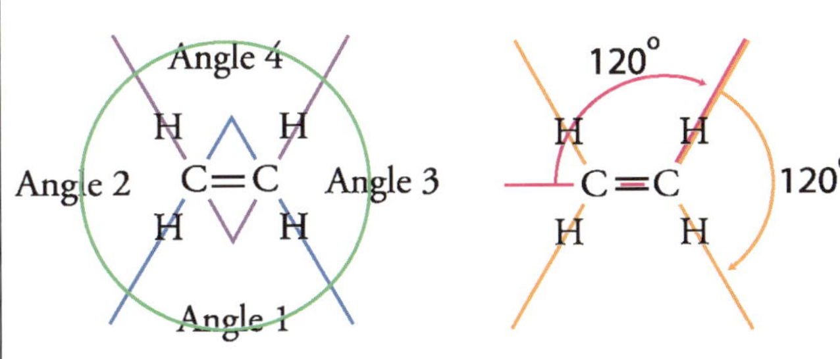


Figure 7. Three different planar molecules are shown. On the left the angles are identified, and on the right, the measurements of those angles are shown. Water (top) is bent, or bigonal; boron trifluoride (center) is trigonal; ethylene (bottom) is tetragonal.

**Exercises**

Please complete the table by drawing the Lewis structure and/or naming the molecular geometry. If we have done them for you in the text, see that you can redo them on your own. If you can't, some review may be necessary.

| Name | Formula | Lewis Structure | Molecular Geometry |
|---|---|---|---|
| hydrogen cyanide | HCN | H – C ≡ N: | linear |
| carbon dioxide | $CO_2$ | | |
| acetylene | $C_2H_2$ | | linear |
| methane | $CH_4$ | | tetrahedral |
| carbon tetrachloride | $CCl_4$ | | tetrahedral |
| ammonia | $NH_3$ | | pyramidal |
| oxygen dichloride | $OCl_2$ | | |
| water | $H_2O$ | H–Ö–H (bent, two lone pairs on O) | biangular planar (bent) |
| boron trifluoride | $BF_3$ | F–B(–F)–F | trigonal planar |
| aluminum chloride | $AlCl_3$ | | |
| ethylene | $C_2H_4$ | | tetragonal planar |

# 20: Quiz 3

1) Understand the particle composition of elements, ions and isotopes. Be able to write and interpret isotope symbols given the use of a periodic table.
2) Be able to calculate the average atomic mass of an element given the masses and relative abundances of naturally occurring isotopes.
3) Understand the relationship between ionization energy and electronegativity.
4) Be able to determine whether a bond is ionic, covalent or polar covalent when given the electronegativities of the bonded elements.
5) Be able to draw Lewis structures for simple covalent compounds. (If you can do the ones in the exercises, you'll be able to do the ones on the test as well.)
6) Know the six common geometries of small molecules (right column, page 84), and be able to draw a stick figure representing each.

Sometimes it's good to be different. The best scientists look at things a little differently.

## 21: What Do You Call This Stuff?

Before I start you out naming chemicals, I'd like you to appreciate what it is you are going to be able to do. Once you learn how to name compounds, no chemist bully can any longer intimidate you with a name like "calcium nitrate" or "ammonium sulfate hydrate."

Perhaps some stranger will approach you and say, "Hey young lady, I have a special deal for you on ammonium nitrate, urea, ammonium polyphosphate and potassium sulfate."

You'll just look him straight in the eye and say, "No thanks, I have all the fertilizer I need!"

Way back in Lesson 2 we introduced you to the building blocks of matter which we call the elements. We also told you that, since elements combine with each other to form a wide variety of chemical compounds, it is convenient to have a shorthand method for referring to them. The symbols for each element are given in the periodic table. Furthermore, through the first nine chapters we have seen the chemical formulas for many different compounds. But how do we name those compounds? Oh, we know that many of them have "common" names, like water, salt, ammonia, etc. But there are literally millions of compounds, and we couldn't possibly hope to memorize common names for them all. Therefore, standard **nomenclatures**—systems of names—have been devised. In this lesson you will begin to learn some of the rules for naming **binary compounds**, that is, compounds that consist of only two elements.

In Lesson 10 we noted that elements can, either through the loss or gain of electrons, form ions. That information will serve you well as you try to name binary compounds. The simplest situation in terms of naming binary compounds is when one of the elements is a metal that can form only one kind of cation.

In earlier studies of chemistry or of physical science, you learned that a metal is described as a substance having a series of characteristics, namely:

1) conduct heat and electricity
2) having surfaces that, when polished, reflect light
3) ductile and malleable

The metallic elements occupy the entire left side of the periodic table (See Figure 1), including group IA, IIA and all the B groups. A few of the post-transition elements behave like metals as well. You also learned that the periodic table is organized into rows called *periods* (from which the name *periodic table* derives) and columns called *groups*. Groups IA and IIA are the *alkali metals* and *alkaline earth metals*, respectively. The B groups are called the *transition metals*.

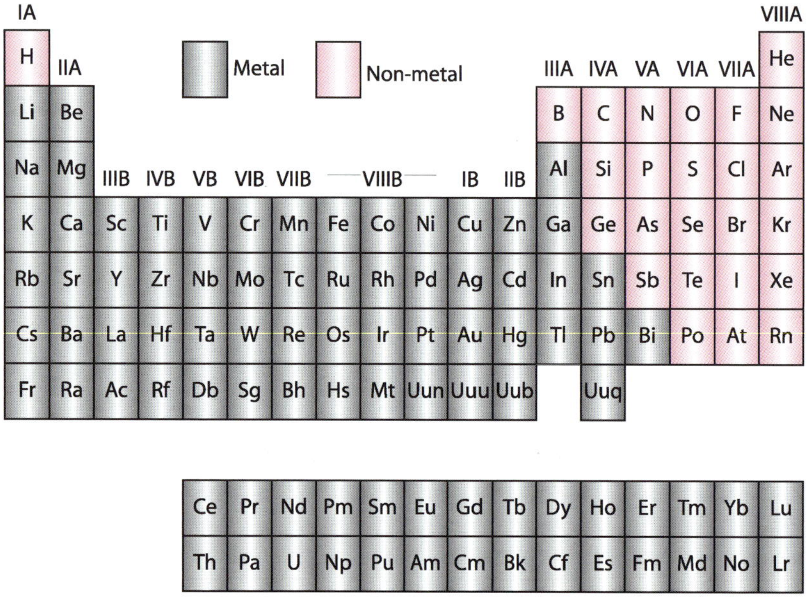

Through studying the detailed electron structures of atoms, you have now learned that groups of elements behave similarly because they have similar electron configurations in their valence shells. There are many

Metal | Non-metal

| IA | IIA | IIIB | IVB | VB | VIB | VIIB | VIIIB | VIIIB | VIIIB | IB | IIB | IIIA | IVA | VA | VIA | VIIA | VIIIA |
|---|---|---|---|---|---|---|---|---|---|---|---|---|---|---|---|---|---|
| H | | | | | | | | | | | | | | | | | He |
| Li | Be | | | | | | | | | | | B | C | N | O | F | Ne |
| Na | Mg | | | | | | | | | | | Al | Si | P | S | Cl | Ar |
| K | Ca | Sc | Ti | V | Cr | Mn | Fe | Co | Ni | Cu | Zn | Ga | Ge | As | Se | Br | Kr |
| Rb | Sr | Y | Zr | Nb | Mo | Tc | Ru | Rh | Pd | Ag | Cd | In | Sn | Sb | Te | I | Xe |
| Cs | Ba | La | Hf | Ta | W | Re | Os | Ir | Pt | Au | Hg | Tl | Pb | Bi | Po | At | Rn |
| Fr | Ra | Ac | Rf | Db | Sg | Bh | Hs | Mt | Uun | Uuu | Uub | | Uuq | | | | |

| Ce | Pr | Nd | Pm | Sm | Eu | Gd | Tb | Dy | Ho | Er | Tm | Yb | Lu |
|---|---|---|---|---|---|---|---|---|---|---|---|---|---|
| Th | Pa | U | Np | Pu | Am | Cm | Bk | Cf | Es | Fm | Md | No | Lr |

**Figure 1.** Everything on the left side of the periodic table (that is, with the exception of hydrogen) and a few of the post-transition elements are metals.

ways in which these grouped elements act alike, but for right now we want to focus on only one of those. All the elements in group IA, the alkali metals, behave similarly in terms of the *kinds of ions they form*. It is easy for each of these elements to lose an electron and form a 1+ cation, because they all have a single electron in a half-empty *s* orbital. When any of these elements forms an ion, it will be a 1+ cation. Hydrogen is not a metal, but it can behave in the same way as every metal in this group. This should be pretty easy to remember—group IA elements form 1+ cations.

Now look at group IIA. The elements in IIA (the alkaline earth metals) all two electrons in their valence shells. (Their *s* sublevels are full.) When they form ions, both of those *s* electrons go away at the same time leaving them with 2+ charges. They always form 2+ cations.

The next group to look at is group IIIA. Yes, I know it's not the next column of elements on the periodic table. The next 10 groups, going left to right, after group IIA are the transition elements, and are given "B" designations. Don't worry about them right now. group IIIA is the first group after the transition elements—you might call them post-transition elements. Again, boron (like hydrogen) is actually a nonmetal, but it behaves the same way as the metals in the group. What kind of ion do you suppose group IIIA elements form? Right, 3+ cations. To summarize then, each of the groups IA to IIIA form the cations that correspond to their group numbers: group IA elements form 1+ ions, group IIA elements form 2+ ions, and group IIIA elements form 3+ ions.

Now move on over to the far right of the periodic table to the noble gases. Remember? Their *s* and *p* sublevels are full. They have eight electrons in their valence shells. The fact is, they just don't tend to form ions at all. They don't even form compounds with other elements very easily. So don't worry about them right now.

Moving one group to the left of the noble gases, we find group VIIA—the halogens. Notice that these are all nonmetals. These elements are alike in the way they form ions. It is easy for each of these elements to gain an electron and form a 1– anion, thereby filling their valence shells with eight electrons.

Now move another group to the left to group VIA. These elements all form 2– anions. And you can probably guess what kind of ions are formed by the elements in group VA. Right, 3– anions. Please remember these tendencies: Groups IA, IIA, and IIIA form 1+, 2+, and 3+ ions, respectively. Groups VIIA, VIA and VA form 1–, 2– and 3– ions, respectively. You'll need this information to name compounds properly.

### Type I Binary Compounds

The first question to ask when you are given a binary compound to name is: "Does this compound contain a metal?" (See Figure 1.) If the answer is "No", then you'll have to wait a few days to learn how to name it. But if the answer is "Yes", then the next question you want to ask is this: "Does the metal in this compound form only one kind of cation?" If the answer is "No", then you'll have to wait a few days to name it. But if the answer is "Yes", then you have what is called a "Type I" binary compound, and by the end of today's lesson you'll know exactly what to call it. Then you can communicate the name of that compound to any other person in the world who knows the language of chemistry.

Of course, you may want to know how to tell if the metal in your compound can form more than one kind of cation. Here's how you tell: does it belong to group IA, IIA or IIIA? If so, then it can form only one kind of cation. The other common metals that can form only one kind of cation are Ni (2+), Zn (2+), Ag (1+) and Cd (2+). (See Figure 2.) You'll just have to remember these. There are some other, less common ones that show up on the Figure, but chemists run across these so infrequently that you needn't bother memorizing them.

Once you have established that the binary compound you are trying to name contains a metal that can form only one kind of cation, then you already have half the name. The name will always begin simply with the name of the metal. The second part of the name will be related to the nonmetal with which the metal

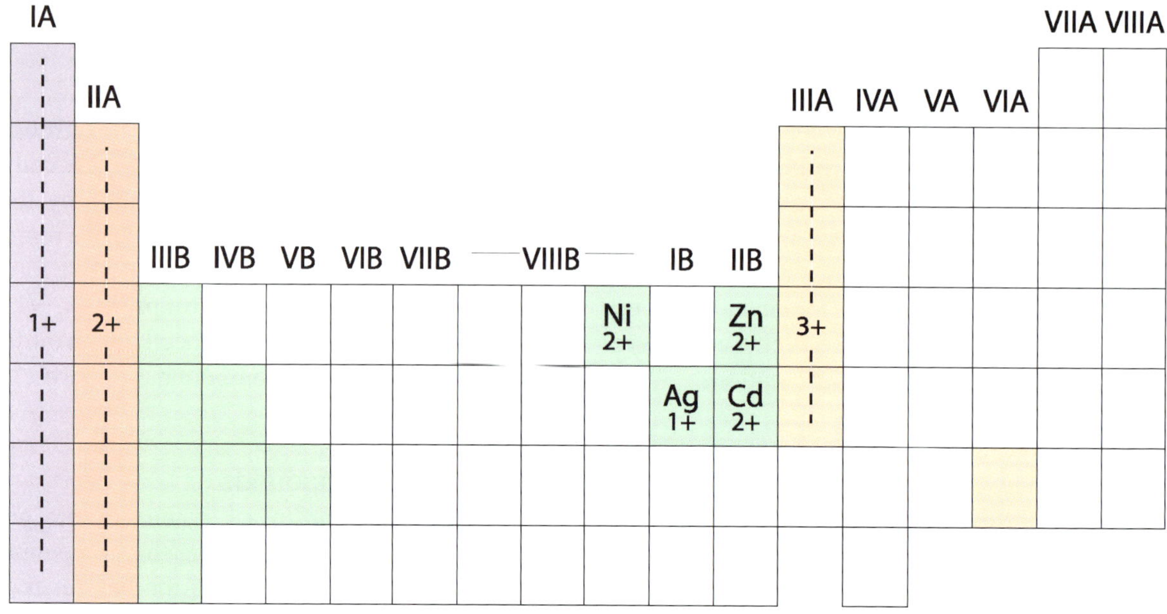


**Figure 2.** These are the elements that form only one kind of cation. However, if you just remember Groups IA (1+), IIA (2+), IIIA (3+), Ni, Zn and Cd (2+) and Ag (1+), you'll probably never run across a need for any of the others.

reacts. Whatever that element is, you take the stem of that element's name and add the ending "ide." A listing of the stems of some common nonmetals is given in the Table.

As I said once before, when hydrogen forms a cation, its charge is always 1+. However, it can also be the nonmetal component of a binary compound—it can also form an anion with a 1– charge. That's why it sometimes appears in two places on a periodic table. It appears in group IA (with other elements that form 1+ ions) and in group VIIA (with other compounds that form 1– ions).

Now you have all the information you need to name a *lot* of compounds. Let's work our way through the naming of some. Just remember that we are covering only Type I compounds so far, so each will contain one of the metals from Figure 2 that forms only one type of cation and one of the nonmetals (Figure 1) that forms an anion.

Your first mission is to name the compound that has the formula $CaCl_2$. First, let's confirm that it is the correct type of compound:

**Table.** Stems that represent common nonmetals in the naming of compounds.

| Element Name | Element Symbol | Stem + "ide" |
|---|---|---|
| hydrogen | H | hydride |
| carbon | C | carbide |
| nitrogen | N | nitride |
| oxygen | O | oxide |
| fluorine | F | fluoride |
| phosphorus | P | phosphide |
| sulfur | S | sulfide |
| chlorine | Cl | chloride |
| selenium | Se | selenide |
| bromine | Br | bromide |
| iodine | I | iodide |

1) Is it binary? That is, does it consist of only two elements? Yes, it consists of only calcium and chlorine.
2) Does the compound contain a metal? Yes, according to Figure 1 its metal is calcium.
3) Does the metal form only one kind of cation? Yes, it's a group IIA metal. Those metals form only 2+ ions.

We can confidently conclude that we are naming a Type I binary compound.

Second, let's name it:

1) The first part of the name is *calcium*.
2) The second part of the name is *chloride*.

Third, put it all together.

The name of the compound is *calcium chloride*.

Instead of going through this process in your mind every time, you should learn to name compounds quickly at a glance. Let's name each of the following Type I binary compounds. (Cover up the names I provided and see if you can name them on your own. Check them first to be sure they are all Type I binary compounds. We wouldn't want to be fooled by one, now would we?):

1) $LiH$ lithium hydride
2) $SrBr_2$ strontium bromide
3) $Na_2O$ sodium oxide
4) $AlI_3$ aluminum iodide
5) $CuCl_2$ copper chloride
6) $Mg_3N_2$ magnesium nitride

Notice that if you don't know which elements are metals and which carry only one kind of charge, you'll never quite know whether you have named the compound properly. Take compound number 5 for instance. If you didn't check carefully, you overlooked the fact that copper is in one of the B groups. It forms more than one type of cation. This compound is improperly named. I hope you caught it! If you didn't, pinch yourself. If you did, give yourself a hug. I'm sorry I can't be there to see it. (As you'll find out in the next lesson, if the metal forms more than one type of cation, you'll have to specify how many atoms are present. At that point you'll learn how to properly name copper chloride.)

## Writing the Formulas of Type I Compounds

You may have noticed that the elements making up binary compounds don't always combine in one-to-one ratios. For example, the formula for magnesium nitride is $Mg_3N_2$. Why is that? If you think about it, it's exactly what you would expect. Remember, magnesium is a group IIA element and therefore forms 2+ ions. Nitrogen is a group VA element and forms 3– ions. So in order for magnesium and nitrogen to combine and produce a neutral compound, there must be 3 magnesium cations (with a total charge of +6) and 2 nitrogen anions (with a total charge of –6). Once we understand this principle, if we are given the name of a Type I binary compound, we can figure out what the formula *must* be.

Let's try a few. What would be the formula for calcium sulfide? Write the symbol for the metal and lightly pencil its charge above it, like this:

$$\overset{2+}{\mathrm{Ca}}$$

Remember, for a Type 1 compound there must be only one possibility. Now we need enough sulfur atoms to neutralize a 2+ charge. Sulfur is in group VIA and forms a 2– anion. Write it:

$$\overset{2+}{\mathrm{Ca}}\ \overset{2-}{\mathrm{S}}$$

The charges are balanced, so we are finished. The formula will simply be CaS.

That was much too easy. Let's do one that requires you to be awake. What would be the formula for calcium fluoride? Write the metal and its charge, then the nonmetal and its charge. Calcium is in group IIA (2+), while fluorine is in VIIA (1–):

$$\overset{2+}{Ca}\quad\overset{1-}{F}$$

Obviously we need two fluorine atoms to balance the charge of calcium:

$$\overset{2+}{Ca}\quad\overset{2(1-)}{F_2}$$

Now the charges are balanced. The formula will be:

$$CaF_2$$

This last example is as tough as it gets. If you can do this you've got it licked. What would be the formula for calcium phosphide?

$$\overset{2+}{Ca}\quad\overset{3-}{P}$$

Since charges of 2+ and 3– do not balance, the lowest common multiple of 2 and 3 is 6. We now need to balance a charge of 6+ with a charge of 6–. Since calcium has a charge of 2+, show 3 calcium ions to give it a total charge of 6+. Likewise, since phosphorus has a charge of 3–, show 2 phosphorus ions to give it a total charge of 6–.

$$\overset{3(2+)}{Ca_3}\quad\overset{2(3-)}{P_2}$$

The charges on the three calcium ions [3(2+), or 6+] and the two phosphorus ions [2(3–), or 6–] are now balanced:

$$\overset{\overset{6+}{3(2+)}}{Ca_3}\overset{\overset{6-}{2(3-)}}{P_2}$$

The formula is $Ca_3P_2$. Now you do some.

**Exercises**

1) Name the following Type I binary compounds:

a) $Li_3N$
b) $AlCl_3$
c) $CaH_2$
d) $BaO$
e) $K_2S$

2) Write the formulas for the following Type I binary compounds:

a) zinc chloride
b) silver oxide
c) magnesium bromide
d) aluminum oxide
e) lithium iodide

## 22: Naming Type II Binary Compounds

Now that you've mastered the naming of Type I binary compounds, let's add a level of complexity. Remember the two questions to ask when confronted with a binary formula:

Does it contain a metal?

If so, can this metal form only one kind of a cation?

If the answer to both questions is "Yes," then you are dealing with a Type I binary compound, and you use the rules we discussed in the last lesson. But what if the answer to the second question is "No?" What if the metal in your binary compound can form more than one kind of cation? Then you are dealing with a Type II binary compound, and you have to do something a little different.

Take iron, for example. Iron can react with chlorine to form two different binary compounds, $FeCl_2$ and $FeCl_3$. Obviously we can't just call both of them *iron chloride*. They're different compounds, having different physical and chemical properties. Somehow we have to distinguish between them. Well, there are two ways that have been devised to do this, but we'll use only one of them in this course—the Stock system. The old, "classical" system is going out of style. In the Text Box on the next page there's more information on the classical system. You will see chemicals labeled using this system from time to time. If you're thinking about a career in science you should definitely become familiar with it.

When you look at the formula, $FeCl_2$, what can you deduce about the kind of cation that the iron must have formed in order to make this compound? Well, since we know that chlorine always forms 1– anions, and since there are two of them, the total negative charge is 2–. Therefore, in order to have an electrically neutral molecule, the iron must be 2+. Likewise, when you see the formula $FeCl_3$, you should be able to deduce that the iron must be 3+ in this case. All those deductions are right on the money—iron can form either a 2+ or 3+ cation. How then do we name the two different iron chlorides? We start the same way as with Type I compounds; we put the name of the metal first. Then we add, in parentheses, the Roman numeral that represents the **valence number** or charge number of the iron. Finally we add the word *chloride*. So:

$FeCl_2$ = iron(II) chloride
$FeCl_3$ = iron(III) chloride

Table 1 shows several common metals and some of the more common cations they form. Armed with this information, you should now be able to name binary compounds for each of the metals listed in the Table 1. Let's see if you can. We'll begin by giving a name to each of several compounds.

**Table 1.** Common metals and their most common cations.

| Metal | Atomic No. | Cations Formed |
|---|---|---|
| Titanium | 22 | 2+, 3+, 4+ |
| Vanadium | 23 | 2+, 3+, 4+, 5+ |
| Chromium | 24 | 2+, 3+, 6+ |
| Manganese | 25 | 2+, 4+ |
| Iron | 26 | 2+, 3+ |
| Cobalt | 27 | 2+, 3+ |
| Nickel | 28 | 2+, 3+ |
| Copper | 29 | 1+, 2+ |
| Arsenic | 33 | 3+, 5+ |
| Tin | 50 | 2+, 4+ |
| Antimony | 51 | 3+, 5+ |
| Mercury | 80 | 1+, 2+ |
| Lead | 82 | 2+, 4+ |
| Bismuth | 83 | 3+, 5+ |

## Names of Metals—Old Style

In the old system of nomenclature, any element that formed two different ions was given a different name for each of those ions. The ion of lower charge was given a name that ended with the *-ous* ending. The ion of higher charge was given the *-ic* ending. For example, iron is a well-loved element that is present in compounds as either iron II or iron III. Iron II (2+) is called *ferrous* iron, while iron III (3+) is called *ferric* iron.

In the names of its compounds, the element would be referred to by its classical name as well as its classical ending. That is, instead of *iron (II) chloride*, the name *ferrous chloride* would be used. Table 2 shows the names and charges of many of the common metal ions in their classical forms.

**Table 2.** Comparison of ion names according to the Stock system and the old system of nomenclature. Ions are listed by their atomic numbers and element names.

| Metal | Atomic No. | Ion Symbol | Stock System | Old System |
|---|---|---|---|---|
| Iron | 26 | $Fe^{2+}$ | iron II | ferrous |
| | | $Fe^{3+}$ | iron III | ferric |
| Copper | 29 | $Cu^{1+}$ | copper I | cuprous |
| | | $Cu^{2+}$ | copper II | cupric |
| Arsenic | 33 | $As^{3+}$ | arsenic III | arsenous |
| | | $As^{5+}$ | arsenic V | arsenic |
| Tin | 50 | $Sn^{2+}$ | tin II | stannous |
| | | $Sn^{4+}$ | tin IV | stannic |
| Mercury | 80 | $Hg^{1+}$ | mercury I | mercurous |
| | | $Hg^{2+}$ | mercury II | mercuric |

This system of naming is still used in the realm of acids. Later we will speak of sulfuric and sulfurous acids, of nitric and of nitrous acids. Although the system is applied somewhat differently to acids than to metals, you'll recognize those endings as being the same.

Let's begin with HgO. From Table 1 we see that mercury commonly forms two ions—$Hg^{1+}$ and $Hg^{2+}$. However, oxygen is in group VIA so it only forms one kind of anion—the $O^{-2}$ ion. (See Lesson 21 if you've forgotten how to decide.) So mercury must be in its 2+ form. The proper name for this compound is mercury(II) oxide. Here are several more. Cover up the answers and see if you can get them right without looking. Then uncover the answers and see how you did.

| | |
|---|---|
| $Cr_2O_3$ | chromium(III) oxide |
| $PbS_2$ | lead(IV) sulfide |
| CuS | copper(II) sulfide |
| $AsF_5$ | arsenic(V) fluoride |
| CuBr | copper(I) bromide |

Please become good at doing this. You'll need it all year.

Of course, just as with the Type I compounds, you also want to be able to write the formula if given the name. And in some ways, this might be a little easier, because in the name you are told the cation value. Suppose, for example, you are asked to write the formula for chromium(III) oxide. The name of the compound tells you that chromium should be considered as a 3+ cation. And you know that oxygen forms a 2– anion. Therefore, finding the lowest common multiple (6), we need two chromium ions (total of 6+) and three oxygen ions (total of 6–). So the formula is $Cr_2O_3$.

Here are some for you to practice on. Cover up the answers again until you see if you can write the formulas properly:

| | |
|---|---|
| nickel(II) sulfide | $NiS$ |
| lead(IV) oxide | $PbO_2$ |
| arsenic(III) chloride | $AsCl_3$ |
| manganese(IV) iodide | $MnI_4$ |
| copper(I) oxide | $Cu_2O$ |

Okay, if you're ready, go ahead and show the world you can name a Type II compound.

**Exercises**

1) Name the following Type I or Type II compounds:

a) $K_2O$
b) $HgCl_2$
c) $TiS_2$
d) $SnF_2$
e) $AsH_3$
f) $Zn_3N_2$

2) Write the formulas for the following Type I or Type II compounds:

a) manganese(II) sulfide
b) antimony(V) chloride
c) calcium hydride
d) bismuth(III) oxide
e) silver oxide
f) nickel(II) oxide

## 23: Polyatomic Ions

Now that you have mastered the art of naming Type I and Type II binary compounds, we want to add another tool to your tool box. In the last few lessons we have observed that many elements can, by the losing or gaining of electrons, form ions. However, those are not the only kinds of ions. There is a wide variety of ions that form from clusters of atoms. Some of them, in fact, are so common that you need to memorize their formulas and charges just like you memorized the symbols of the more common elements. (You have done that, haven't you?!) The Table lists the more common **polyatomic ions.** As I said, you need to memorize the names, formulas and charges of each of these ions. I suggest you make another set of flash cards to help you.

**Table.** The common polyatomic ions, with their formulas and charges (or "valences").

| Valence | Name | Formula |
|---|---|---|
| 1+ | ammonium | $NH_4^+$ |
| 1- | acetate | $C_2H_3O_2^-$ |
| | bicarbonate* | $HCO_3^-$ |
| | bisulfate* | $HSO_4^-$ |
| | chlorate | $ClO_3^-$ |
| | cyanate | $OCN^-$ |
| | cyanide | $CN^-$ |
| | hydroxide | $OH^-$ |
| | hypochlorite | $OCl^-$ |
| | iodate | $IO_3^-$ |
| | nitrate | $NO_3^-$ |
| | nitrite | $NO_2^-$ |
| | perchlorate | $ClO_4^-$ |
| | permanganate | $MnO_4^-$ |
| 2- | carbonate | $CO_3^{2-}$ |
| | chromate | $CrO_4^{2-}$ |
| | dichromate | $Cr_2O_7^{2-}$ |
| | sulfate | $SO_4^{2-}$ |
| | sulfite | $SO_3^{2-}$ |
| 3- | phosphate | $PO_4^{3-}$ |

* The bicarbonate and bisulfate ions may also be called by the names *hydrogen carbonate* and *hydrogen sulfate*, respectively.

Why are we talking about these now? Because these polyatomic ions can combine with the ions of elements (or with each other) to form compounds which are named in a similar way to the binary compounds. For the purposes of naming compounds these ions should be regarded as individual entities. So when you see a formula like $NaHSO_4$, you should recognize that this is a compound formed by the combination of a sodium cation and a bisulfate anion.

Let's learn how to name these compounds. There is only one polyatomic ion in your Table that is a cation: the ammonium ion, $NH_4^+$. Since it is a cation, we would expect it to act like the metal cations do, and combine with anions to produce electrically neutral compounds—and that is just what it does. Also, it always has a 1+ charge. When naming compounds in which the ammonium ion is combined with a single nonmetal, treat it like a Type I compound. The first part of the name will simply be "ammonium." The second part of the compound is then named by taking the stem of the element and adding "ide" as before.

Here are a couple of examples:

| | |
|---|---|
| $NH_4Cl$ | ammonium chloride |
| $(NH_4)_2S$ | ammonium sulfide |

Ammonium **halides** (halides are compounds containing halogen anions: fluoride, chloride, bromide and iodide) and ammonium sulfide are the only compounds you will encounter that fall into this category.

The ammonium ion will also combine with many of the polyatomic anions listed in our Table. In those cases, the name is simply the word "ammonium" followed by the name of the appropriate anion. For example:

| | |
|---|---|
| $(NH_4)_2SO_4$ | ammonium sulfate (Sulfate's charge is 2–) |
| $NH_4ClO_3$ | ammonium chlorate |
| $NH_4NO_3$ | ammonium nitrate |

Of course, in addition to reacting with the ammonium ion, the polyatomic anions in our Table combine with cations (often metal cations). If they combine with a metal that can form only one kind of cation, then the name of the resulting compound is simply the name of the metal followed by the name of the polyatomic anion. Here are some examples:

| | |
|---|---|
| $Zn(NO_3)_2$ | zinc nitrate |
| $KOH$ | potassium hydroxide |
| $Mg(C_2H_3O_2)$ | magnesium acetate |
| $AlPO_4$ | aluminum phosphate |

(No, this last one is not Alpo—the dog food. Let me reemphasize the importance of proper capitalization in your atomic symbols. "Mom left a note. I'm not sure, but I think she wants me to feed the dog some aluminum phosphate.")

But what if a polyatomic anion combines with a metal that can form more than one kind of cation? Then you use the Stock system, just like before:

| | |
|---|---|
| $SnBr_4$ | tin (IV) bromide |
| $Fe_2(CO_3)_3$ | iron (III) carbonate |
| $Pb(CN)_2$ | lead (II) cyanide |

Let me emphasize again how important it is that you memorize the name, formula *and* charge of each of these polyatomic ions. Knowing the charges of these ions is essential in order to figure out the formula when given the name of a compound that contains one. For example, what is the formula for aluminum sulfate? It is *not* simply $AlSO_4$. Aluminum forms a 3+ cation and sulfate is a 2– anion, so the charges wouldn't balance. The only formula that makes sense is $Al_2(SO_4)_3$. (Two 3+ cations with a total charge of 6+, and three 2– anions with a total charge of 6– ...ah, neutrality!) Here are more of them. Make sure you understand why each of these formulas is correct before going on to work on the exercises. I've explained the first two examples. You should be able to explain the others to yourself:

| | |
|---|---|
| copper (I) sulfate | $Cu_2SO_4$ |

(We need two 1+ Cu ions to offset the 2– charge of the sulfate.)

| | |
|---|---|
| titanium (IV) sulfide | $TiS_2$ |

(We need two 2– sulfur anions to offset the 4+ charge of the titanium.)

| | |
|---|---|
| potassium permanganate | $KMnO_4$ |
| ammonium carbonate | $(NH_4)_2CO_3$ |
| sodium chromate | $Na_2CrO_4$ |
| iron (II) acetate | $Fe(C_2H_3O_2)_2$ |

Ready when you are!

**Exercises**

1) Name the following compounds:

a) $Ba(NO_3)_2$
b) $CaCrO_4$
c) $Al_2O_3$
d) $Sn(NO_3)_4$
e) $Mg(MnO_4)_2$
f) $SrO$
g) $HgCl_2$

2) Write the formulas for the following compounds:

a) barium bicarbonate
b) manganese(II) hydroxide
c) cobalt(II) cyanide
d) sodium chromate
e) antimony(III) sulfate
f) sodium dichromate
g) chromium(III) chlorate

## 24: Naming Type III Binary Compounds

Well, just when you thought you knew everything you needed to know with regard to naming binary compounds, we're going to add yet another level of complexity. Again, let's remind ourselves of the two questions to ask when confronted with a binary formula:

Does it contain a metal?
If so, can this metal form only one kind of a cation?

So far, we've looked at naming compounds for which the answer to the first question is "Yes." But what if the answer to the first question is "No?" The fact is that some binary compounds don't contain a metal at all; they are made from two nonmetals. And, naturally, there is a naming system that is designed for just this case.

When you have a binary compound containing only nonmetals (called a **Type III binary** compound), the name of the compound will start with the element that occurs first in this series:

Si, B, P, H, C, S, I, Br, N, Cl, O, F

The reason? This is the order of increasing *electronegativity*. This is the order in which all elements are named in compounds. Nonmetals are more electronegative than metals so we are accustomed to putting the metal first. Now, when there is no metal the less electronegative nonmetal goes first.

So then, if you have a compound containing carbon and fluorine, the carbon will be named first. A compound containing phosphorus and nitrogen will be named by putting the phosphorus first. (Don't try to say "phosphorus first" five times in a row or you'll get your tang all tonguled up.) In a compound containing silicon and phosphorus, the silicon will be named first. The second element, as in previous lessons, is given an "ide" ending.

The problem with this kind of compound is that it can be formed from so many different ratios of the elements. For example, nitrogen and oxygen can combine as $N_2O$, NO, $NO_2$ or $N_2O_5$. Therefore, we have to have a way to know how many atoms of each element is contained within a specific compound. This is done by putting a Greek prefix in front of each element's name that corresponds to the number of atoms of that element in the compound. You probably already know these prefixes which are listed in the Table.

For example, what is the standardized name for $H_2O$? Since H comes before O in our series, the hydrogen is named first. Since there are two hydrogen atoms, then the name will begin with "dihydrogen." Since there is only one oxygen atom, then it will have the prefix "mono." Since oxygen starts with the letter "o" we change "mono" to "mon." Therefore, the standardized name for water is **dihydrogen monoxide**. (We might drop the vowel from the end of the prefix if it would cause two vowels to be pronounced together.)

For $H_2O_2$, the hydrogen will be named first. The number of hydrogen atoms is the same as in the previous compound, so the name starts with "dihydrogen" again. But there are two oxygen atoms in this compound, so the correct name is **dihydrogen dioxide**.

**Table.** Prefixes used to designate the number of atoms of a particular element appearing in a Type III compound.

| No. Atoms | Prefix |
|---|---|
| 1 | mono |
| 2 | di |
| 3 | tri |
| 4 | tetra |
| 5 | penta |
| 6 | hexa |
| 7 | hepta |
| 8 | octa |
| 9 | nona |
| 10 | deca |

Of course, as with many compounds, the two examples that have been given have **common names** as well as their proper, standardized names. The common names of these compounds are *water* and *hydrogen peroxide*. (The $O_2^{2-}$ group is commonly called *peroxide*.)

If only one atom of the first element in the formula is present, the prefix is dropped. For example, the compound corresponding to the formula $N_2O$ is called *dinitrogen monoxide*. However, when naming the compound corresponding to the formula NO, we do not say "mononitrogen monoxide." We just say "nitrogen monoxide."

Here are several other examples to show you how to name Type III compounds before you start naming them on your own. You might want to test yourself on these before proceeding to the exercises.

$CCl_4$

Carbon comes before chlorine in our priority list, so the name starts with "carbon." (Since carbon is named first, and there is only one atom of it, we drop the "mono" prefix.) There are four chlorine atoms, so the complete name is carbon tetrachloride.

$Cl_2O$

Chlorine comes before oxygen, and there are two chlorine atoms. So we start with "dichlorine" (not "dichloride"). There is only one oxygen atom, so we have dichlorine monoxide.

$N_2O_5$

Two nitrogens, five oxygens—dinitrogen pentoxide

| | |
|---|---|
| ICl | iodine chloride |
| $S_4N_4$ | tetrasulfur tetranitride |

Okay, you know the routine by now. Let's see how to write formulas for the following compounds. Since we specify exactly how many of each atom is present in the name, it's no trouble figuring out how to write a formula:

| | |
|---|---|
| phosphorus pentachloride | $PCl_5$ |
| diarsenic pentoxide | $As_2O_5$ |
| disulfur dichloride | $S_2Cl_2$ |
| carbon disulfide | $CS_2$ |
| sulfur trioxide | $SO_3$ |

(Notice, we don't *always* drop the vowel from the end of the prefix when it puts two vowels together.)

Your turn!

**Exercises**

1) Name the following compounds:

a) $As_4O_6$ (don't drop the "a" from "hexa")
b) $N_2O$
c) $O_2F_2$
d) $NCl_3$
e) $SiCl_4$
f) $NH_3$
g) $P_4O_{10}$

2) Write the formulas for the following compounds:

a) iodine monobromide
b) diphosphorus pentoxide
c) diboron trioxide
d) carbon tetraiodide
e) diboron hexahydride
f) dichlorine heptoxide
g) dihydrogen sulfide

## 25: The Mole

In this lesson we want to consider the concept of the **mole**. When chemists talk about a mole, they are not referring to a little brown spot on the skin, nor are they talking about a small animal that burrows in the ground. Chemists use the term "mole" in an entirely different way. Let's introduce the concept by considering the reaction of magnesium and sulfur to form magnesium sulfide.

| Mg | + | S | → | MgS |
|---|---|---|---|---|
| 24.31 g | + | 32.06 g | = | 56.37 g |

It can be shown experimentally that 24.31 g of magnesium will react with 32.06 g of sulfur to make 56.37 g of magnesium sulfide. Of course, if 24.31 g of Mg and 32.06 g of S react together completely, then we'd better get 56.37 g of product. That's in keeping with the **Law of Conservation of Mass**—"atoms in equals atoms out"; "mass in equals mass out." But we also know that when elements react to form a specific compound, the elements combine in simple numeric ratios (**Law of Definite Composition**). That also holds true in this case; 1 magnesium atom reacts with 1 sulfur atom to form 1 magnesium sulfide molecule. What if we had two magnesium atoms and two sulfur atoms. How many molecules of MgS could we make? Right—two. What if we had three atoms of Mg and three atoms of S? We could make three molecules of MgS. You get the idea.

Well, obviously we have a lot more than one, two or three atoms of magnesium if we have 24.31 grams of it. And there are a lot more than one, two or three atoms of sulfur contained in 32.06 grams of sulfur. Actually, we have sextillions (lots) of atoms in these mass quantities. But, since the Mg and the S were both completely consumed in this reaction, and since the compound formed requires one atom of Mg for every atom of S, then we must conclude that, however many atoms we had of magnesium, in this case, we must have had the same *number of atoms* of sulfur. And we do. Furthermore, however many atoms we have of each of these two elements, when they react, they must produce the same number of molecules of magnesium sulfide. And they do.

Well, it turns out that 24.31 g of Mg contain $6.022 \times 10^{23}$ atoms of Mg, and 32.06 g of sulfur contain $6.022 \times 10^{23}$ atoms of S. That means, of course, that 56.37 g of MgS contain $6.022 \times 10^{23}$ molecules of MgS. This is where the term "mole" comes in. You see, the word "mole" is like the word "dozen." It stands for a number of something. If I said to you, "I have a dozen apples," you'd know I had twelve apples. I could have used the word "twelve", but I used the word "dozen". It doesn't matter which of the two words I choose to use because they both send exactly the same message.

Well, in the same way, the word "mole" as a chemistry term represents a number, and that number is $6.022 \times 10^{23}$. That number is called Avogadro's number, named for a fellow named Amadeo Avogadro, who determined many quantitative concepts in chemistry. (You'll recognize that the word *quantitative* comes from the same root as the word *quantity*. It conveys the idea that something is being measured, counted or *quantified*. It has to do with "putting a number on" something, just as Avogadro "put a number on" the molar quantity of pure substances.) So that's how many somethings you have when you have a mole of something: $6.022 \times 10^{23}$. Now, it could be something besides atoms or molecules. It could be ions, for example. Just remember that if you have $6.022 \times 10^{23}$ of them, you've got a mole of them. The mole is abbreviated "mol."

I said earlier that 24.31 g of Mg contain $6.022 \times 10^{23}$ atoms of Mg. Therefore, 24.31 g of Mg is equivalent to one mole of Mg. We also observed that 32.06 g of S contain $6.022 \times 10^{23}$ atoms of S. Therefore, this is one mole of S. Notice that one mole of Mg does not have the same mass as one mole of S. *They have the same number of atoms, but not the same mass.* Why? Because the individual atoms don't have the same mass. A sulfur atom has a greater mass than a magnesium atom, so one mole of sulfur must have greater mass than one mole of magnesium.

There is a basic difference in the ways the words *dozen* and *mole* came about. The word *dozen* is rather arbitrary. Twelve of something is called a dozen, but anything could have been called by that word. Avogadro's number has much greater underlying significance. Look at your periodic table. The atomic mass of Mg is 24.31 amu, while the atomic mass of S is 32.06 amu. Do you recognize these numbers? Can you see that if you took the atomic mass of Mg and changed the units from amu to grams, you'd have the mass of one mole of Mg? Similarly, if you took the atomic mass of S and did the same, you'd have the mass of one mole of S. This is universally true. To determine the mass of one mole of an element, just look up the atomic mass of that element and change the units from amu to grams. In fact, a mole may be defined as the **gram atomic weight** of a chemical entity (element, ion, compound, etc.).

You might also observe that if we added the atomic masses of one atom of Mg and one atom of S we would obtain the mass of one molecule of MgS—56.37 amu. Likewise, the gram atomic weight of MgS is 56.37 g. As with atoms, the same is true with molecules—one gram atomic weight is equal to one mole. We may confidently say that there are 56.37 g of MgS in one mole of MgS. So this method of determining how many grams there are in one mole of an element works for compounds, too. *To determine the mass of one mole of a compound (its "**molar mass**"), add up the atomic masses for each atom contained in one molecule of the compound and change the units from amu to grams.*

For example, if you were asked to calculate the molar mass of NaCl, you would proceed as follows. Since one molecule of NaCl contains one atom of sodium and one atom of chlorine, we simply add the atomic masses of these two elements as illustrated in the Figure.

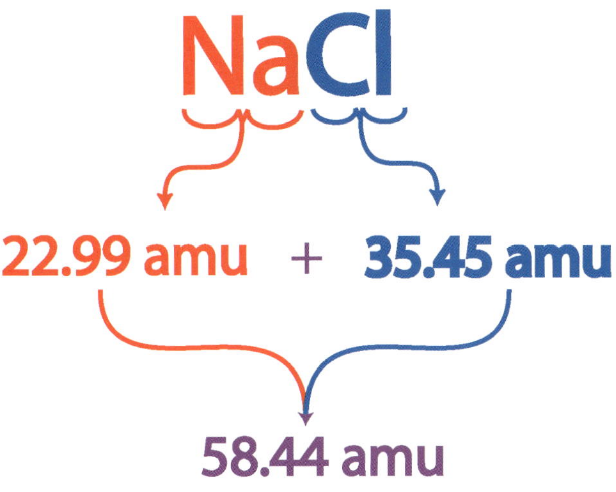


**Figure.** Calculating the molar mass of a compound is just a matter of adding up the masses of the atoms making it up.

This gives us the mass of one molecule of NaCl, which is equal to 58.44 amu. We might call this the **molecular mass**, **molecular weight** or **formula weight** of NaCl. They all mean exactly the same thing, and all would be expressed in amu (atomic mass units).

Now, if the mass of one molecule of NaCl is equal to 58.44 amu, then one *mole* of NaCl equals 58.44 g. We might say that "the *molar mass* of NaCl is equal to 58.44 g," that "the *gram molecular weight* of NaCl is equal to 58.44 g" or that the "*gram formula weight* of NaCl is equal to 58.44 g." All three of these mean exactly the same thing.

Let's summarize what we have learned so far using MgS as our example:

| | |
|---|---|
| Atomic mass of Mg: | 24.31 amu |
| Atomic mass of S: | 32.06 amu |
| Molecular mass of MgS: | 56.37 amu |

| | |
|---|---|
| Molar mass of Mg: | 24.31 g |
| Molar mass of S: | 32.06 g |
| Molar mass of MgS: | 56.37 g |

Now, going back to our original example reaction then, we could say about it that 1 mole of Mg reacts with 1 mole of S to produce 1 mole of MgS.

| Mg | + | S | → | MgS |
|---|---|---|---|---|
| 1 mol | | 1 mol | | 1 mol |

Also, please notice that you've now been given two equivalencies that can be used as unit factors. The first equivalency is:

$$1 \text{ mol} = 6.022 \times 10^{23} \text{ atom}$$

From this we get:

$$\frac{1\,\text{mol}}{6.022 \times 10^{23}} = 1 \quad \text{and} \quad \frac{6.022 \times 10^{23}}{1\,\text{mol}} = 1$$

The second equivalency is, for an element:

$$1 \text{ mol} = 1 \text{ gram atomic weight (GAW)}$$

From this we get:

$$\frac{1\,\text{mol}}{1\,\text{GAW}} = 1 \quad \text{or} \quad \frac{1\,\text{GAW}}{1\,\text{mol}} = 1$$

The idea of a mole will be very important for you for the rest of this course because it will greatly simplify your life. Suppose, for example, you were given a mass of 15 g of magnesium metal and asked how much sulfur would be needed to completely react with it. Without knowledge of moles, this would be a very difficult problem. Magnesium and sulfur don't react with each other in gram-for-gram quantities, but they do react with each other mole for mole. So the question is how many *moles* are there in 15 g of magnesium. Because we know that one mole of sulfur will react with one mole of magnesium to give us one mole of magnesium sulfide, we can solve the problem.

Do you realize how powerful this is? If you know the formula, you can determine the amounts required for one substance to react with another. You can also determine the quantity of product you would expect from the reaction if it went to completion.

In order to do this we must first become comfortable with converting gram quantities to molar quantities and vice versa. For example, how many moles of silver are present in 36.3 g of the element? Remember your problem solving method. Start with whatever you're given. In this case:

$$36.3 \text{ g of Ag}$$

What are you asked to figure out? The moles of silver:

$$? \text{ mol Ag} = 36.3 \text{ g Ag}$$

Now the question arises, "Do you have a unit factor that relates moles of Ag to grams of Ag?" Yes, you do. How many grams of Ag are in one mole of Ag? Look at the periodic table. The atomic mass of Ag is 107.9 amu. Therefore, there are 107.9 g of Ag in one mole of Ag. So set up the problem this way:

$$? \text{ mol Ag} = \frac{36.3 \text{ g Ag} \mid 1 \text{ mol Ag}}{\mid 107.9 \text{ g Ag}} = 0.336 \text{ mol Ag}$$

(Remember, the one mole under consideration in this problem is not an approximation. It is *precisely one mole*. Don't let that "1" limit the precision of your answer!)

Suppose, instead of being asked for the *moles* of silver, you had been asked how many *atoms* of silver were contained in this same 36.3 g of Ag? Again, you would start with the 36.3 g of Ag (what you're given) and use the unit factor to convert to moles—but then you would need that second unit factor to convert moles to atoms:

$$? \text{ atom Ag} = \frac{36.3 \text{ g Ag}}{} \left| \frac{1 \text{ mol Ag}}{107.9 \text{ g Ag}} \right| \frac{6.022 \times 10^{23} \text{ atoms Ag}}{1 \text{ mol Ag}} = 2.06 \times 10^{23} \text{ atoms Ag}$$

Now, then, what is the mass (g), of a single titanium atom? Set it up using information on Ti from the periodic table:

$$? \text{ g Ti} = \frac{1 \text{ atom Ti}}{} \left| \frac{1 \text{ mol Ti}}{6.022 \times 10^{23} \text{ atoms Ti}} \right| \frac{47.87 \text{ g Ti}}{\text{mol Ti}} = 7.949 \times 10^{-23} \text{ g Ti}$$

As you continue in this course you will find that many problems in chemistry require you to "go through moles" in order to solve them. Since the only way for you to get comfortable doing these kinds of problems is to do them, and since we wouldn't want to disappoint you, here are some exercises for you.

**Exercises**

1) Determine the molar masses of the following compounds:

a) $CaCl_2$
b) $BaSO_4$
c) $NaOH$
d) $C_6H_5OH$
e) $Mg(NO_3)_2$

2) For future reference, make a table for yourself like the one on page 94. Calculate the gram formula weight of each of the polyatomic ions and add those weights into a new column on the right of the formulas. If you try, you will find many uses for this table as you complete the rest of the course. (If you don't try, you will find yourself calculating these masses over and over and over again!) Preserve as many significant digits as the periodic table will allow. You don't want the precision of future calculations to be limited by the number of significant digits you carry in your calculations. Flag this page in your notebook with a paper clip, sticky note, etc.

3) How many moles of atoms are contained in the following?

a) 5.60 g S
b) 432 g Co
c) $6.022 \times 10^{23}$ atoms Ti
d) $7.654 \times 10^{32}$ atoms Ga

4) Calculate the mass, in grams, in each of the following:

a) 0.390 mol Cu
b) $7.654 \times 10^{22}$ atoms Ti
c) $5.30 \times 10^{22}$ molecules $CH_3OH$
d) 5.130 mol $NH_4NO_3$

## 26: Percent Composition

Now that you are adept in calculating the molar mass of a given compound, it will be simple for you to also calculate its **mass percent composition**. The mass percent composition, sometimes just called the **percent composition**, is the percentage contribution of each element to the total molar mass.

But why would you ever want to calculate the percent composition of anything? Will it add one cubit to your stature? Will it change one hair of your head from white to black? Well, I hope not. But what it might do is allow you to appreciate what portion of a compound is contributed by each of the elements that make it up. For example, suppose you were instructed by your doctor or nutritionist to take 500 mg of zinc every day. But when you looked at the label on your favorite supplement, it said, "Every teaspoon contains 1,200 mg zinc sulfate." Terrific. Zinc sulfate is $ZnSO_4$. Well, if there are 1,200 mg of zinc sulfate, then how much zinc is there? You're smart, you figure it out.

You can best solve this problem if you understand percent composition. Every compound is made up of more than one atom. Percent composition tells what portion of a compound by mass is contributed by each of its elements. To figure it, you start by dividing the mass of the element you are concerned about by the mass of the entire molecule. That gives you the *mass proportion* of the molecule made up by that element, right? Then, as you well know, you can express any fraction as a percent by multiplying it by 100. So, if the element we are concerned with is element A, then the mass percent composition of A in a compound is:

$$\text{mass \% A} = \frac{\text{mass A}}{\text{molar mass}} \times 100$$

Don't let this phrase *molar mass* throw you. Remember, from the last lesson, it's simply the mass of the entire molecule. (See the Figure.)

Here's our first example. Let's calculate the percent of the composition of KCl that is contributed by each of the elements in it. In short, let's calculate the "percent composition of KCl." First, we'll calculate the molar mass of KCl, just as we did all of those molecules in the last lesson:

Gram Atomic Weight K
+ Gram Atomic Weight Cl
Molar Mass KCl

So:

| | |
|---|---|
| 39.10 g | K |
| + 35.45 g | Cl |
| 74.55 g | KCl |

**Figure.** The mass percent of nitrogen in magnesium nitrate is given by the mass of all the nitrogen atoms divided by the molar mass of the compound times 100.

$$\% N = \frac{\text{wt N}}{\text{wt } Mg(NO_3)_2} \times 100$$

Now we will determine the portion of the entire molecule contributed by each of its elements, each expressed as a percentage:

$$\text{mass \% K} = \frac{\text{mass K}}{\text{molar mass}} \times 100 = \frac{39.10}{74.55} \times 100 = 52.45\%\ \text{K}$$

Likewise:

$$\text{mass \% Cl} = \frac{\text{mass Cl}}{\text{molar mass}} \times 100 = \frac{35.45}{74.55} \times 100 = 47.55\%\ \text{K}$$

Of course, if we've done our work correctly, the individual elements should make up 100% of the molecule, so let's check our work. Do 52.45% and 47.55% add up to 100%? Yep! Now, I'm not going to remind you every time that you should perform this simple check. After all, you're way up into high school and have your permanent teeth and everything. But if you know what's good for you, you'll do it without being told.

Let's try a second example. Let's calculate the percent composition of calcium nitrate. Hopefully, by now, it's easy for you to figure out that the formula for calcium nitrate is $Ca(NO_3)_2$. So this is the formula for which we must calculate the molar mass. (Please note that if you use the wrong molecular formula, then your calculation will be wrong too.) This will add insult to injury. Begin as before by calculating the molar mass:

$$\text{mass } Ca(NO_3)_2 \text{ g} = \text{mass Ca g} + 2 \text{ mass } NO_3 \text{ g}$$

Remember that you calculated all the molar masses of those polyatomic ions just for this occasion. So use that mass of nitrate now in your equation:

$$\text{mass } Ca(NO_3)_2 \text{ g} = 40.08 \text{ g} + 2(62.01) \text{ g} = 164.10 \text{ g}$$

And then calculate the percent of this molar mass made up by each of the elements:

$$\text{mass \% Ca} = \frac{\text{mass Ca}}{\text{molar mass}} \times 100 = \frac{40.08}{164.10} \times 100 = 24.42\%\ \text{Ca}$$

$$\text{mass \% N} = \frac{\text{mass N}}{\text{molar mass}} \times 100 = \frac{2(14.01)}{164.10} \times 100 = 17.07\%\ \text{N}$$

$$\text{mass \% O} = \frac{\text{mass O}}{\text{molar mass}} \times 100 = \frac{6(16.00)}{164.10} \times 100 = 58.50\%\ \text{O}$$

The most common mistake students make on these problems is to use the mass of an element in place of the total mass of all atoms of that element. Remember, we are not trying to calculate the portion of the compound made up by one atom of oxygen, but rather the *total contribution* of all the oxygen atoms to that molecule's mass.

If you did your cross-check to be sure the percentages added up, you found that they added up to 99.99%. That's within an acceptable margin of error. If the numbers don't add up to within two digits of 100% in the last decimal place, then you have probably made a mistake somewhere.

### Percent Composition from Experimental Data

As long as you are given the formula of a particular compound, you should be able to calculate its percent composition. But sometimes, rather than the formula, you might have experimental data. For example, a strip of copper metal having a mass of 4.767 g was heated in air until all of it was converted to its oxide. The oxide had a mass of 5.967 g. What is its percent composition?

The reaction takes place between copper and oxygen. Since we know how much copper reacted (4.767 g) and how much oxide was produced (5.967 g), the Law of Conservation of Mass tells us that the difference between these two masses will be the mass of the oxygen that reacted. So 5.967 g – 4.767 g = 1.200 g oxygen.

To calculate the percent composition we just divide the mass of each reactant by the mass of the product and multiply by 100:

$$\text{mass \% Cu} = \frac{\text{mass Cu}}{\text{mass product}} \times 100 = \frac{4.767\text{ g}}{5.967\text{ g}} \times 100 = 79.89\%\text{ Cu}$$

$$\text{mass \% O} = \frac{\text{mass O}}{\text{mass product}} \times 100 = \frac{1.200\text{ g}}{5.967\text{ g}} \times 100 = 20.11\%\text{ O}$$

Finally, suppose a 1.300-g sample of a hydrocarbon (a compound made up entirely of hydrogen, carbon and sometimes oxygen) was found to contain 0.678 g of carbon, 0.171 g of hydrogen and 0.451 g of oxygen. What is its percent composition? Simple stuff:

$$\text{mass \% C} = \frac{\text{mass C}}{\text{mass sample}} \times 100 = \frac{0.678\text{ g}}{1.300\text{ g}} \times 100 = 52.2\%\text{ C}$$

$$\text{mass \% H} = \frac{\text{mass H}}{\text{mass sample}} \times 100 = \frac{0.171\text{ g}}{1.300\text{ g}} \times 100 = 13.2\%\text{ H}$$

$$\text{mass \% O} = \frac{\text{mass O}}{\text{mass sample}} \times 100 = \frac{0.451\text{ g}}{1.300\text{ g}} \times 100 = 34.7\%\text{ C}$$

Don't quit yet! There is one more important application of this new tool you have learned. Be strong and endure.

### Percent Water in a Hydrate (not a Hydrant!)

Some ionic compounds, when they crystallize in the presence of water, or when their crystals are exposed to moist air, trap water in their crystal structures. The water that binds with the crystalline matrix is always present in stoichiometric amounts with the molecules of the compound itself. Such a compound is known as a **hydrate**, and the water is called its **water of hydration** or **water of crystallization**. The formulas of hydrates are made by writing: (1) the formula of the ionic compound, (2) a dot, (3) the number of water molecules present and (4) the formula for water itself.

The names of hydrates are made from (1) the name of the ionic compound, (2) the number of water molecules (expressed using the Greek prefixes), and (3) the word *hydrate*. Some examples of the formulas and names of some common hydrates are given in the Table.

The reason we take the time to bring up hydrates here is that you use the same method to calculate the portion of water associated with a hydrate that you used to calculate percent composition. For example, if you are asked what percent of the mass of cobalt(II) chloride hexahydrate is water of hydration, you would follow the same procedure as before. The first thing to do is to calculate the total molar mass. It'll be easier for you if you calculate the mass of the water separately from the mass of the ionic molecule. First calculate the mass of cobalt chloride:

$$\text{mass CoCl}_2\text{ g} = 58.93\text{ g} + 2(35.45)\text{ g} = 129.83\text{ g}$$

Then the mass of the water of hydration:

$$\text{mass 6H}_2\text{O g} = 6(18.02)\text{ g} = 108.12\text{ g}$$

Then use these quantities to calculate the percent water in the hydrate:

$$\text{mass \% H}_2\text{O} = \frac{\text{mass H}_2\text{O}}{\text{mass CoCl}_2\cdot 6\text{H}_2\text{O}} \times 100$$

$$= \frac{108.1\text{ g}}{129.8\text{ g} + 108.1\text{ g}} \times 100 = 45.44\%\text{ H}_2\text{O}$$

**Table.** Common hydrates.

| Formula | Name |
|---|---|
| $CoCl_2 \cdot 6H_2O$ | cobalt(II) chloride hexahydrate |
| $CuSO_4 \cdot 5H_2O$ | copper(II) sulfate pentahydrate |
| $MgSO_4 \cdot 7H_2O$ | magnesium sulfate heptahydrate |
| $BaCl_2 \cdot 2H_2O$ | barium chloride dihydrate |
| $NiSO_4 \cdot 6H_2O$ | nickel(II) sulfate hexahydrate |
| $Na_2CO_3 \cdot 10H_2O$ | sodium carbonate decahydrate |

Okay, we can finally get you to your exercises so you can be done for the day. Way to hang in there!

**Exercises**

1) Calculate the percent composition of the following compounds:

   a) $Ba(ClO_3)_2$
   b) $Na_2CO_3$

2) An ionic compound was prepared by reacting 12.1 g Al with 47.9 g Cl. What is the percent composition of the ionic compound?

3) An enterprising chemistry student took 10.0 g of an unknown hydrate and heated it in order to drive off the water of hydration. The mass of the **anhydrous salt** (that's what is left after the water is gone) was 5.82 g. What was the percent water in the original hydrate?

4) The student in question #3 was then told by his chemistry instructor that his hydrate was one of the following three compounds: $BaCl_2 \cdot 2H_2O$; $MgSO_4 \cdot 7H_2O$; or $NiSO_4 \cdot 6H_2O$. Which of these three hydrates was probably the one the student used? Why?

## 27: Quiz 4

From now on, you'll be allowed to use the periodic table on all your quizzes.

1) Be able to name any binary or polyatomic compound. (Names such as Helga, Alfonzo and Beatrice will not be accepted.)
2) Be able to write the formula of any binary or polyatomic compound.
3) Know the formulas and charges of all the polyatomic ions listed in the Table in Lesson 23. (Make flashcards for yourself if needed.)
4) Be able to calculate the molar mass of any compound for which you are given either a name or a molecular formula.
5) Have a thorough understanding of moles.
6) Memorize Avogadro's number, and know what it represents. Be able to calculate the number of moles of a substance from its mass or its number of atoms.
7) Be able to calculate the mass percent composition of any compound, given either the mass of each element or the molecular formula.
8) Be able to calculate the percent water in any hydrate of which you are given the formula.

Some people are happiest when they are thinking or exploring. They would probably be good scientists.

## 28: Empirical Formulas

In the last lesson you learned how to calculate the mass percent composition of a compound. Way back in Lesson 8 we observed that elements combine in simple numeric ratios to form compounds. Well, what if you were given the percent composition of a compound, but you didn't know the formula? Could you determine anything about the ratios of the elements with regard to each other? I don't mean to answer for you, but yes, you can. It's a simple matter of converting the percentages from percent composition into small, whole numbers.

Once the percent composition of a chemical compound is known, then its **empirical formula** can be determined. The empirical formula is defined as *the simplest formula of a compound that still gives the correct whole-number ratios of the atoms*. However, keep in mind that the compound does not actually exist with those numbers of atoms. For example, the formula for hydrogen peroxide is $H_2O_2$. Its *empirical formula* is HO. You see, it is the simplest formula that gives the correct whole-number ratio of hydrogen to oxygen—1:1. Another example would be benzene, $C_6H_6$. (See Figure 1.) The empirical formula? CH. What about cyclopropane—$C_3H_6$? The empirical formula is $CH_2$. This is the simplest formula that still expresses the correct carbon-to-hydrogen ratio—1:2.

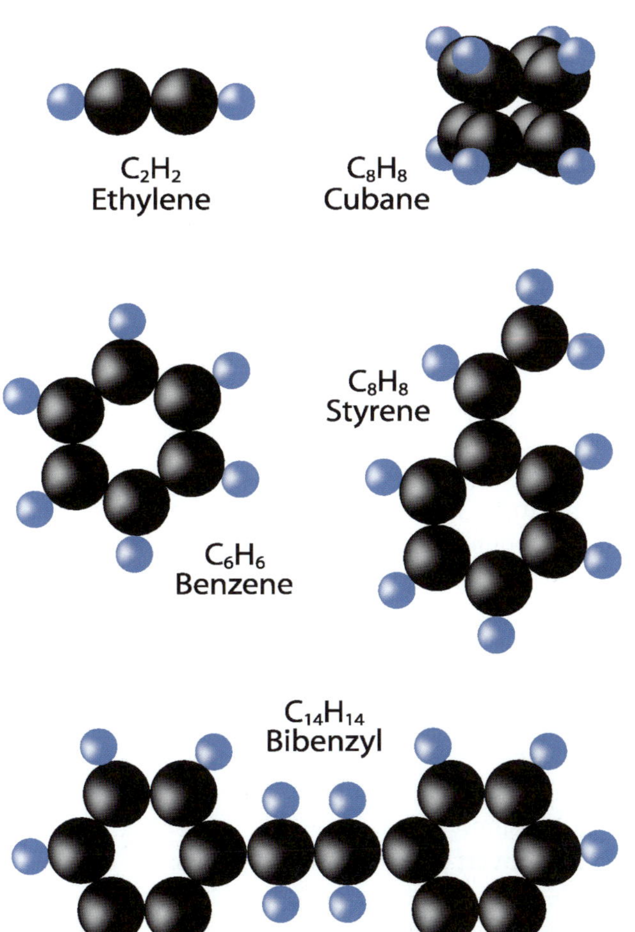


And *why* would anyone ever want to convert the percent composition into a small whole-number ratio? Well, let's say you went to a party where the percent composition of adults, teenagers and small children was 50%:25%:25%, respectively. What is the simple, whole-number ratio of adults to teenagers to small children? Clearly, it's 2:1:1 (by numbers, not by mass). Sometimes we just like to say things like, "The teenagers were outnumbered at the party by adults 2:1." Or, "There were as many little munchkins running around as there were teenagers." But sometimes small-number ratios can reveal hidden secrets that percentages keep covered up. You'll just have to read through this lesson to see what I mean. (But you can start by looking at Figure 2.)

**Figure 1.** Knowing the empirical formula of an unknown compound narrows the possible molecular formulas it can have. If the empirical formula is not the actual molecular formula itself (which it often is) you can obtain the molecular formula by multiplying the number of atoms of each element by a constant. Here are the structures having the empirical formula CH up through those having the molecular formula $C_{14}H_{14}$. Other CH compounds exist having more than 14 carbon atoms. However, because of the God-given uniqueness of carbon, it forms far more compounds than any other element. Without the variety of structures that carbon takes on, life (as we know it) could not exist.

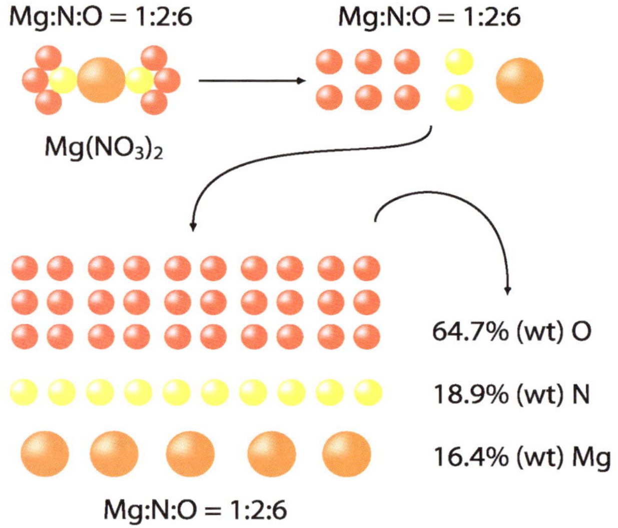


**Figure 2.** The ratio of atoms present in a single molecule dictates the ratio of atoms present in any amount of that substance. The percent composition of a compound is specific to that ratio of atoms.

How do we calculate the empirical formula from percent composition? Here are the steps:

1) Assume a given amount of the unknown compound. Usually 100 g is assumed so that the percentages can be easily converted into grams.
2) Write down the number of grams of each element that would be present in the assumed amount of the compound.
3) Convert the grams of each element into moles.
4) Divide each of the numbers determined in step 3 by the smallest of them. This will usually yield whole-number ratios of the different atoms.
5) If necessary, multiply each of the values obtained in step 4 by the smallest number possible that will convert them into whole-number ratios.

Let's do one. Then you'll understand exactly what all that gobbledygook means. For starters, what is the empirical formula of a compound that has the following percent composition: 43.4% Na, 11.3% C, and 45.3% O? (Look at these elements. Would you like to venture a guess as to what it might be?)

Step 1. Assume 100 g.
Step 2. Convert percentages into grams.

Since we assumed 100 g, then we would have 43.4 g Na, 11.3 g C, and 45.3 g O.

Step 3. Convert the grams of each element into moles.

$$? \text{ mol Na} = \frac{43.4 \text{ g Na}}{} \left| \frac{1 \text{ mol Na}}{22.99 \text{ g Na}} \right. = 1.89 \text{ mol Na}$$

$$? \text{ mol C} = \frac{11.3 \text{ g C}}{} \left| \frac{1 \text{ mol C}}{12.01 \text{ g C}} \right. = 0.941 \text{ mol C}$$

$$? \text{ mol O} = \frac{45.3 \text{ g O}}{} \left| \frac{1 \text{ mol O}}{16.00 \text{ g O}} \right. = 2.83 \text{ mol O}$$

Step 4. Divide all three of the molar quantities you just calculated by the smallest of the three.

$$\frac{1.89 \text{ mol Na}}{0.941 \text{ mol C}} \approx 2 \frac{\text{mol Na}}{\text{mol C}}$$

(The symbol used in place of the equal sign means "approximately equal to.")

$$\frac{0.941 \text{ mol C}}{0.941 \text{ mol C}} = 1 \frac{\text{mol C}}{\text{mol C}}$$

$$\frac{2.83 \text{ mol O}}{0.941 \text{ mol C}} \approx 3 \frac{\text{mol O}}{\text{mol C}}$$

Therefore, the empirical formula is $Na_2CO_3$. This also happens to be the molecular formula of a very common compound. Notice that step 5 wasn't needed in this case. We already have a ratio of small whole numbers. Let's try one for which we will need it.

Derive the empirical formula from a compound whose percent composition is: 29.08% Na, 40.58% S, and 30.34% O.

Step 1. Assume 100 g.
Step 2. Convert percentages into grams.

Since we assumed 100 g, then we would have 29.08 g Na, 40.58 g S, and 30.34 g O.

Step 3. Convert the grams of each element into moles:

$$? \text{ mol Na} = \frac{29.08 \text{ g Na}}{} \bigg| \frac{1 \text{ mol Na}}{22.99 \text{ g Na}} = 1.265 \text{ mol Na}$$

$$? \text{ mol S} = \frac{40.58 \text{ g S}}{} \bigg| \frac{1 \text{ mol S}}{32.07 \text{ g S}} = 1.265 \text{ mol S}$$

$$? \text{ mol O} = \frac{30.34 \text{ g O}}{} \bigg| \frac{1 \text{ mol O}}{16.00 \text{ g O}} = 1.896 \text{ mol O}$$

Step 4. Divide the calculated values by the smallest calculated value.

$$\frac{1.265 \text{ mol Na}}{1.265 \text{ mol Na}} = 1 \frac{\text{mol Na}}{\text{mol Na}}$$

$$\frac{1.265 \text{ mol S}}{1.265 \text{ mol Na}} = 1 \frac{\text{mol S}}{\text{mol Na}}$$

$$\frac{1.896 \text{ mol O}}{1.265 \text{ mol Na}} = 1.499 \frac{\text{mol O}}{\text{mol Na}} \approx 1.5 \frac{\text{mol O}}{\text{mol Na}}$$

Here is where we obviously need step 5.

Step 5. If necessary, multiply each of the values obtained in step 4 by the smallest number possible that will convert them into whole-number ratios.

Clearly, multiplication by 2 will result in whole numbers. Therefore:

$1 \text{ Na} \times 2 = 2 \text{ Na}$    $1 \text{ S} \times 2 = 2 \text{ S}$    $1.5 \text{ O} \times 2 = 3 \text{ O}$

The empirical formula is $Na_2S_2O_3$. This also happens to be the formula of sodium thiosulfate. (Thiosulfate is a divalent polyatomic anion we haven't shown you before.)

Okay, one more problem before we move on to the Exercises. This one is actually easier than the others (if you can imagine it). But it's in a slightly different form. Just in case you should happen to see one of these on a quiz, I'd like you to see the logic of it.

A 35.5-g sample of an unknown compound was analyzed and found to contain 8.26 g Mg, 10.91 g S, and 16.33 g O. What is its empirical formula? In this case, since we are given actual masses of the elements present, we don't need to assume a fictitious mass of 100 g of starting material. We can just use those actual masses to calculate the whole-number ratios. So:

$$? \text{ mol Mg} = \frac{8.26 \text{ g Mg}}{} \bigg| \frac{1 \text{ mol Mg}}{24.31 \text{ g Mg}} = 0.340 \text{ mol Mg}$$

$$? \text{ mol S} = \frac{10.91 \text{ g S}}{} \bigg| \frac{1 \text{ mol S}}{32.07 \text{ g S}} = 0.3402 \text{ mol S}$$

$$? \text{ mol O} = \frac{16.33 \text{ g O}}{} \bigg| \frac{1 \text{ mol O}}{16.00 \text{ g O}} = 1.021 \text{ mol O}$$

Then we divide by the smallest, as previously:

$$\frac{0.340 \text{ mol Mg}}{0.340 \text{ mol Mg}} = 1 \frac{\text{mol Mg}}{\text{mol Mg}}$$

$$\frac{0.3402 \text{ mol S}}{0.340 \text{ mol Mg}} \approx 1 \frac{\text{mol S}}{\text{mol Mg}}$$

$$\frac{1.021 \text{ mol O}}{0.340 \text{ mol Mg}} \approx 3 \frac{\text{mol O}}{\text{mol Mg}}$$

Hey! The empirical formula is $MgSO_3$. There happens to be a compound having that empirical formula, and it's called magnesium sulfite. I'll bet that's what we have here. Do some chemistry on your own!

**Exercises**

1) Calculate the empirical formula for a compound that has the following percent composition: 12.26% Mg, 31.24% P, 56.48% O.

2) A hydrocarbon was analyzed and found to contain 18.6 g C, 3.1 g H, and 55.02 g Cl. What is its empirical formula?

3) A 4.43-g sample of P was burned in air to produce 10.15 g of an oxide. What's the empirical formula of this compound?

## 29: How Many Empiricals in a Molecular?

As we saw in the last lesson, empirical formulas for compounds are not that difficult to determine, as long as we know the mass percent composition of the compound. But empirical formulas do not always tell us everything we would like to know about a compound. Now sometimes they do. For example, the empirical formula for water is $H_2O$. That is also its *molecular formula*—the *actual* formula of the compound. The same is true for ammonia, $NH_3$; table salt, NaCl; hydrochloric acid, HCl; and a host of other compounds. But, as we have already observed, the empirical formula for other compounds will not be the same as their molecular formulas.

Like empirical formulas, molecular formulas can be determined if we have sufficient information. In addition to the percent composition (which enables you to find the empirical formula) you also need the molar mass of a compound. And there are analytical techniques, such as mass spectrometry, that enable the determination of molar mass for a wide variety of compounds. (See the Figure.)

Remember, the empirical formula represents the simplest whole-number formula for a compound. Therefore, the molecular formula will be a whole-number multiple of the empirical formula. Consider our old example, hydrogen peroxide. The molecular formula, $H_2O_2$, is the empirical formula, HO, times 2. The molecular formula of benzene, $C_6H_6$, is its empirical formula, CH, times 6. When we have the empirical formula of a compound, and its molar mass, then we can calculate that multiplier. The multiplier tells us how many empirical formula "units" make up a whole molecular formula for that compound. So then, if we have an empirical formula (or a percent composition from which we can calculate it) and the molar mass of an unknown compound, we can calculate the multiplier that will get us the molecular formula. If you think about it for a moment, you will realize that the multiplier is

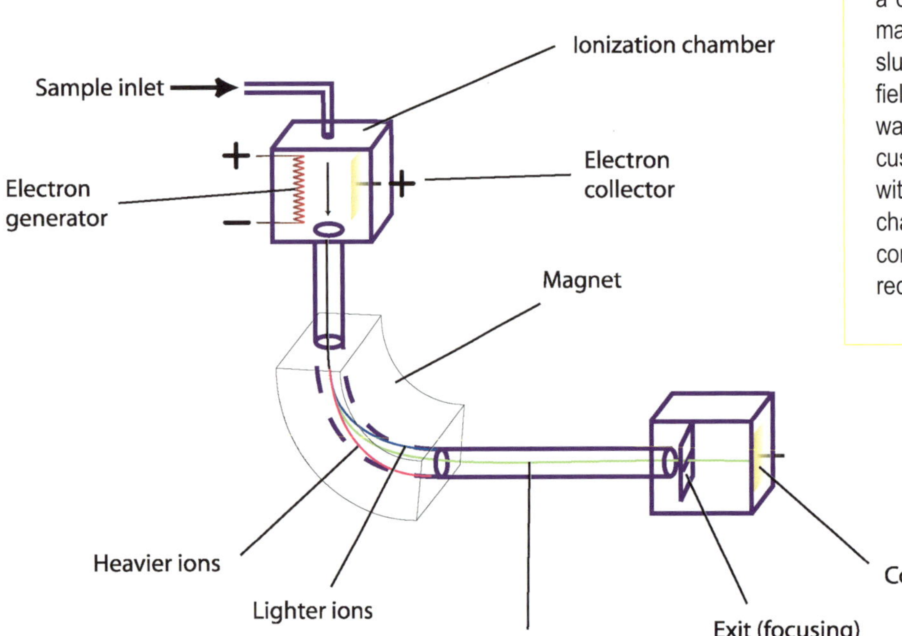


**Figure.** The mass spectrometer separates particles based on their velocity, mass and charge. A tiny amount of sample is bombarded with electrons to form ions. The particles are then accelerated through a curved tube surrounded by a powerful magnet. Heavy particles will respond more sluggishly to the effects of the magnetic field compared to the lighter ions. In this way the magnetic field can be used to focus ions on the detector that have masses within a certain range. The detector is a charged plate that attracts the ions and is connected to a device that amplifies and records their impact.

not only the ratio of atoms in the two formulas, it's also the ratio of molar masses of the two formulas:

$$\text{multiplier} = \frac{\text{molar mass of compound}}{\text{molar mass of empirical formula}}$$

We can then multiply each of the elements of the empirical formula by the multiplier to get the molecular formula.

Let's do it. Suppose the FBI had found an unknown compound having an empirical formula of $P_2O_5$. The substance was subjected to mass spectroscopy and it was determined that it had a molar mass of 283.88 g. They are willing to pay you $1 million worth of Milk Duds to tell them the molecular formula. Can you do it?

Once again, the first thing to do is to calculate the molar mass of the empirical formula:

$$\text{mass } P_2O_5 \text{ g} = 2\,(30.97)\text{ g} + 5\,(16.00)\text{ g}$$

$$= 141.94 \text{ g } P_2O_5$$

Now we simply divide the molar mass of the unknown compound by the molar mass of the empirical formula:

$$\text{multiplier} = \frac{283.88}{141.94} = 2.0000 \approx 2$$

Therefore, to express the molecular formula we write the empirical formula but multiply each of the elements in that empirical formula by 2. The answer then is $P_4O_{10}$.

Now that you've gotten your Duds, let's give you a slightly different problem. Let's combine what you learned in the previous lesson with what you have learned today. If 7.0 g of nitrogen combine with 19.98 g of oxygen to produce a compound that has a molar mass of 108 g, what is the molecular formula of that compound?

The first step is to determine the empirical formula as we did in the previous lesson:

$$?\text{ mol N} = \frac{7.0\text{ g N}}{} \left|\frac{1\text{ mol N}}{14.01\text{ g N}}\right. = 0.50\text{ mol N}$$

$$?\text{ mol O} = \frac{19.98\text{ g O}}{} \left|\frac{1\text{ mol O}}{16.00\text{ g O}}\right. = 1.248\text{ mol O}$$

$$\frac{0.50\text{ mol N}}{0.50\text{ mol N}} = 1\frac{\text{mol N}}{\text{mol N}}$$

$$\frac{1.248\text{ mol O}}{0.50\text{ mol N}} = 2.5\frac{\text{mol O}}{\text{mol N}}$$

$$1\text{ mol N} \times 2 = 2\text{ mol N}$$
$$2.5\text{ mol O} \times 2 = 5\text{ mol O}$$

So the empirical formula is $N_2O_5$.

The next step is to determine the molar mass of the empirical formula. For $N_2O_5$ the molar mass is:

$$\text{molar mass } N_2O_5\text{ (g)} = 2(14.01)\text{ g} + 5\,(16.00)\text{ g}$$

$$= 108.02\text{ g } N_2O_5$$

We were told that this is also the molar mass of the compound. Therefore, the molecular formula is the same as the empirical formula—$N_2O_5$.

Finally, the molar mass of dibromocyclohexane is 241.94 g. It contains 29.78% C, 4.17% H and 66.05% Br. What is its molecular formula?

First, determine the empirical formula. Assume 100 g:

$$?\text{ mol C} = \frac{29.78\text{ g C}}{} \left|\frac{1\text{ mol C}}{12.01\text{ g C}}\right. = 2.480\text{ mol C}$$

$$?\text{ mol H} = \frac{4.17\text{ g H}}{} \left|\frac{1\text{ mol H}}{1.008\text{ g H}}\right. = 4.14\text{ mol H}$$

$$?\text{ mol Br} = \frac{66.05\text{ g Br}}{} \left|\frac{1\text{ mol Br}}{79.90\text{ g Br}}\right. = 0.8267\text{ mol Br}$$

$$\frac{2.480 \text{ mol C}}{0.8267 \text{ mol Br}} \approx 3\,\frac{\text{mol C}}{\text{mol Br}}$$

$$\frac{4.14 \text{ mol H}}{0.8267 \text{ mol Br}} \approx 5\,\frac{\text{mol H}}{\text{mol Br}}$$

$$\frac{0.8267 \text{ mol Br}}{0.8267 \text{ mol Br}} = 1\,\frac{\text{mol Br}}{\text{mol Br}}$$

So, the empirical formula is $C_3H_5Br$.

Next, we'll need the molar mass of the empirical formula:

$$C_3H_5Br\ (g) = 3\,(12.01)\text{ g} + 5\,(1.008)\text{ g} + 79.90\text{ g}$$

$$= 120.97\text{ g } C_3H_5Br$$

So we can divide the molar mass of the unknown compound by the molar mass of the empirical formula:

$$\text{multiplier} = \frac{241.94}{120.97} = 2.0000 \approx 2$$

In this case, we find that it takes 2 empirical formula units to account for the total molar mass of the compound. So we then multiply each element of the empirical formula by 2. This gives the molecular formula, $C_6H_{10}Br_2$.

**Exercises**

1) A compound having an empirical formula of $CH_2O$ was found to have a molar mass of approximately 90 g. What is its molecular formula?

2) Caffeine, an additive found in many soft drinks, coffee and teas, has a molar mass of 194.1 g. Its percent composition is 49.5% C, 5.20% H, 28.9% N and 16.5 % O. Determine the molecular formula of caffeine.

3) Glucose, a common sugar, is composed of 40.0% C, 6.72% H and 53.3% O. Its molar mass is 180 g. What is the molecular formula of glucose?

# Chemical Reactions

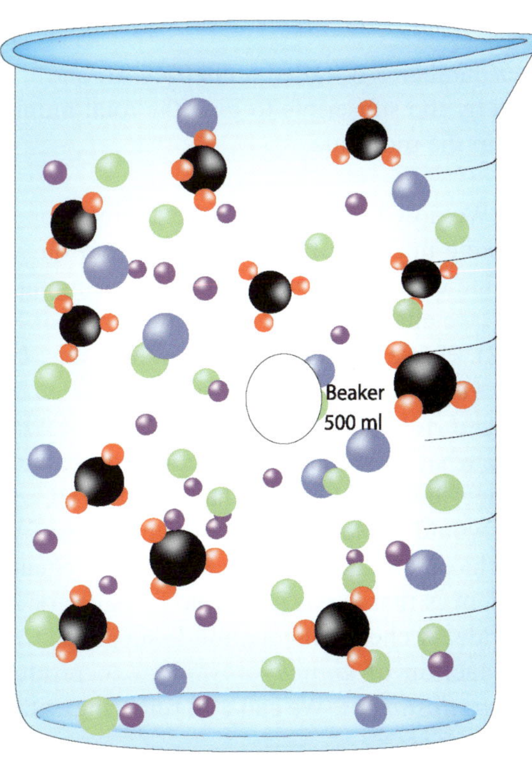


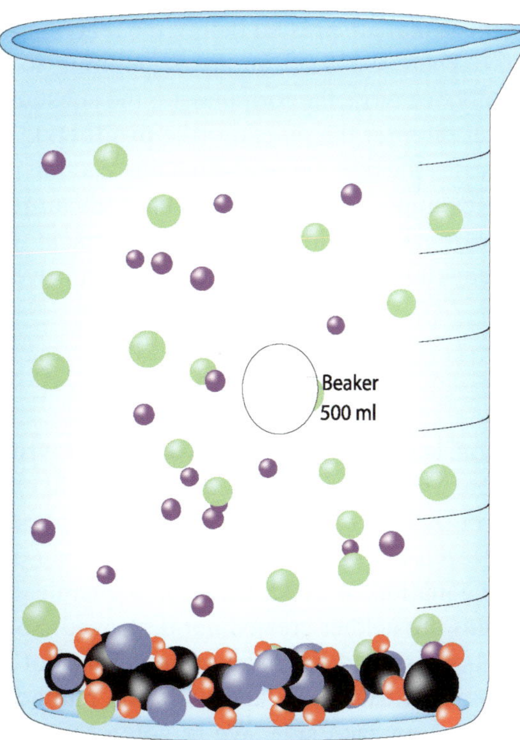

# 30: Teetering on the Brink of Reaction Chemistry

Thus far in our studies we have talked a lot about the shorthand methods used to describe elements, ions and isotopes, and the formulas for chemical compounds. We have also mentioned the fact that substances can undergo chemical changes by "reacting" with other substances to form products that have different physical and chemical properties than the initial substances. We have even talked about the charge environments at the surfaces of atoms and molecules (e.g., electronegativity, polarity, the octet rule and electron configurations) that might cause them to interact with one another. Now it's time to put all this together.

When substances react to form products, the shorthand method that is used to describe that reaction is called a **chemical equation**. Perhaps you have seen these before and even have some basic understanding of how to read or even balance them. To write a chemical equation we first write the chemical formulas for the substances that are present before the reaction (the **reactants**). Then, we draw an arrow pointing to the right. Finally, we write the formulas for the substances that are present after the reaction (the **products**). The arrow indicates the direction in which the reaction proceeds. For example, hydrogen and chlorine can react to form hydrogen chloride or hydrochloric acid. The chemical equation that describes that reaction would be:

$$H_2 + Cl_2 \rightarrow HCl$$

Well, at least that's how we would *start* writing the equation. You see, chemical equations are like teeter totters in that they should be **balanced**. The same number of each kind of atom should appear on each side of the equation. Why? Because in a chemical reaction, atoms are neither created nor destroyed. So the equation above should have the same number of hydrogen atoms on each side, and the same number of chlorine atoms on each side. We're going to balance this equation in a bit, but first let's notice how we began. When we wrote the formulas for the reactants, we didn't just write H and Cl. We wrote $H_2$ and $Cl_2$. Why? Because both of these are elements that exist in nature as diatomic molecules. We told you about seven elements that are diatomic all the way back in Lesson 2. In case you forgot, they are: $H_2$, $N_2$ , $O_2$, $F_2$, $Cl_2$, $Br_2$ , and $I_2$. Therefore, when any of these substances is written in a chemical equation, it must be written in its diatomic form.

Now we are ready to talk about *balancing* the equation. How many hydrogen atoms are on the left side of the equation? Two. Yes, I know there is only one hydrogen *molecule*, but since it is diatomic, there are two hydrogen *atoms* present. How many hydrogen atoms on the right side? Only one. What about chlorine? There are two atoms on the left side of the equation and only one on the right side. To balance this equation, then, we would write a 2 in front of the formula for hydrogen chloride, like this:

$$H_2 + Cl_2 \rightarrow 2\ HCl$$

That number *2* is called a **coefficient**, and it is multiplied by the subscripts to give the total number of atoms in the molecule.

We now have a balanced equation—one that has the same number of each kind of atom on each side. And the way we balanced it was by changing a coefficient (we changed the coefficient in front of HCl from 1 to 2). Please make sure that you learn this. *In balancing chemical equations, only coefficients can be changed, not subscripts.* You see, someone who doesn't know what he's doing might come along and want to balance this equation by changing the HCl to $H_2Cl_2$. That would indeed put the same number of hydrogen and chlorine atoms on each side of the equation. The problem is that $H_2Cl_2$ is not the formula for hydrogen chloride. A chemical equation is supposed to describe the reaction of *known substances* to produce other *known substances*. We can't go tinkering with molecular formulas.

So what *is* a coefficient? It is an indication of the *number of molecules or moles* of a substance required to balance the reaction between known reactants and products. Looking back at our example equation, you can see that the reaction of a single molecule of $H_2$ with a single molecule of $Cl_2$ will make *two molecules* of HCl. Whenever more than a single molecule of a substance is represented in an equation, a coefficient appears before it to indicate the number of those molecules needed for the equation to be balanced. It's not necessary to write a coefficient of 1, since the absence of a coefficient is assumed to indicate a 1.

Also notice, as I already hinted, that a coefficient can also represent a number of moles as well as a number of molecules. This reaction:

$$H_2 + Cl_2 \rightarrow 2\ HCl$$

may be thought of as a reaction between *molecules* of $H_2$ and $Cl_2$ or as a reaction between *gram atomic masses* of these two substances. In either case, the balanced equation accurately represents the real reaction. That's because gram atomic masses (or moles) represent the exact proportions of atoms in the reactants and products. There'll be more on this subject as we continue.

To balance an equation is simply to recognize and record what happens in nature. It's not that we can't talk about solitary hydrogen atoms, but we must recognize that such atoms are absent from nature for most practical purposes. We express entities on paper that are true to nature. Practically all free hydrogen is present in the form of $H_2$, so that's what we put in our equations. The same is true of $H_2Cl_2$. We don't write $H_2Cl_2$ into our equations because, for most practical purposes, it can be considered nonexistent. (See Figure 1.)

So back to reality. Suppose we started out with a single hydrogen molecule consisting of two hydrogen atoms bound together by the "covalent" sharing of a pair of electrons. Now a second molecule enters the picture—a diatomic chlorine molecule. What will happen when these two different kinds of structures come into close contact with one another? The forces that hold those molecules together will begin to interact. (See Figure 2.) What happens from that point depends on the forces.

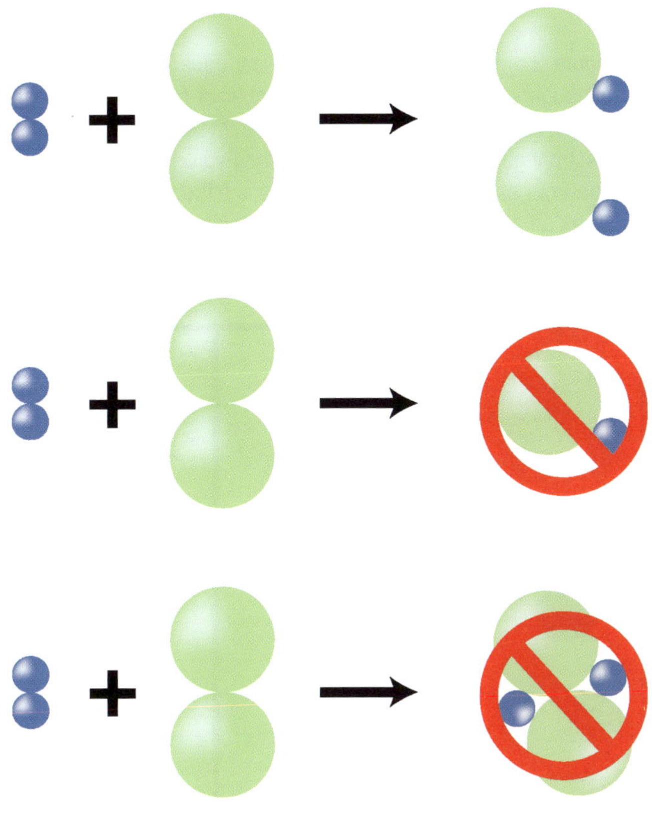

**Figure 1.** When one molecule of $H_2$ and one molecule of $Cl_2$ react, the product is always two molecules of HCl. The atoms on the right side of the equation are the exact same atoms as those on the left side of the equation. They have just rearranged under their natural forces.

So then, the 2 HCl is not just *necessary* to make the formula balance; it's the *only* way that this reaction can take place. In our equations we must represent *real* situations. Now I realize that you do not yet have all the information you need to start deciding what molecules will exist in nature and which will not. But, for the present, just be aware that chemical equations are written to represent real situations.

Let's look at another simple example—the reaction of hydrogen and oxygen to produce water. You know how all three of these molecules exist in nature.

**Figure 2.** When we write a chemical equation we are trying to communicate what happens in a real chemical reaction. Although we might imagine any number of products, the real products will be those that result from the real forces among atoms. **A:** Here we see an encounter between a hydrogen molecule and a chlorine molecule. **B:** Because chlorine is strongly electronegative, the electrons from the hydrogen atoms are pulled toward separate chlorine atoms. Likewise, one outer electron from each of the chlorine atoms is attracted to the corresponding hydrogen atom. **C:** The chlorine and hydrogen atoms break off attractions with their previous partners to join with new partners. The result is two new molecules of HCl.

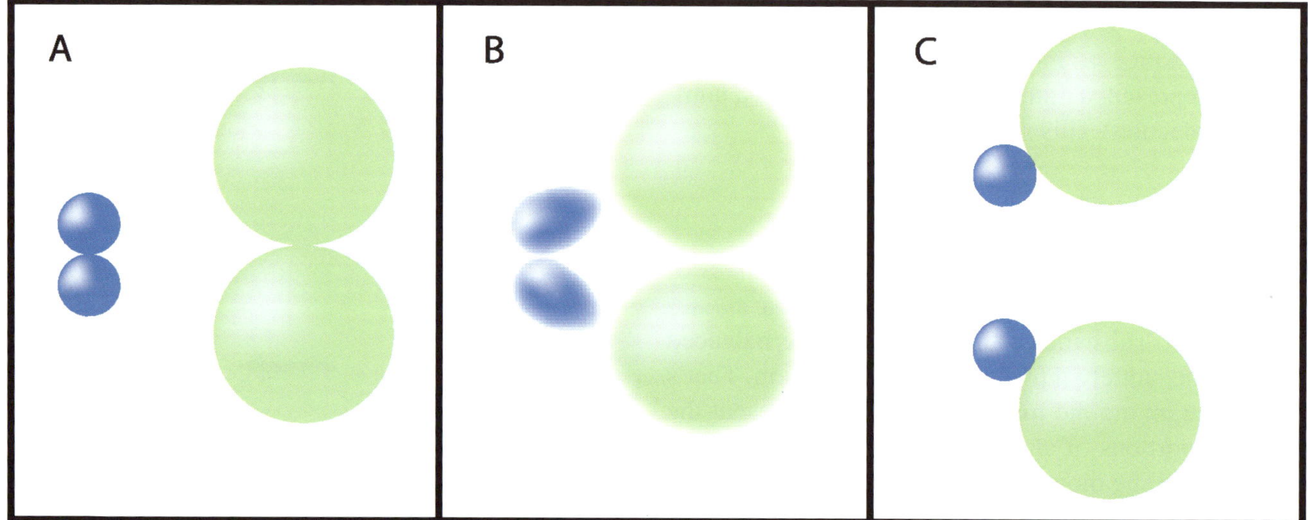


Both hydrogen and oxygen are diatomic. Water is water—$H_2O$. Our initial equation would look like this:

$$H_2 + O_2 \rightarrow H_2O$$

A quick look at this equation indicates that the hydrogen atoms are balanced (two on each side). But the oxygen atoms aren't—two on the left side, one on the right. The first thing we would do to clear up this problem is to put a coefficient of 2 in front of the formula for water to get the same number of oxygen atoms on each side:

$$H_2 + O_2 \rightarrow 2\ H_2O$$

However, after doing so, the hydrogen atoms are no longer balanced—now we have two H atoms on the left side, and four on the right side. But we can fix this by placing a coefficient of 2 in front of $H_2$. So we get:

$$2\ H_2 + O_2 \rightarrow 2\ H_2O$$

We now have 4 hydrogen atoms and 2 oxygen atoms on each side of the equation. (Please notice that we didn't monkey with any molecular formulas – only with coefficients.)

Figure 3 on the following page shows you this same process using molecules instead of molecular formulas. If you read the caption and follow the logic, you'll see that balancing a reaction equation is an attempt to mimic what happens to real molecules using the smallest number of molecules possible to get the job done. It's important to realize that, when you are finished, the numbers of atoms (and therefore, the masses of the atoms) must be identical on both sides of the equation. We can't have atoms appearing and disappearing into thin air. (See Figure 4.) This is the well-known concept of **mass balance**. Chemists (and scientists in general) are cautious to be sure that they can account for all the mass of reactants by showing that it ended up in the products. If they can't it's very difficult for them to convince an audience of careful people that their research is worthwhile.

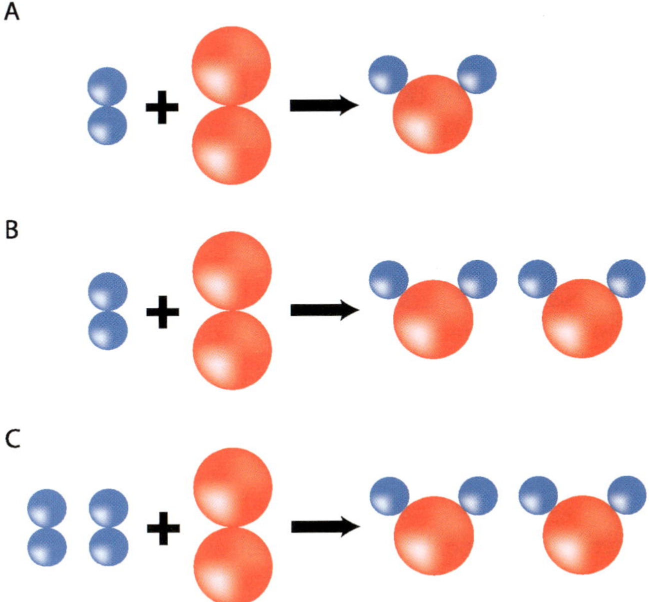


**Figure 3.** When we begin writing an equation of a reaction, we start out with the molecular formulas. We represent only one molecule of each reactant and product as in **A.** However, you'll quickly notice that there are four atoms on the left and only three on the right. You can't react four and only get three. What really happens? **B.** We begin trying to reconcile this problem by adding atoms to the right. Since the only product of this reaction is water, all we can do is add more water molecules. Here we've added a second water molecule. That's the equivalent of placing a "2" in front of the molecular formula while balancing the equation. Now we have six atoms on the right and still have only four on the left. There are two oxygen atoms on each side, but the hydrogen atoms don't balance. **C.** If we double the number of hydrogen molecules on the left (just like putting a "2" in front of the molecular formula) the equation is balanced in terms of oxygen and in terms of hydrogen. And we balanced it without changing any molecular formulas. Only $H_2$ and $O_2$ are present before the reaction and only $H_2O$ is present afterward.

**Figure 4.** If you were able to put the molecules of the reactants and the molecules of the products in a pan balance, it would balance perfectly. The masses on the left are equal to the masses on the right. This called the concept of "mass balance."

Let's do one more simple reaction—the reaction of hydrogen and nitrogen to produce ammonia ($NH_3$). Our initial equation would be:

$$H_2 + N_2 \rightarrow NH_3$$

Balancing nitrogen first we would discover that we need a "2" in front of the formula for ammonia so that we would have two nitrogen atoms on each side:

$$H_2 + N_2 \rightarrow 2\ NH_3$$

Then, in order to balance the hydrogen, we would need to put a "3" in front of $H_2$:

$$3\ H_2 + N_2 \rightarrow 2\ NH_3$$

Now there are six hydrogen molecules on each side. The equation is balanced.

Often chemical equations are more complex than these examples. For that reason, some guidelines have been developed in order to make the balancing of equa-

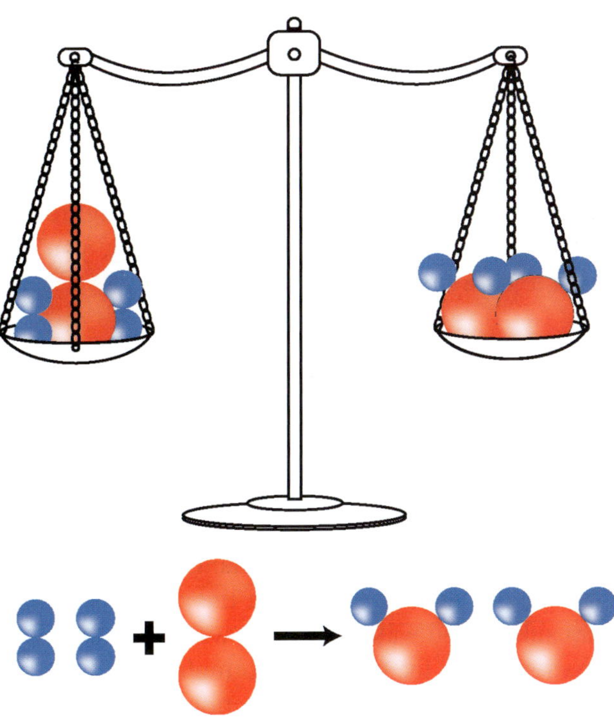

tions a little more structured. Here are the steps to follow:

1) **Identify the reactants and products.** Somehow, information will be given to you concerning the chemical reaction that is taking place. The substances involved may be given as formulas or they may be identified by name. In any event, the first step is to identify what the reactants are and what the products are.
2) **Write an unbalanced equation.** Once you know which substances are reactants and which ones are products, write an equation showing the molecular formula for each substance involved in the reaction.
3) **Balance the equation.** In general, it is a good practice to balance metals first, then nonmetals, then hydrogen, then oxygen. Occasionally, by the time you get to hydrogen and oxygen, they are already balanced.
4) **Do a final check.** When changing the coefficients of reactants and products, it is easy to find that making a change in a coefficient to balance one element will put another element out of balance. So it is highly recommended that you make sure each element is balanced before thinking you're done.

One more point–don't be a slave to the balancing rules. If you can balance an equation by inspection, then by all means, go ahead. The structured method is given only to provide a way for you to approach equations that appear to be too complex to just "figure out." But, however you approach the balancing, you'd be smart to do a final check.

Let's look at a couple of other examples, then you can do some on your own. Aluminum hydroxide will react with $H_2SO_4$ (sulfuric acid) to form aluminum sulfate and water. Show the balanced equation for this reaction.

First, we identify the reactants—aluminum hydroxide and sulfuric acid—and the products—aluminum sulfate and water. Then we write the unbalanced equation, using the molecular formulas for all the substances participating in the reaction. We were given the formula for sulfuric acid; it's $H_2SO_4$. Using our nomenclature rules we determine that the formula for aluminum hydroxide is $Al(OH)_3$, and the formula for aluminum sulfate is $Al_2(SO_4)_3$. So the unbalanced equation is written as:

$$Al(OH)_3 + H_2SO_4 \rightarrow Al_2(SO_4)_3 + H_2O$$

Next, we balance the equation. Starting with aluminum, we note that there are is one aluminum atom on the left side and two on the right side. So we put a "2" in front of $Al(OH)_3$. That gives us:

$$2\ Al(OH)_3 + H_2SO_4 \rightarrow Al_2(SO_4)_3 + H_2O$$

Then we balance the nonmetal, sulfur. At this point we have one sulfur atom on the left side and three on the right. So we put a "3" in front of the $H_2SO_4$. That gives us:

$$2\ Al(OH)_3 + 3\ H_2SO_4 \rightarrow Al_2(SO_4)_3 + H_2O$$

Then we balance oxygen. At this point we have 18 oxygen atoms on the left side (make sure you see this), and 13 on the right side. Well, there are two compounds on the right side of the equation whose coefficients we can increase, but if we increase the coefficient of the $Al_2(SO_4)_3$ by only 1 unit, we increase the number of oxygen atoms by 12. Besides that, we would unbalance the aluminum and the sulfur which we worked so hard to balance. The only logical thing to do is to raise the number of water molecules to "6." That gives us 18 oxygen atoms on each side, and it takes us to this point:

$$2\ Al(OH)_3 + 3\ H_2SO_4 \rightarrow Al_2(SO_4)_3 + 6\ H_2O$$

Lastly, we need to balance hydrogen. But upon checking we find that there are now 12 hydrogen atoms on the left and 12 on the right. The equation is balanced. Check it to be sure. It isn't that there aren't tougher equations to balance, but most of the equations you run across in practice are no tougher than that one.

Aluminum metal will react with nickel(II) sulfate to produce aluminum sulfate and elemental nickel. Let's write the balanced equation for this reaction. First, we must identify the reactants and products. Metallic elements exist and are written into equations as individual atoms of that element. So aluminum is written simply as Al; the other reactant is $NiSO_4$. The products are $Al_2(SO_4)_3$ and Ni. So we write the unbalanced equation as:

$$Al + NiSO_4 \rightarrow Al_2(SO_4)_3 + Ni$$

Then we balance the equation. Since we see that there is only one Ni on each side of the equation, we'll start with aluminum

$$2\ Al + NiSO_4 \rightarrow Al_2(SO_4)_3 + Ni$$

Next we'll balance sulfur:

$$2\ Al + 3\ NiSO_4 \rightarrow Al_2(SO_4)_3 + Ni$$

Then oxygen. Hey! Oxygen is already balanced! But when we recheck we notice that the nickel is no longer balanced. You see, we threw it out of whack when we balanced sulfur. We need to put a 3 in front of Ni on the right side to throw it *back in* whack:

$$2\ Al + 3\ NiSO_4 \rightarrow Al_2(SO_4)_3 + 3\ Ni$$

Now we're finished.

**Exercises**

Balance the following equations:

1) $Ba(OH)_2 + HCl \rightarrow BaCl_2 + H_2O$

2) $Fe + FeCl_3 \rightarrow FeCl_2$

3) $La(OH)_3 + HClO_3 \rightarrow La(ClO_3)_3 + H_2O$

4) $Sb_2S_3 + HCl \rightarrow H_3SbCl_6 + H_2S$

## 31: Chemical Equations: The Dark Side

It's probably occurred to you as you have been practicing the balancing of chemical equations that some of these might get pretty complex. Right you are. Although you might not think so, the ones you've been given thus far are fairly straightforward. In this lesson we want to show you what to do when a certain kind of situation is encountered.

Let's ease into this by balancing the equation that shows the complete oxidation of glucose to form carbon dioxide and water. In case you don't remember, the formula for glucose is $C_6H_{12}O_6$. (See Figure 1.) So our initial equation would look like this:

$$C_6H_{12}O_6 + O_2 \rightarrow CO_2 + H_2O$$

**Figure 1.** Glucose is the most used fuel molecule in the plant and animal kingdoms. Plants can produce it using energy from sunlight and carbon dioxide from the atmosphere. Animals produce it from a variety of source materials or consume it directly. It can be stored in a variety of forms and broken down quickly to obtain needed energy. (Black is used for carbon, red for oxygen and blue for hydrogen.)

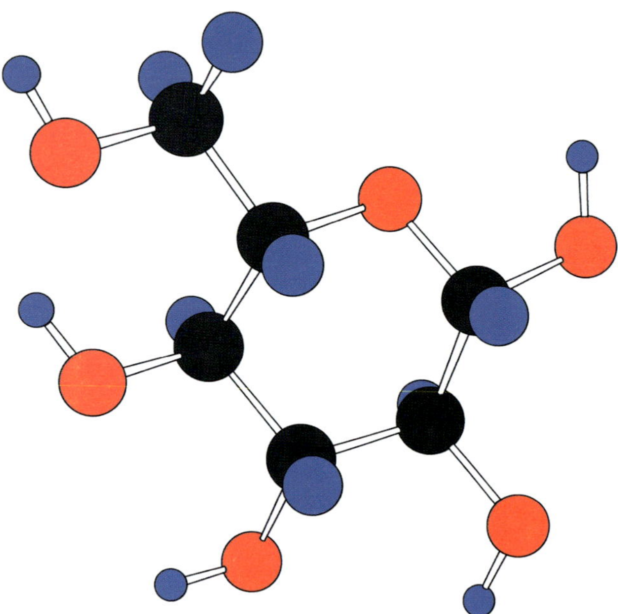

Starting with carbon, we notice that there are 6 carbon atoms on the left and only one on the right, so we put a "6" in front of the carbon dioxide:

$$C_6H_{12}O_6 + O_2 \rightarrow 6\ CO_2 + H_2O$$

Moving on to hydrogen we observe that there are 12 hydrogen atoms on the left and 2 on the right. So we put a 6 in front of water to give us 12 hydrogen atoms on each side:

$$C_6H_{12}O_6 + O_2 \rightarrow 6\ CO_2 + 6\ H_2O$$

Finally, we observe that there are now 8 oxygen atoms on the left and 18 on the right. So if we put a "6" in front of $O_2$ then we will have 18 on each side. This gives us the balanced equation:

$$C_6H_{12}O_6 + 6\ O_2 \rightarrow 6\ CO_2 + 6\ H_2O$$

Okay, I know this was pretty easy—this was just to give you more practice and warm you up. But this example also shows something that you need to remember. *The complete oxidation of any compound that has the general formula of $C_xH_yO_z$ or $C_xH_y$ will always produce $CO_2$ and $H_2O$ as the final products.* As I said, you should remember this. It will come in handy later.

Okay, so armed with that information, if I were to tell you to write the balanced equation for the complete oxidation of propane, $C_3H_8$, then you'd be set. First you'd write the unbalanced equation:

$$C_3H_8 + O_2 \rightarrow CO_2 + H_2O$$

Then you'd balance carbon:

$$C_3H_8 + O_2 \rightarrow 3\ CO_2 + H_2O$$

Then you'd balance hydrogen:

$$C_3H_8 + O_2 \rightarrow 3\ CO_2 + 4\ H_2O$$

## Chemical Oxidation as a Source of Energy

Complete oxidation reactions have a special place in the heart of every chemist and every biologist. Let me appeal to your sense of loyalty to this noble process. Hydrocarbons (chemicals consisting of hydrogen, carbon and sometimes oxygen), are highly reduced chemicals. In an atmosphere of oxygen, to be "reduced" means that there is considerable potential energy bound up in their molecules. That makes these molecules excellent sources of usable energy.

It wouldn't be useful to know that these chemicals supply considerable energy if the total world store of that energy were small. However, consider that these are the compounds that make up petroleum fuels and coal, and the wood in trees. Oxidation of these chemicals is responsible for most of the mechanical energy that runs automobiles and heavy equipment. It's responsible for a great deal of the energy that heats our homes and businesses. Since the fuel is only one of the reactants in this equation, it is also important to know that oxygen, the other reactant, is also abundant.

Of course, anytime two such abundant reactants are allowed to combine in such high quantities, the products of that reaction will also be abundant. That's why there is concern among some in the scientific community, industry, government and (especially) the media that the burning of fossil fuels will cause an overabundance of carbon dioxide in our atmosphere.

The other point of discussion about fueling reactions is that the more oxidized hydrocarbons ($C_xH_yO_z$) are not petroleum products. They are largely carbohydrates (including simple sugars) and fats. Glucose, the most common sugar in living things, is $C_6H_{12}O_6$. Energy storage fats are of the structure $C_3H_5O_3R_3$, where each of the R units is one of many fatty acids. These are the fuels of plants and animals.

In brief, plants and animals live off the same chemical processes (although using different source materials) that cars, boats and furnaces use. Incidentally, some bacteria can actually feed on petroleum.

These "machines" get their energy from fuel molecules: the machines on the left from glucose; the machines on the right from a variety of carbohydrates and fats. Some bacteria even use gasoline as though they were little automobiles.

Then you'd balance oxygen:

$$C_3H_8 + 5\ O_2 \rightarrow 3\ CO_2 + 4\ H_2O$$

Again, no problem.

But what if I asked you to write the balanced equation for the complete oxidation of butane, $C_4H_{10}$? Well, it looks pretty much like the other one, doesn't it? So let's write the unbalanced equation:

$$C_4H_{10} + O_2 \rightarrow CO_2 + H_2O$$

Then let's balance it around carbon:

$$C_4H_{10} + O_2 \rightarrow 4\ CO_2 + H_2O$$

Then let's balance it with respect to hydrogen:

$$C_4H_{10} + O_2 \rightarrow 4\ CO_2 + 5\ H_2O$$

Then let's balance oxygen. Well, what are you waiting for. Balance those oxygen atoms!

Houston, we have a problem! At this point you have 13 oxygen atoms on the right side of the equation and two oxygen atoms on the left. But the only oxygen-containing substance on the left side is $O_2$, and no matter how much you increase its coefficient you will still have an even number of oxygen atoms on the left. Alas, what to do?

Here's what you do. You need 13 oxygen atoms on each side of the equation, but the only way that you can get 13 oxygen atoms from $O_2$ is if you have 13/2 of $O_2$ molecules. Okay, so do that. Write an *interim* equation like this:

$$C_4H_{10} + \tfrac{13}{2}O_2 \rightarrow 4\,CO_2 + 5\,H_2O$$

This equation is balanced—it just doesn't express the number of molecules of each substance in whole numbers. But you can make it do that. Just multiply everything on both sides of the equation by 2:

$$2\ C_4H_{10} + 13\ O_2 \rightarrow 8\ CO_2 + 10\ H_2O$$

In this regard, chemical equations are just like mathematical equations—if you multiply everything by the same number, you haven't changed the balance.

Let's look at another example—the reaction that occurs when aluminum metal reacts with nitric acid ($HNO_3$) to produce aluminum nitrate and hydrogen gas. The initial unbalanced equation would be:

$$Al + HNO_3 \rightarrow Al(NO_3)_3 + H_2$$

The aluminum atoms are balanced, so we move on to nitrogen. It can be balanced by placing a "3" in front of the nitric acid:

$$Al + 3\ HNO_3 \rightarrow Al(NO_3)_3 + H_2$$

Then we try to balance hydrogen. Again, there is an odd number of hydrogen atoms on the left side and an even number on the right. There is no single coefficient we can increase to make things balance. So we put 3/2 in front of $H_2$:

$$Al + 3\,HNO_3 \rightarrow Al(NO_3)_3 + \tfrac{3}{2}\,H_2$$

That gives us an equation that balances, although not everything is expressed in whole numbers. Then again we multiply through by 2 to get the final balanced equation.

$$2\ Al + 6\ HNO_3 \rightarrow 2\ Al(NO_3)_3 + 3\ H_2$$

If multiplying through by the same number doesn't affect the balance of a chemical equation, it should follow that dividing through by the same number doesn't affect the balance. Check your final equation to be sure that you cannot simplify it further by dividing the coefficients by a common factor. For example, a chemistry student might come up with the following balanced equation for the complete oxidation of ethanol:

$$2\ C_2H_5OH + 6\ O_2 \rightarrow 4\ CO_2 + 6\ H_2O$$

Although this equation is balanced, it is not as simplified as it could be. Each of the coefficients is divisible by 2. The following balanced equation is easier to look at and understand:

$$C_2H_5OH + 3\ O_2 \rightarrow 2\ CO_2 + 3\ H_2O$$

Because this equation is equivalent to the previous one, but simpler, it is accepted among friendly chemists that we will always simplify when we can. If you don't join the ranks of friendly chemists, friendly chemists will not give you full credit for your answers on friendly tests.

**Exercises**

Please balance the following equations:

1) $KO_2 + H_2O \rightarrow KOH + O_2$

2) $Co + O_2 \rightarrow Co_2O_3$

3) $P + I_2 \rightarrow PI_3$

4) $C_2H_6 + O_2 \rightarrow CO_2 + H_2O$

# 32: Reaction Types

You've probably been able to figure out by now that there are zillions of chemical reactions. However, as is true with most things, they can be grouped into a much smaller number of categories. In this lesson we want to begin considering briefly a couple of different kinds of reactions, and to consider one particular type in some detail.

## Combination Reactions

One of the most common types of chemical reactions, called **combination reactions**, occurs when one substance combines with another substance to produce a third substance. Because the third substance is "synthesized", or formed, during these reactions, they are sometimes called **synthesis reactions**. Without specifying the elements, we can write a "**general form**" of this type of reaction as follows:

$$A + B \rightarrow AB$$

AB is a single compound that is *synthesized* by the *combination* of A and B.

We have already seen many examples of this kind of reaction. Some chemical reactions that fall into this category are shown here:

$$2\ H_2 + O_2 \rightarrow 2\ H_2O$$

$$2\ Na + Cl_2 \rightarrow NaCl$$

$$CaO + H_2O \rightarrow Ca(OH)_2$$

## Decomposition Reactions

A second (opposite) category of reactions includes those in which a compound *decomposed* or "fell apart." Unsurprisingly, that reaction is called a **decomposition reaction**. The general form of this reaction is:

$$AB \rightarrow A + B$$

Here are some examples of decomposition reactions.

$$CaCO_3 \xrightarrow{\Delta} CaO + CO_2$$

(The Greek letter *delta* is often used in this fashion to show that heat is required to make the reaction proceed in the direction of the arrow.)

$$2\ H_2O \xrightarrow{\text{electric current}} 2\ H_2 + O_2$$

$$N_2H_4 \rightarrow N_2 + 2\ H_2$$

## Single Displacement Reactions

Another type of reaction, and one on which I want us to spend some time, is that type that can be characterized in this way:

$$A + BC \rightarrow B + AC$$

There are three different subtypes of this general reaction I want you to know. They are:

1) A is either a solid metal or hydrogen gas, and BC is an ionic compound.
2) A is a metal and BC is $H_2O$.
3) A is a halogen and BC is a metal halide.

**Displacement of a Metal Ion by a Metallic Solid.** Zn metal will react with an aqueous solution of hydrochloric acid according to the following balanced equation:

$$Zn\ (s) + 2\ HCl\ (aq) \rightarrow H_2\ (g) + ZnCl_2\ (aq)$$

Can you see that the Zn has "displaced" the hydrogen? Before the reaction it was the hydrogen that was bonded to the chlorine. Afterwards it is the zinc that is bonded to the chlorine. Of course, it is not simply a one-on-one kind of thing. We don't get ZnCl because there is no such thing—it's not charge balanced. And, when hydrogen is liberated, it exists in its normal diatomic form, $H_2$. But the element "partners" have switched.

$$Zn + 2\,HCl \rightarrow H_2 + ZnCl_2$$

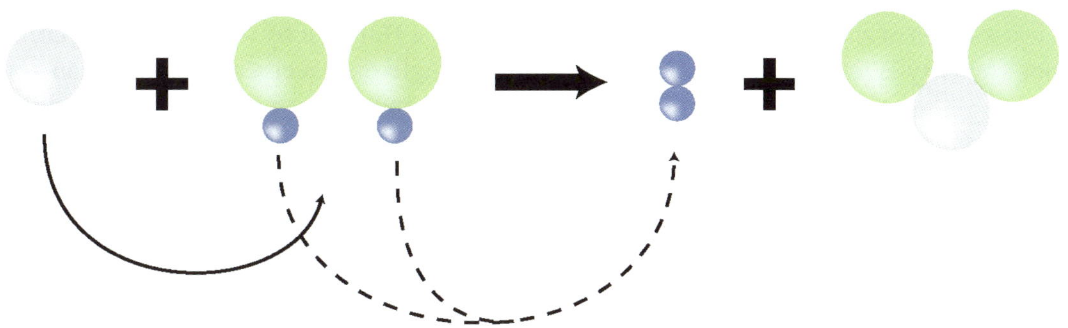

**Figure.** This is a single displacement reaction. A single, lone element (zinc) displaces another single element (hydrogen) in a compound. Notice that a single displacement reaction may involve more than one atom of the elements involved. In this Figure, the element *hydrogen* is being displaced in two molecules at the same time.

| Activity Series |
|---|
| Li |
| K |
| Ca |
| Na |
| Mg |
| Al |
| Mn |
| Zn |
| Cr |
| Fe |
| Co |
| Ni |
| Sn |
| Pb |
| H |
| Cu |
| Ag |
| Hg |
| Pt |
| Au |

(See the Figure.) This kind of reaction is very common with metals and acids.

Aluminum, manganese, chromium, iron and many other metals will displace hydrogen from HCl just like zinc will. However, copper, silver, platinum and some other metals won't. Is there any way that you can predict which will and which won't, so that you don't have to memorize zillions of reactions? Yes, what you need is an **activity series**. This is a list of elements in order of their "chemical activities." Chemical activity is an indication of how readily a metal will participate in a chemical reaction by giving up its valence electrons—those outer, most loosely held, most energetic electrons that tend to be vulnerable to an electron thief.

The Table on the upper right of the page shows an activity series of selected metals (the most common ones encountered in everyday chemistry) and of hydrogen (because hydrogen so often acts like a metal).

As long as you have access to this list, then you should be able to figure out which metals will displace hydrogen from HCl and which ones won't. More than that, you will also know which metals will displace hydrogen from sulfuric acid ($H_2SO_4$) nitric acid, ($HNO_3$), or any other acid. For example,

$$Zn\ (s) + H_2SO_4\ (aq) \rightarrow H_2\ (g) + ZnSO_4\ (aq)$$
$$Zn\ (s) + HNO_3\ (aq) \rightarrow H_2\ (g) + Zn(NO_3)_2\ (aq)$$

As you can see, the list also indicates that copper, silver, mercury, platinum and gold are less active than hydrogen, so they will not displace hydrogen from HCl or any other acid. The following reaction will not go:

$$Cu\ (s) + HCl\ (aq) \rightarrow H_2\ (g) + CuCl_2$$

The activity series can be used for other purposes besides determining which metals will displace hydrogen. Remember, we said that this list is given in order of descending activity. Any element found in the list will displace anything below it, but nothing above it. Since Zn is listed above hydrogen, we can know ahead of time that it will displace hydrogen. What would happen if you put a piece of Zn metal in an aqueous solution of $CuSO_4$? Would anything happen? Yes, indeed.

Since Zn is higher than Cu in the activity series, then Zn would displace Cu, too:

$$Zn\ (s) + CuSO_4\ (aq) \rightarrow Cu\ (s) + ZnSO_4\ (aq)$$

What about the converse? That is, what if you put a piece of copper metal in a solution of zinc sulfate? Nothing. Nada. Zip. Copper's activity is lower than zinc's.

Perhaps you can now see how useful the activity series is. It can guide you in deciding the results of a large number of chemical reactions. This is pretty easy stuff–the more active metal displaces the less active metal. Just don't forget that hydrogen is in this list, too.

**Displacement of Hydrogen from Water.** Metals having a higher activity than hydrogen can displace hydrogen from water, giving a metal hydroxide or metal oxide as the second product. Some metals, like sodium or potassium, do it easily. Others might require more energetic conditions for the reaction to take place at a reasonable rate.

Here are some examples:

$$2\ K\ (s) + 2\ H_2O \rightarrow H_2\ (g) + KOH\ (aq)$$

$$3\ Fe(s) + 4\ H_2O \xrightarrow{\Delta} 4\ H_2\ (g) + Fe_3O_4\ (s)$$

**Displacement of a Halogen by Another Halogen.** It turns out that single displacement reactions can also take place where one halogen displaces another. For example, when an aqueous solution of chlorine ($Cl_2$), which is slightly yellow-green, is mixed with a colorless, aqueous solution of sodium bromide, the chlorine will displace the bromine. The color of the solution turns deep red—the color of bromine ($Br_2$). The reaction takes place according to this equation:

$$Cl_2\ (g) + 2\ NaBr\ (aq) \rightarrow Br_2\ (l) + 2\ NaCl\ (aq)$$

And, just like the reaction between the metals and hydrogen, there is an activity series for the halogens. The order of their activity, from greatest to least, is $F_2 > Cl_2 > Br_2 > I_2$. That should be easy to remember; it's the same as their order on the periodic table.

**Exercises**

Write balanced equations for the reaction of each of the following pairs. If the reaction will not occur, strike through the arrow to indicate "no reaction."

1) magnesium metal and an aqueous nitric acid to form hydrogen gas and aqueous magnesium nitrate

2) chlorine gas and aqueous sodium iodide to form iodine gas and aqueous sodium chloride

3) copper metal and an aqueous iron(II) nitrate to form aqueous copper(II) nitrate and iron metal

4) calcium and water to form hydrogen gas and aqueous calcium hydroxide

5) solid iodine aqueous hydrochloric acid to form chlorine gas and aqueous hydrogen iodide

6) iron metal and aqueous of magnesium chloride to form solid magnesium and iron(II) chloride

7) zinc and water to form hydrogen gas and aqueous zinc hydroxide

8) zinc metal and an aqueous lead(II) sulfate

## 33: The Double Switch

Another prominent category of reactions is that of **double displacement** reactions or **double replacement** reactions. The general form of this type of reaction can be expressed as:

$$AB + CD \rightarrow AD + BC$$

As you can see, there is a little similarity between these reactions and the single displacement reactions we looked at in the last lesson. But in this case, *two* elements exchange partners with each other. Most of the time, reactions of this sort occur between two ionic compounds in aqueous solution. In only a few cases will one or both of the reactants not be in solution. Here are some examples of double displacement reactions:

$$AgNO_3\ (aq) + NaCl\ (aq) \rightarrow AgCl\ (s) + NaNO_3\ (aq)$$
$$Na_2CO_3\ (aq) + CaCl_2\ (aq) \rightarrow CaCO_3 + 2\ NaCl\ (aq)$$

These reactions are depicted in Figure 1.

Water is one of the few solvents that has enough polar character to it to wedge its way between the ions in an ionic compound. So, when you dissolve ionic compounds, the respective ions are "**solvated**" by the solvent molecules, and the compound "**dissociates**" into its ions. In other words, when you dissolve NaCl in water, you don't have NaCl molecules moving about the water. Instead, you have free sodium cations and free chloride anions. The water molecules surround the ions and satisfy their charges.

That helps to explain how the compounds in double displacement reactions "exchange partners." Dissociated ions float around in solution, free to interact with each other. And frankly, sometimes that's all that happens when aqueous solutions of two ionic compounds are mixed together. If the dissolved ions have no great attraction for one another, they just remain separate in solution. On the other hand, when there is an attraction among ions that is stronger than their at-

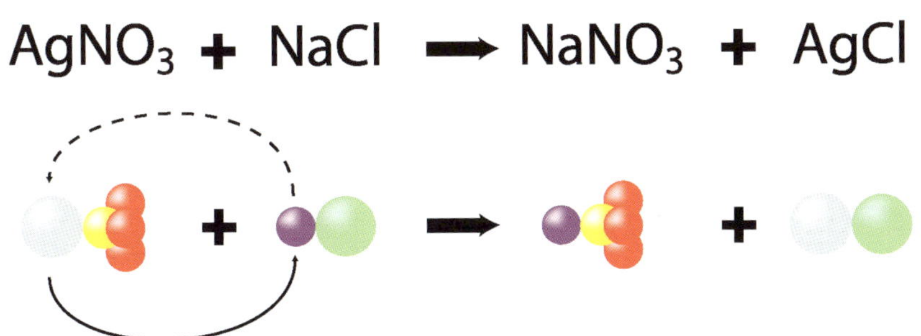


**Figure 1.** In a double displacement reaction, both pairs of elements swap bonding partners, forming two new compounds.

$Na_2CO_3 + CaCl_2 \rightarrow CaCO_3 + 2\ NaCl$

tractions to the water molecules, ions combine to form a new compound. Now it could be that one pair of ions forms a new compound that leaves the solution. Afterward, the solution would be made up of water and the ions that are left over. Or, it could be that both pairs of ions will form new compounds, leaving no ions in solution. Any ions that do not form an insoluble compound will remain in solution until the water is evaporated, leaving behind a soluble ionic compound that forms from the leftover ions. (See Figure 2.)

Some of the more obvious signs that indicate a change in bonding partners are: formation of a precipitate, indicating that one or both of the products is insoluble in water, evolution (giving off) of a gaseous product or formation of a molecular compound, such as water.

You see, when any of these three things happens, ions are removed from the solution. Let's look at some examples. First, what happens when aqueous solutions

**Figure 2.** **(A)** When two compounds dissolve, four kinds of ions will be present in solution. **(B)** If two of those four kinds of ions have forces of attraction that are strong enough to overcome their separation by water molecules, they will combine and form a compound. This leaves the other two kinds of ions remaining in solution **(C)** unless their forces are strong enough to bring them together as well. When ions combine they may form an ionic compound or a molecular compound. The product may be a solid (which will precipitate), a liquid or a gas (which will bubble from the beaker). Ions remaining unreacted in solution will come together as compounds if the water is evaporated. In these diagrams, the reaction between sodium carbonate and calcium chloride is depicted. The condition represented in A takes place for only a fraction of a second. The attraction between the calcium and carbonate ions is very strong. These ions will fall out of solution immediately as solid calcium carbonate. The ions left behind are sodium and chloride. Their attractions are not strong enough to pull them together as long as water is present. If the sodium chloride solution were decanted from the calcium carbonate precipitate, the water could be evaporated to provide pure sodium chloride as the second reaction product.

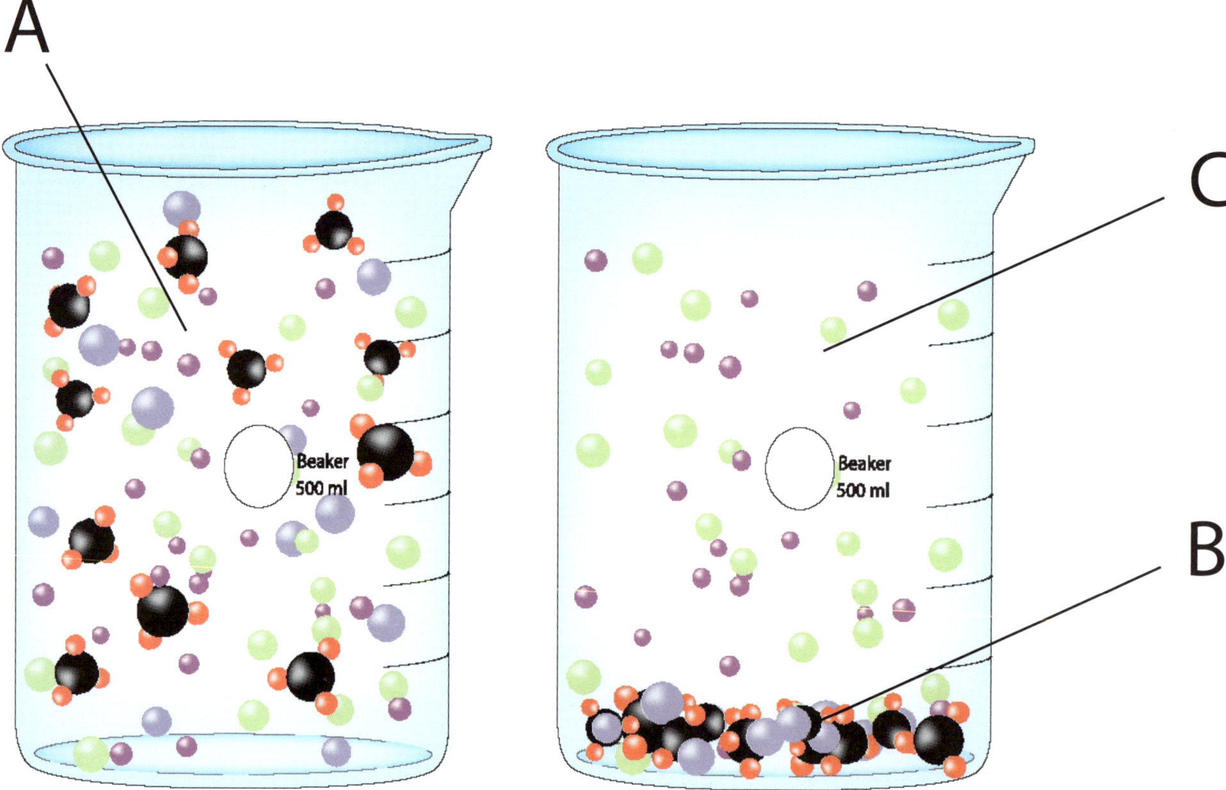

of barium chloride and sodium sulfate are combined? Let's begin by writing the first half of the equation:

$$BaCl_2 + NaSO_4 \rightarrow$$

Anytime you have two ionic compounds as reactants, the only way you can get a reaction without leaving ions stranded is to swap the bonding partners. The products would have to be barium sulfate and sodium chloride. The completed equation, when balanced, would look like this:

$$BaCl_2 + Na_2SO_4 \rightarrow BaSO_4 + 2\ NaCl$$

We know that sodium chloride, if formed, would remain in aqueous solution. But what about the barium sulfate? This is where it helps to know what makes a compound soluble. After the hundreds of years of exploration in chemistry, the best way to know is still by experience. That's because there are many complex factors that decide solubility. There is a wealth of information on the subject, and there are extensive tables that tell what every imaginable compound is soluble in under a variety of conditions. However, there are some rules of thumb that come about through experience. The most critical ones are listed in Table 1.

Well, if we look up barium sulfate in Table 1, we find that it is one of the few insoluble sulfates. So, as the two solutions were mixed together, a precipitate of barium sulfate would form. Reactions of this sort are also called **precipitation reactions**. Another way compounds leave solution is via gas bubbles. There are only a handful of gases that commonly evolve from chemical reactions. These are shown in Table 2.

**Table 1.** Solubilities of ionic compounds in water.

| Anion | Soluble | Insoluble |
|---|---|---|
| Nitrate | All | None |
| Chloride | All except | Ag, $Hg_2Cl_2$, Pb(II)* |
| Sulfate | All except | Ca*, Sr, Ba, Hg, Pb, Ag* |
| Sulfide | IA, IIA, $NH_4^+$ | All other |
| Carbonate | IA, $NH_4^+$ | All other |
| Hydroxide | IA, Sr, Ba, Ca* | All other |

* Somewhat soluble.
Modified from Masterton and Slowinski, 1978

As a second example, what do you suppose you would see if solid iron(II) sulfide were added to an aqueous solution of hydrochloric acid and a double displacement reaction took place? The balanced equation would be as follows:

$$FeS + 2\ HCl \rightarrow FeCl_2\ (aq) + H_2S\ (g)$$

Thus, we would expect to see bubbles, as hydrogen sulfide gas is formed. In most cases, by writing the balanced equations of proposed double displacement reactions, you should be able to determine if a gas will be evolved. Let me mention to you, though, that sometimes an unstable compound will be formed as a product, which will decompose to form water and a gas. Compounds of this sort to look out for are $NH_4OH$, $H_2CO_3$ and $H_2SO_3$. Here are those reactions:

$$NH_4OH\ (aq) \rightarrow H_2O\ (l) + NH_3\ (g)$$
$$H_2CO_3\ (aq) \rightarrow H_2O\ (l) + CO_2\ (g)$$
$$H_2SO_3\ (aq) \rightarrow H_2O\ (l) + SO_2\ (g)$$

Here is a third example of double displacement. Both sodium hydroxide and hydrochloric acid completely dissociate in water. But when 1 mole of NaOH is dissolved in water and added to an aqueous solution containing 1 mole of hydrochloric acid, the resulting solution contains no appreciable amount of either hydrogen or hydroxide ions. Why? You can figure out the answer to this question if you write the balanced equation for the reaction. Start with the reactants:

$$NaOH + HCl \rightarrow$$

Do the swap:

$$NaOH + HCl \rightarrow NaCl + H_2O$$

Since this equation is already balanced, no further work is necessary. Notice that of these two products, one (NaCl) is an ionic compound and the other ($H_2O$) is a molecular compound. If you recall, a molecular compound is one in which the attractions between atoms are not primarily ionic, but covalent. Hydrogen ions and hydroxide ions form water molecules, so that the ions no longer exist in their ionic forms. Once ions bind one another as molecules, they are no longer available in solution to combine with other ions.

This reaction is an example of an acid-base neutralization reaction in which a strong acid (HCl) reacts with a strong base (NaOH). The products of these reactions always include a salt (an ionic compound, such as NaCl) and water.

**Table 2.** Gases commonly evolved from chemical reactions.

| | |
|---|---|
| Diatomic gases | $H_2$, $Cl_2$*, $O_2$ |
| Nitrogenous gases | $N_2$, $NH_3$*, $N_2O$, $NO_2$, NO |
| Other | HCl*, $CO_2$*, $H_2S$ |

* These gases are soluble in water. When small amounts are formed they may remain in solution.

**Exercises**

For each of the following pairs, predict whether a double displacement reaction would take place by determining whether a precipitate, a gas or water would be formed. If no reaction would occur, just write "no reaction." Balance all equations where a reaction would take place.

1) $Pb(NO_3)_2$ (aq) + KOH (aq) →

2) $Na_2CO_3$ (aq) + HCl (aq) →

3) $Na_2SO_4$ (aq) + $Cu(NO_3)_2$ (aq) →

4) CaO (s) + HCl →

5) $Na_2CO_3$ (aq) + $NH_4Cl$ (aq) →

6) $(NH_4)_2S$ (aq) + HCl (aq) →

7) $CuSO_4$ (aq) + $Na_2S$ (aq) →

## 34: Oxidation-Reduction Reactions

This lesson is on a particular type of chemical reaction called an **oxidation-reduction** reaction or "**redox**" reaction for short. Please don't misunderstand, we don't mean to say "oxidation reaction *and* reduction reaction." What we mean to say is a *single reaction* that involves *both* an oxidation and a reduction. You see, you can't have one without the other. Any time one atom or ion gains an electron (**reduction**), another atom or ion must have lost one (**oxidation**).

Where do these terms—oxidation and reduction—come from? When a substance gains an electron, its charge is made more negative, or reduced. Hence, the term *reduction*. Going through this course, we'll mention the fact that oxygen and a few other atoms tend to hog any electrons in their neighborhood. When an atom steals an electron from another, it is acting like oxygen—hence, the term *oxidation*. (Although oxygen is the most common example of an oxidizer, there are stronger oxidizers than $O_2$.)

When will an electron change hands? Anytime the atom that possesses it encounters an atom that has a greater attraction to it. *Now* you have a basis for understanding the activity series we learned about a few lessons ago. The activity series is simply a listing of elements according to the ease with which they give up electrons to others. The activity series will approximately correspond to the ionization energies and electronegativities we saw in Lesson 13. Elements at the top of the activity series are those which give up their electrons most easily (have the lowest ionization energies). Those at the bottom of the list are those that hold their own electrons most tightly and accept electrons from other elements most readily (have the highest ionization energies).

### Why Do I Care?

- When bread browns in the oven, it is incomplete oxidation of the chemicals making up the bread that causes it to change color.
- When sliced apples have been exposed to the air for a period of time, oxidation makes their surfaces turn brown.
- When molds and other fungi grow on the surfaces of leaves and decompose them, it's oxidation of the carbon structure of the leaf that is responsible for its decay. The leaf is converted to carbon dioxide and literally becomes part of the atmosphere.
- Fire results from the oxidation of fuels: coal, petroleum and firewood.
- Laundry bleach cleans clothes by oxidizing the chemicals that stain the fabric. (It may also oxidize the chemicals that *color* the fabric. That's why bleach is known for "taking the color out.")
- Iron changes to rust when it is oxidized by oxygen.
- Certain chemicals, called antioxidants, protect your body from damage by oxidation. These reducing agents include vitamins C and A.
- Metal ores usually come out of the earth as oxidized minerals. A "reduction plant" (a factory that processes these ores) applies electrical current to reduce these oxidized metals to their metallic forms.
- All batteries use oxidation-reduction reactions to generate electrical current.
- You need redox reactions to fuel your body's activities.

I could go on but I'm guessing I've made my point!

Oxidation-reduction reactions have historically fueled most human activities. (Although lately, many of those activities are being given over to nuclear energy.) The burning of coal, petroleum and wood and the usage of electrochemical reactions in batteries are examples of oxidation-reduction reactions. The burning of carbon fuels results in a change in the oxidation state of carbon from its 4– state (in $H_4C$, for example) to its 4+ state (in $CO_2$). That represents an enormous release of energy. In the case of fuel oxidation, that energy is released in the form of heat and light (fire).

## Understanding Electron Transfer

What could the browning of bread and the operation of a battery possibly have in common? In a chemical sense, they are from the same class of reactions that take place when one substance has a greater affinity for electrons than another. When they come into contact with one another a redox reaction takes place. In the case of browning bread, it's the carbon in the dough coming into contact with oxygen at high temperature. In the case of an alkaline battery, it's manganese ions coming into electrical contact with metallic zinc. (See Figure 1.)

The simplest redox reactions involve an ion in solution passing an electron to a more positive ion. For example:

$$Zn^0 + Cu^{2+} \rightarrow Zn^{2+} + Cu^0$$

In this reaction, zinc metal (all metals are neutral in their "**metallic**" forms) passes two electrons to copper. You may think of this reaction as taking place in two separate steps even though those two steps take place

**Figure 1.** The alkaline batteries used in most electrical devices today are made from a manganese oxide core which serves as a source of electrons (at the cathode or negative terminal), and a zinc wrapper forming the anode (or positive terminal). The zinc anode is prevented from contact with the pasty core by an insulating layer. These batteries are called "alkaline" because the zinc is in an alkaline (or basic) paste made of potassium hydroxide. The potassium hydroxide serves as an "electrolyte" conducting medium for the chemical reaction between the manganese oxide oxidizer and the zinc reductant.

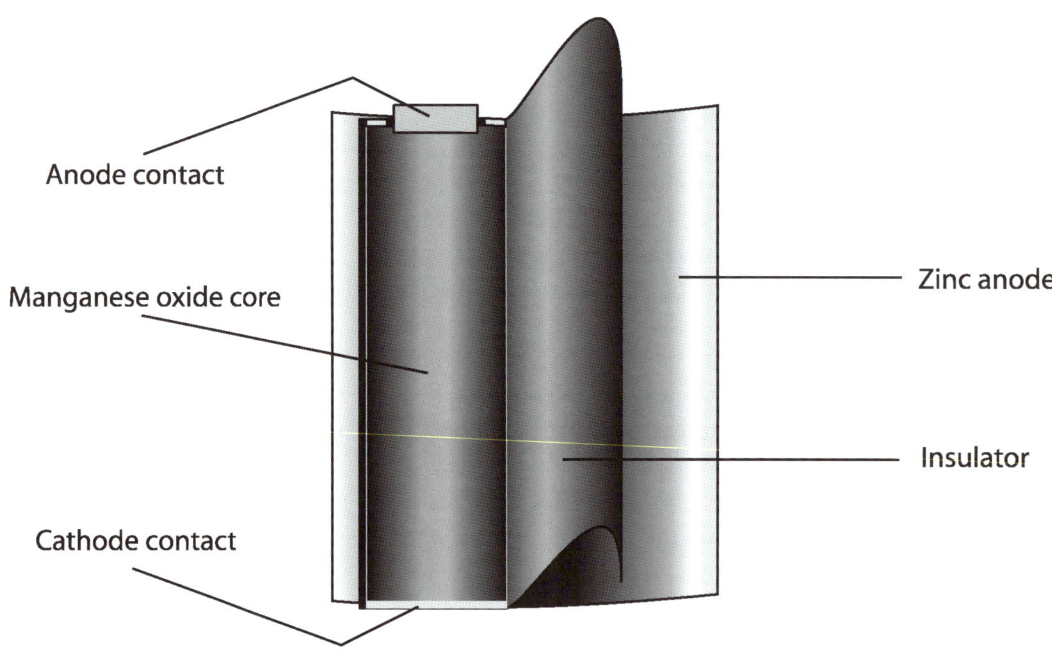

at the same time. First, zinc loses its electrons (under pull from the positive copper ions):

$Zn^0 \rightarrow 2e^- + Zn^{2+}$ (oxidation of metallic zinc)

and the positive copper ions receive them:

$Cu^{2+} + 2e^- \rightarrow Cu^0$ (reduction of copper ions)

These two expressions are referred to as **half reactions**. One of these half reactions describes the oxidation of zinc while the other describes the reduction of copper ions. You will refer to these as the **oxidation half reaction** and the **reduction half reaction**.

## Oxidation State and Oxidation Number

A given element may exist in a number of **oxidation states**, depending on its own atomic structure. It's easy to understand how two ions in an ionic compound react with one another based on their attractive charges—one positive for one negative. However, the attraction between two covalently bound atoms is not as easy to visualize.

It turns out that the same groups of atoms that react (because of their opposite charges) to make ionic compounds will also react to form covalent compounds in order to satisfy the octet rule. For example, you know that sodium chloride is an ionic compound that forms because of the tendency that group IA atoms have to become 1+ ions, and the tendency that group VIIA atoms have to become 1– ions. When sodium meets chlorine, the sodium atom gives up its extra electron. Chlorine accepts it, and the two become ionically bound because of the attraction of their new charges.

Hydrogen bromide (HBr) is an example of a *covalent* compound formed from the attraction of a group IA atom to a group VIIA atom. They bond, not because of ionic attraction, but because of the need that all atoms have to fill their valence shell with eight electrons. However, if hydrogen *were* to form an ion, then, it *would* form a 1+ ion. If bromine *were* to form an ion it would most likely form a 1– ion. For the sake of convenience, then, why don't we just assign them numbers that suggest how they would bond if they were ionic. This will be an easy way to identify potential compounds from among combinations of elements. In the compound hydrogen bromide, we would assign hydrogen an **oxidation number** of 1+ and bromine an oxidation number of 1–. *The oxidation number is a number that chemists made up to describe the ways in which elements have been seen to react.*

Most elements can take on more than one oxidation state. For example, the element bromine can exist as a 1– ion in solution, or it can covalently bind to another bromine atom and be neutral. We can assign a separate oxidation number to bromine for each oxidation state it is known to take on. We assign it an oxidation number of 1– when it acts in the manner of a monovalent, negative ion (as in HBr), or of 0 when it is neutral (as in $Br_2$). In any diatomic molecule, the bond connecting the two atoms has to be nonpolar. Because the compound is molecular and not ionic, we know that each atom is acting as a neutral atom.

Iron, having electrons that readily pass to other substances in its surroundings (as is true with many of the transition elements), can take on a number of different oxidation states, so it can be assigned corresponding oxidation numbers. Metallic iron is neutral and has an oxidation number of 0; iron (II) is a divalent cation (as in $FeCl_2$) and is assigned an oxidation number of 2+; iron (III) is a trivalent cation (as in $FeCl_3$), assigned an oxidation number of 3+.

In a neutral compound all the oxidation numbers will add up to zero. Iron(III) chloride is the same as $Fe^{3+}Cl^-_3$. Notice that the valence of iron complements the valence of the three chlorine atoms. Here, we have a single iron atom that is in its 3+ oxidation state bound to three chlorine atoms, each in its 1– oxidation state. All of the group IA elements can take on a 1+ valence, so they will certainly react with any of the halogens, all of which can take on a 1– valence. In fact, we know that HF, HCl, HBr, and HI are compounds that exist in reality. LiF, LiCl, LiBr and LiI are also realities. Any element with an oxidation number of 2+ (such as $Ca^{2+}$) will combine in a 1:1 molar relationship with any 2– (such as $O^{2-}$, to form CaO), or in a 1:2 ratio with any 1– element (such as $Cl^-$, to form $CaCl_2$), and so on.

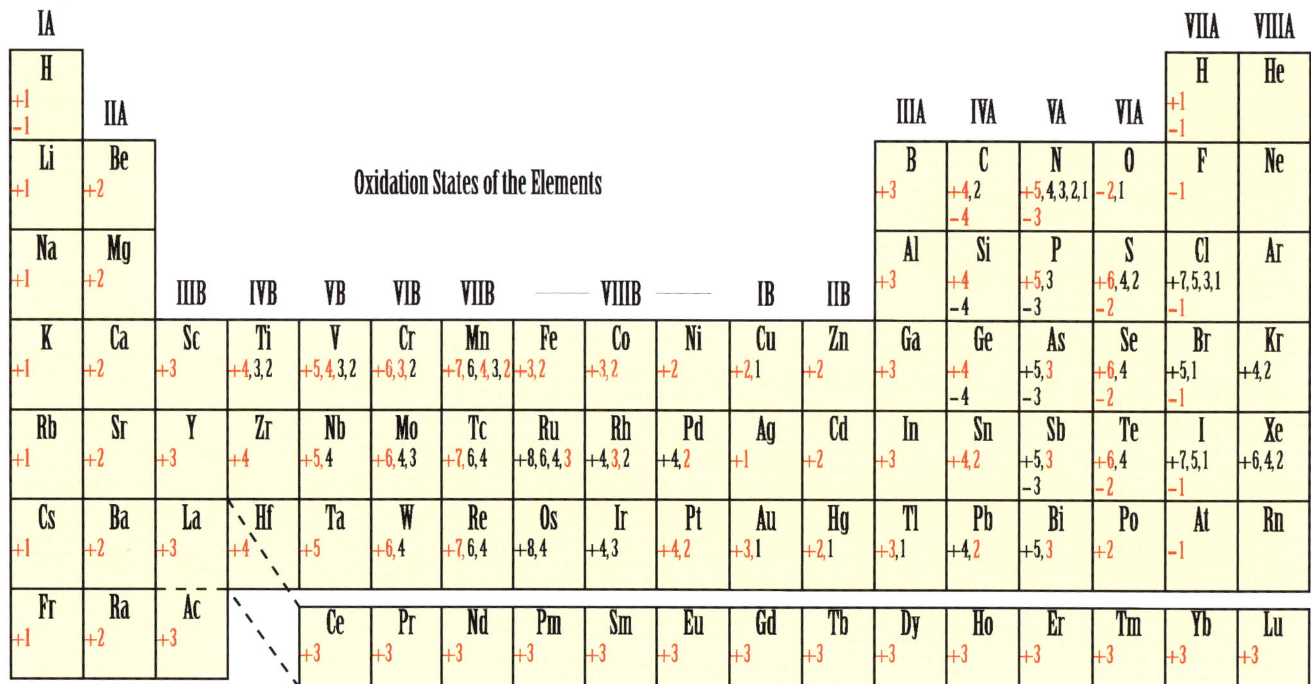


**Figure 2.** These are the common oxidation states (other than neutral) of the elements. Groups IV-VIA and the transition metals of groups IV-VIIIB are the ones that are most frequently seen to take on more than one oxidation state. To master oxidation numbers is to master the natural occurrence of compounds. Compounds whose oxidation numbers cancel are likely to be found bound together in a compound.

For a complete listing of oxidation numbers of the elements, see the periodic table in Figure 2. There the most common oxidation numbers are shown in red, while the ones that have been observed in less usual situations are shown in black. You can pretty much make any existing compound by placing together elements in combinations of oxidation numbers which produce a net oxidation number of zero. However, if you mix and match the less common oxidation numbers, you are unlikely to get anything ordinary.

While different elements may have several oxidation states that they can assume, most have one or two favorites in which they are found most of the time. For example, because the noble gases (group VIIIA) have outer *p* suborbitals that are filled, they are neutral. It's rare to find any of them existing in any other form without blasting them with energy. Every halogen (group VIIA) is missing an electron from its outer *p* suborbital, so they tend to gain an electron from a partnering element. This gain of an electron means that they will have negative oxidation numbers in the compounds they form. Every alkali metal (group IA) has only one electron in its outer shell, so they tend to lose that electron to a partnering element. This loss of an electron means that they will have positive oxidation numbers in the compounds they form.

Nitrogen is the extreme case. It has been found to assume six different oxidation states. This stands to reason because it is among the group VA elements. These elements can accept or lose any number of electrons from their outer shells without being less satisfied with their electrical state than they are when neutral. In other words, there are lots of compounds of nitrogen that provide more satisfactory electron configurations than does its neutral state. Nitrogen's favorite states will be those that result in a full valence shell. For that to happen it would have to gain three electrons (oxidation number = 3–) or lose five electrons (oxidation number 5+).

Notice that, in our discussion, we have used interchangeably the terms *oxidation state*, *valence* and *charge*. On a practical level they all mean the same thing. If we talk about an atom having an oxidation number of 3+, don't freak out. All we are saying is that it is carrying a 3+ charge or is reacting as an ion that would like to gain three electrons. The only difference to keep in mind is that compounds may be molecular (rather than ionic) in nature. For example, water ($H_2O$) is a neutral molecule with little ionic character. However, we can still assign oxidation numbers to its atoms. It has one oxygen atom that we might think of as being in a 2– oxidation state and two hydrogen atoms, each in a 1+ oxidation state.

Oxidation results in the increase in the oxidation number. For example, in the oxidation half reaction:

$$Zn^0 \rightarrow 2e^- + Zn^{2+}$$

(**Oxidation** = **increase** in oxidation number)

the oxidation number of zinc changes from 0 (neutral) to 2+. In the reduction half reaction, copper, being reduced, undergoes a decrease in its oxidation number from 2+ to 0.

$$Cu^{2+} + 2e^- \rightarrow Cu^0$$

(**Reduction** = **reduction** in oxidation number)

## What State?

If you memorize the oxidation numbers of a few elements, you can nearly always tell by looking at a compound what oxidation state its elements are in. The group IA, IIA, IIIA and IVA elements tend to lose their valence electrons, so they have oxidation numbers of 1+, 2+, 3+ and 4+, respectively. The group VIA and VIIA elements tend to gain their complement of electrons to fill their *p* suborbitals, so they tend to take on oxidation numbers of 2– and 1– respectively.

The group VA elements are right in the middle. They might just as soon either lose or gain their complement of valence electrons. For this reason they tend to take on oxidation numbers of either 5+ or 3–. That is to say, they might just as easily lose five electrons to obtain a noble gas configuration as to gain 3 and take on the next noble gas configuration. However, the more protons these atoms have in their nuclei, the greater will be their tendency to gain electrons rather than lose them. For this reason, the small N and P atoms might lose their 5 electrons (and take on 5+ oxidation numbers), while As, Sb (antimony) and Bi will more likely gain electrons and take on 3+ oxidation numbers.

Here are some rules that will also help you in determining the oxidation state of a given atom in a compound. Follow along on the periodic table as you go over these, and you should memorize them.

1) Every element (e.g., $Fe^0$ or $Cu^0$), while in its elemental state, will be neutral.
2) Every diatomic, triatomic (etc.) element (e.g., $H_2$, $P_4$, $Cl_2$, $O_2$, $N_3$, $O_3$) will be neutral.
3) When it is not in its elemental (neutral) form, every group IA, IIA or IIIA element takes on *only* a 1+, 2+ or 3+ oxidation state, respectively (except Tl, which you may ignore. See the Figure in the previous lesson if you're curious).
4) When it is not in its elemental (neutral) form, group VIIA elements *usually* form 1– ions. (When it isn't in its 1– state, you can tell what state it's in by looking at its bonding partner.)
5) Hydrogen is always in a 1+ oxidation state, except when it is either neutral (as in $H_2$) or in the form of a metal hydride (1–), such as NaH.
6) Oxygen is always in a 2– oxidation state, except when it is either neutral (as in $O_2$) or in a peroxide ($O_2^{2-}$: two oxygen atoms, each in a 1– state).

What practical difference does it make whether an element has gone through an oxidation or reduction? The chief difference is that, during redox reactions, molecules find more stable arrangements (lower potential energy states), and electrons find paths that allow them to give up some of their potential energy. This can result in a tremendous energy release. That's what makes these reactions so useful as fueling reactions.

Every oxidation results in an increase or decrease in potential energy from reactants to products. Every reduction results in a similar change in potential energy, but in the opposite direction. When the energy changes from the two half reactions are combined, the difference in total potential energy is the energy change for the entire reaction. It represents the energy that the reaction either gives up or absorbs. Some redox reactions require a continuous input of energy to drive them—they are **endothermic**. The reverse reactions give off energy—they are **exothermic**. The amount of energy given off by an exothermic reaction can be pretty significant.

**Exercises**

1) Please write and balance half reactions for each. Tell which is the oxidation half reaction and which is the reduction half reaction.

a) $Mg^0$ (s) + $Ni^{2+}$ (aq) → $Mg^{2+}$ (aq) + $Ni^0$ (s)
b) $Zn^0$ (s) + $Cu^{2+}$ (aq) → $Zn^{2+}$ (aq) + $Cu^0$
c) $Fe^0$ (s) + $Pb^{2+}$ (aq) → $Fe^{2+}$ + $Pb^0$

2) Please write redox reactions for each of the following pairs of half reactions.

a) $Cr^0 \rightarrow 3\ e^- + Cr^{3+}$ and $As^{3+} + 3\ e^- \rightarrow As^0$
b) $Sn^0 \rightarrow 2\ e^- + Sn^{2+}$ and $Pb^{2+} + 2\ e^- \rightarrow Pb^0$
c) $Cu^0 \rightarrow 2\ e^- + Cu^{2+}$ and $Hg^{2+} + 2\ e^- \rightarrow Hg^0$

3) Please give the oxidation number of each element in the following compounds.

a) NaCl
b) CrO
c) $H_2S$
d) $NH_3$
e) $Al_2O_3$
f) $H_2SO_4$
g) $Mg(OH)_2$
h) $K_2Cr_2O_7$

## 35: Whose Electron Is This?

Once you can determine the oxidation numbers for the elements in a compound, you are prepared to determine which of the elements making up that compound has undergone an oxidation or a reduction during a reaction. Now remember, in the case of molecular (nonionic) compounds, there are no ions formed, so electrons don't truly transfer from one atom to another. It's only through the oxidation numbers that we can appreciate whether an element has undergone a change with respect to its oxidation state. (See the Figure.)

For example, in the following reaction:

$$CH_4 \quad + \quad 2\,O_2 \quad \rightarrow \quad CO_2 \quad + \quad 2\,H_2O$$

how would you know whether or not there has been any oxidation or reduction? Start by assigning an oxidation number to each element. Let's begin with $CH_4$. We learn from rule 5 in the previous lesson that hydrogen is in its 1+ state. Because there are four hydrogen atoms bound to carbon, carbon must be in its 4– state.

**Figure.** In a reaction involving ions where electrons are swapped from one atom to another it's easy to tell when oxidation is occurring. In the reaction shown below between calcium and oxygen, oxygen receives both of the valence electrons that previously belonged to calcium. In reactions involving covalent molecules, such as the second reaction written below, you can't tell by looking at the Lewis structures which atoms are being oxidized and which are being reduced. To assign oxidation states to atoms, we make believe that the electrons transferred to the atom that would most likely receive them if the transfer should actually occur. Although you can't see it in the written equation, that's oxygen.

$$\cdot Ca \cdot \ + \ [:\ddot{O}:] \longrightarrow Ca^{2+} + [:\ddot{O}:]^{2-}$$

$$H:\!C\!:\!H \ (\text{with } H \text{ above and below } C) \ + \ 2\ O::O \longrightarrow O::C::O \ + \ 2\ H:\!O\!:\!H$$

So we write these oxidation numbers above their symbols:

$$\begin{array}{ccccccccc} 4-\ 1(4) & & & & & & \\ CH_4 & + & 2\,O_2 & \rightarrow & CO_2 & + & 2\,H_2O \end{array}$$

Next, look at oxygen. There are four O atoms represented, and each is in its neutral state (according to rule 2), so we write 4(0) above 2 $O_2$.

$$\begin{array}{ccccccccc} 4-\ 1(4) & & 4(0) & & & & \\ CH_4 & + & 2\,O_2 & \rightarrow & CO_2 & + & 2\,H_2O \end{array}$$

Moving to the right side of the equation, we find a single carbon atom bound to two oxygen atoms. Because oxygen is no longer neutral, it is in its 2– state (rule 6). This means that the carbon atom must be in its 4+ state. Likewise, in water, the oxygen atom is in its 2– state. Hydrogen must be in its 1+ state.

$$\begin{array}{ccccccccc} 4-\ 1(4) & & 4(0) & & 4+\ 2(-2) & & 4(1+)\ 2(2-) \\ CH_4 & + & 2\,O_2 & \rightarrow & CO_2 & + & 2\,H_2O \end{array}$$

Now let's compare the oxidation numbers on the left to the ones on the right. It's clear that carbon has gone from its 4– oxidation state (in $CH_4$) to its 4+ state (in $CO_2$). We can write an equation that shows what has happened to carbon like this:

$$C^{4-} \rightarrow C^{4+} + 8\ e^-$$

(One carbon atom in its 4– state was oxidized to its 4+ state.)

This is an example of oxidation. Carbon gave up eight electrons and in doing so its oxidation number increased by 8.

For every oxidation, there must be a reduction, so what was reduced? Well, hydrogen is in its 1+ state at the beginning of the reaction (in $CH_4$). At the end of the reaction it is still in its 1+ state (in $H_2O$). It was neither oxidized nor reduced. The only other element to consider is oxygen. Two oxygen atoms went from their neutral state in $O_2$ to their 2– state in $CO_2$. The other two have gone from their neutral state in $O_2$ to their 2– state in $H_2O$. So the reduction reaction may be written:

$$4\ O^0 + 8\ e^- \rightarrow 4\ O^{2-}$$

(Four neutral oxygen atoms were reduced to their 2– states.)

Notice that the number of electrons given up to oxidation will always balance the number of electrons used for reduction. If they don't balance, something is amiss—try again.

While the previous redox reaction involved only one element being oxidized and one being reduced, there are many cases where one of these processes affects more than one element simultaneously, as in:

$$KIO_3 + NH_3 \rightarrow KNO_3 + HI + H_2$$

Writing the oxidation number of each element, we get:

| 1+5+3(2–) | | 3– 3(1+) | | 1+5+2(3–) | | 1+ 1– | | 2(0) |
|---|---|---|---|---|---|---|---|---|
| $KIO_3$ | + | $NH_3$ | → | $KNO_3$ | + | HI | + | $H_2$ |

Taking these one element at a time, we see:

| | |
|---|---|
| $K^+ \rightarrow K^+$ | No change |
| $I^{5+} + 6\ e^- \rightarrow I^-$ | Reduction, 6 electrons |
| $3\ O^{2-} \rightarrow 3\ O^{2-}$ | No change |
| $N^{3-} \rightarrow N^{5+} + 8\ e^-$ | Oxidation, 8 electrons |

Notice that hydrogen has two different fates. One of the original 3 atoms ends up in HI, while the other two are used in producing hydrogen gas. These different fates relate to different final oxidation states:

$3\ H^+ + 2\ e^- \rightarrow 2\ H^0 + H^+$ Reduction, 2 electrons

At the end of the day, the number of electrons transferred in the oxidation half of the reaction (8) is the same as the number of electrons transferred in the reduction half (6+2).

Now you try. Practice by covering up the information above the reaction we just completed to see if you can reproduce my answers. When you get it right, go on to the exercises.

**Exercises**

As you just did in the example in the lesson, please break these redox reactions down to show what happens to the individual elements. Tell which are oxidized, which are reduced and which do not change states. Be sure the equations are balanced before you begin, and be sure the transferred electrons balance when you are finished.

1) $Zn + H_2SO_4 \rightarrow ZnSO_4 + H_2$
2) $H_2S + Fe^0 \rightarrow FeS + H_2$
3) $CuO + NH_3 \rightarrow N_2 + Cu^0 + H_2O$
4) $Al^0 + HgCl_2 \rightarrow AlCl_3 + Hg^0$

## 36: Quiz 5

1) Be able to identify the reaction type (combination, decomposition, single displacement, double displacement or oxidation-reduction) when given a generic equation of the form:

$$A + B \rightarrow AB$$
$$AB \rightarrow A + B$$
$$A + BC \rightarrow AB + C$$
$$AB + CD \rightarrow AC + BD$$
$$A^0 \rightarrow A^+ + e^-$$
$$B^+ + e^- \rightarrow B^0$$

2) Be able to predict whether single and double replacement reactions will occur given an activity series, a table of solubility rules and a table of common gases evolved from reactions.
3) Be able to write redox half reactions from redox reaction equations. Also be able to do the converse.
4) Be able to assign oxidation numbers to all the elements in a compound. You will need to know the rules. For example, if I told you to assign oxidation numbers to $CO_2$, you would need to know that oxygen is in its 2– state and what you could conclude about carbon.
5) Know how to write a balanced redox equation for each individual element that takes part in a redox reaction.

Thinking chemistry changes the way you look at everything.

## 37: Stoichiometry

Chemistry is a true "building block" science. Certain concepts cannot really be understood until more elementary ones have been mastered. This lesson will illustrate that very well. Way back in Lesson 3 we talked about the unit factor method. Then there were some lessons on the naming of chemical compounds. In Lesson 14 you were introduced to the concept of the mole. Most recently there have been lessons on setting up and balancing chemical equations. Well, now we are going to put all of these things together and talk about one of the most important skills that chemists need to master—the ability to determine quantities of one substance shown in a chemical equation when the quantity of one or more other substances is known. These quantitative relationships are called **stoichiometry.**

You see, one of the most useful things about balanced equations is that they show the **molar ratios** of reactants and products. For example, back in the lesson on the mole, we observed that magnesium and sulfur can react to form magnesium sulfide according to the following equation:

| Mg | + | S | → | MgS |
|---|---|---|---|---|
| 1 mol | | 1 mol | | 1 mol |

Since one atom of Mg reacts with one atom of S to form one molecule of MgS, then it would also be true that $6.022 \times 10^{23}$ atoms of Mg would react with $6.022 \times 10^{23}$ atoms of S to produce $6.022 \times 10^{23}$ molecules of MgS. Therefore, as we observed back in Lesson 14, *one mole of Mg can react with one mole of S to produce one mole of MgS.* This relationship of moles to moles makes up the **molar stoichiometry** of this reaction. Based on that stoichiometry, which can be obtained from the coefficients in the balanced chemical equation, we may calculate the amount of any substance in the equation, as long as we know how much of one of them is present. For example, suppose we are starting with 1.41 mol of Mg. How many mol of S do we need to completely react with it? And how many mol of MgS will be formed?

All we have to do is to look at the chemical equation. If 1 mol Mg reacts with 1 mol S, how many mol of S will react with 1.41 mol Mg? Right—1.41 mol of S. However many atoms, molecules or moles of Mg we have, we must have the same number of S atoms, molecules or moles to react with them because the molar ratio of Mg to S is 1:1. And this molar ratio information is available in every balanced equation. (See the Figure.)

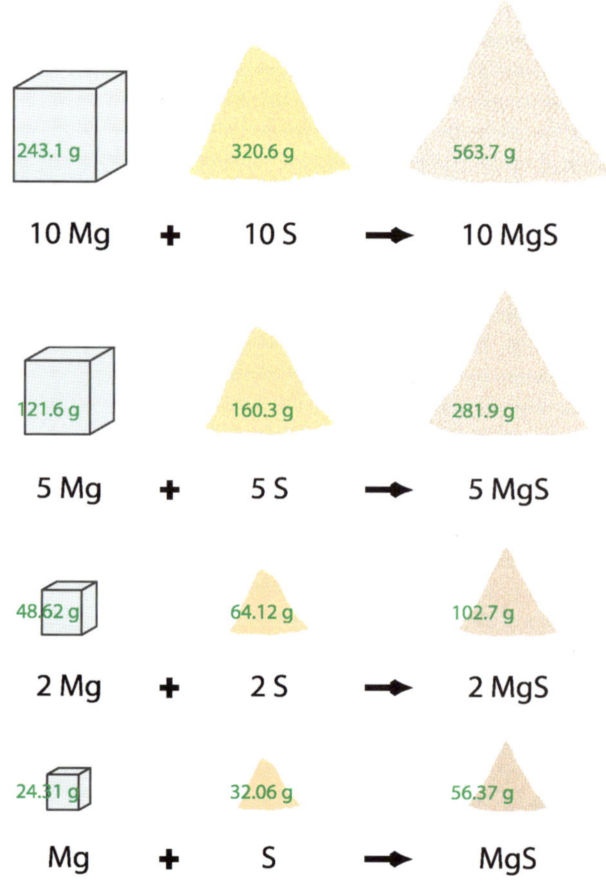


**Figure.** A chemical equation may be thought of as shorthand for the proportions of reacting substances and their products. No matter how much of those substances are present, the proportions are always the same. Small whole-number ratios of moles are used, but any whole-number or fractional number of moles is equivalent as long as the proportions are the same.

Here, let me show you what I mean. If elemental potassium reacts with water according to the following equation,

$$\begin{array}{ccccccccc} 2\ K & + & 2\ H_2O & \rightarrow & H_2 & + & 2\ KOH \\ 2\ mol & & 2\ mol & & 1\ mol & & 2\ mol \end{array}$$

how many moles of hydrogen gas would be produced if you started with 1.7 mol K?

From the coefficients in the balanced equation we find that the molar ratio of potassium to hydrogen is 2:1. Therefore, if we are starting with 1.7 mol K, the calculation would look like this:

$$?\ mol\ H_2 = \frac{1.7\ mol\ K}{} \left| \frac{1\ mol\ H_2}{2\ mol\ K} \right. = 0.85\ mol\ H_2$$

The reason this works is that, for every 1 mol of hydrogen produced, 2 mol of potassium have reacted. Based on the relationship stated in the equation, these are equivalent statements and may be treated as mathematical equivalents:

$$1\ mol\ H_2\ produced = 2\ mol\ K\ reacted$$

Therefore, we can treat the molar ratio of potassium to hydrogen like a unit factor in our calculation:

$$\frac{1\ mol\ H_2}{2\ mol\ K} = 1 \quad or \quad \frac{2\ mol\ K}{1\ mol\ H_2} = 1$$

Let's do another. We'll write the balanced equation for the complete combustion of glucose, $C_6H_{12}O_6$, and then calculate how many moles of oxygen would be required to oxidize 8.3 mol of glucose. First, we write the balanced equation. Remember, the products from the complete combustion of any compound having the general formula $C_xH_yO_z$ are $CO_2$ and $H_2O$. Therefore, the balanced equation would look like this:

$$C_6H_{12}O_6 + 6\ O_2 \rightarrow 6\ CO_2 + 6\ H_2O$$

Now notice the molar ratios. Six moles of $O_2$ are required for every mole of glucose. Our problem says we are starting with 8.3 mol of glucose, so the calculation goes like this:

$$?\ mol\ O_2 = \frac{8.3\ mol\ glucose}{} \left| \frac{6\ mol\ O_2}{1\ mol\ glucose} \right. = 50.\ mol\ O_2$$

(Note: The decimal point after the zero indicates that it is intended to be a significant digit.)

Sometimes the amount of one of the substances in an equation might be given in grams rather than moles. In that event you will have to convert the grams to moles, then solve as before. For example, suppose we had been given the complete equation for the combustion of glucose and were told that we wanted to oxidize just enough glucose to produce 75 g of $CO_2$. How much glucose should we use? The first thing we'd need to know is how many moles of $CO_2$ are contained in 75 g. The molar mass of $CO_2$ is 44.01 g/mol. Therefore,

$$? \text{ mol glucose} = \frac{75 \text{ g } CO_2}{} \left| \frac{1 \text{ mol } CO_2}{44.01 \text{ g } CO_2} \right.$$

Since we are given the amount of $CO_2$ that would be produced, and asked for how much glucose would be consumed, we introduce our unit factor to convert from $CO_2$ produced to glucose consumed. According to the equation, every mole of glucose that reacts produces six moles of $CO_2$. So our equation becomes:

$$? \text{ mol glucose} = \frac{75 \text{ g } CO_2}{} \left| \frac{1 \text{ mol } CO_2}{44.01 \text{ g } CO_2} \right| \frac{1 \text{ mol glucose}}{6 \text{ mol } CO_2} = 0.28 \text{ mol glucose}$$

So 0.28 mol of glucose, when completely oxidized, would produce 1.70 mol of $CO_2$, which is equivalent to 75g of $CO_2$. And what if, rather than *moles* of glucose, you had been asked for the *grams* of glucose? You simply convert the moles to grams. You can do by simply adding that conversion factor to your equation. Since there are 180.16 g glucose/mol glucose:

$$? \text{ g glucose} = \frac{75 \text{ g } CO_2}{} \left| \frac{1 \text{ mol } CO_2}{44.01 \text{ g } CO_2} \right| \frac{1 \text{ mol glucose}}{6 \text{ mol } CO_2} \left| \frac{180.16 \text{ g glucose}}{1 \text{ mol glucose}} \right. = 51. \text{ mol glucose}$$

Okay, this is the kind of thing that you learn by doing. Go ahead and plow straight into the exercises. Just remember: Start off every problem by writing the balanced equation. We have been providing you with molar masses, but to do the exercises you will have to calculate them yourself.

**Exercises**

1) Given the equation:

$$2\ MnO + 5\ PbO_2 + 10\ HNO_3 \rightarrow 2\ HMnO_4 + 5\ Pb(NO_3)_2 + 4\ H_2O$$

a) How many moles of nitric acid ($HNO_3$) are required to react with 110.0 g of MnO?
b) How many moles of water would be produced if 13.2 mol of lead(II) nitrate were formed?

2) Benzene is $C_6H_6$.

a) If 22 mol of benzene are completely reacted with molecular oxygen, how many moles of carbon dioxide will be produced?
b) If 150 g of water were produced, how many moles of oxygen must have reacted?

3) Calcium carbonate, when used as an antacid, neutralizes stomach acid (which is hydrochloric acid). If an antacid tablet contains 750 mg of calcium carbonate, then how many moles of acid can be neutralized by taking two tablets? (The word *neutralized*, when speaking of acids, means reacted with a base—here, calcium carbonate.)

4) Sodium nitride can be formed by reacting elemental sodium with nitrogen gas ($N_2$). Write the balanced equation for this reaction, then calculate how many moles of sodium nitride can be produced by reacting 25.0 g of sodium metal.

5) Hydrogen cyanide, HCN, is a poisonous chemical that is used in the extraction of gold. It is produced by the reaction of ammonia and methane, and hydrogen gas is a by-product of the reaction. Write the balanced equation for this reaction, then calculate how many grams of HCN will be produced if 15.0 mol of hydrogen are generated.

6) How many moles of aluminum sulfate can be produced from 8 mol of sulfuric acid if sulfuric acid reacts with aluminum hydroxide according to the following equation?

$$3\ H_2SO_4 + 2\ Al(OH)_3 \rightarrow Al_2(SO_4)_3 + 6\ H_2O$$

7) Since aluminum is higher on the activity chart than nickel, solid aluminum will displace nickel in solution. Therefore, how many grams of Ni can be produced if a 10.0-g sample of aluminum is placed in an aqueous solution of $NiSO_4$, and if the Al reacts completely?

## 38: Son of Stoichiometry

Stoichiometry is one of those things that is so important in the study of chemistry that I thought it might be a good idea to give you a further opportunity to work on it. So there is no new instruction in this lesson, just extra problems on which you can practice.

**Exercises**

1) Given the equation:

$$Fe_2(CO_3)_3 + 6\ HCl \rightarrow 2\ FeCl_3 + 3\ CO_2 + 3\ H_2O$$

a) How many grams of HCl are required to produce 56.2 g of $FeCl_3$?

b) How many moles of water would be produced if 131.9 g of iron(III) carbonate completely reacted?

2) Given the following reaction:

$$2\ KMnO_4 + 5\ H_2C_2O_4 + 6\ HCl \rightarrow 2\ MnCl_2 + 10\ CO_2 + 2\ KCl + 8\ H_2O$$

a) If 200.0 g of potassium permanganate are completely reacted, what mass of water will be produced?

b) If 38.0 mol of HCl react, how many grams of $MnCl_2$ will be formed?

3) Bismuth(III) chloride will react with hydrogen sulfide to form bismuth(III) sulfide and hydrochloric acid.

a) Write a balanced equation for this reaction.

b) How many mol of acid will be formed if 15 mol of hydrogen sulfide react?

4) Nitrobenzene, $C_6H_5NO_2$, can be "reduced" to form aniline, $C_6H_5NH_2$, according to this equation:

$$C_6H_5NO_2 + 2\ CO + H_2 \rightarrow C_6H_5NH_2 + 2\ CO_2$$

How many mol of gas would be required to react with 75.0 g of nitrobenzene?

5) If you'll recall, deuterium (abbreviated "D") is heavy hydrogen. It's the isotope of hydrogen having an extra neutron in its nucleus. Deuterated ammonia, $ND_3$, can be easily prepared via the following reaction:

$$Mg_3N_2 + D_2O \rightarrow Mg(OD)_2 + ND_3$$

a) Balance this equation.

b) Calculate the mass of $ND_3$ that can be prepared from 25 mol of $D_2O$. (The atomic mass of deuterium is 2.016 amu.)

6) Arsenic is an element that can exist as a multiatomic molecule. In fact, $As_4$ can be prepared via the following reaction:

$$As_4O_6 + 6\ C \rightarrow 6\ CO + As_4$$

If you started with 150 g of carbon, how many molecules of $As_4$ could you produce?

7) Given the following reaction:

$$Fe_2O_3 + 3\ CO \rightarrow 2\ Fe + 3\ CO_2$$

How many moles of each product can be formed by the reaction of 46 g of CO?

**Note:** For the rest of the course, you will be exposed to problems of this nature, so you might as well make sure you understand it well before you go on. If you need to get more practice, you'll have to spend some extra hours in the evening to get it so that you can continue your lessons on schedule.

## 39: How Much Can You Really Get?

All of the problems you've been working in the last two chapters have one thing in common—they assume that the reactants are present in sufficient amounts so that they are completely consumed and the maximum amount of products is obtained. In the real world, of course, chemistry doesn't always work that way. Sometimes in a chemical reaction one of the reactants is present "**in excess**." In other words, more of that reactant is present than is really required to completely react with the other reactant. Consider the following...

Suppose you have five hungry cats. From previous experience you know that if you give a hungry cat a mouse to eat, then he will no longer be a hungry cat, he will be a happy cat. If we were to write this like a chemical equation we might write it like this:

Well, you have five hungry cats. But suppose you only have three mice. You have enough mice to make three of the cats happy, but two will remain hungry. Now our "equation" will look like this:

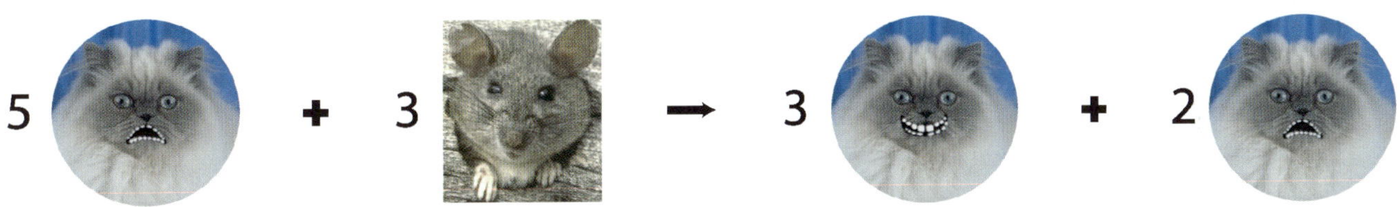


Therefore, you have an excess of two cats. They would be happy to "react" with mice—there just aren't mice for them. After the reaction is over, you still have two unreacted cats. Figure 1 illustrates this point using our simple reaction between magnesium and sulfur.

In a similar way, when one reactant of a chemical reaction is present in excess, that means that there isn't enough of the other reactant to completely react with it. Consider again the reaction:

$$Mg + S \rightarrow MgS$$

This balanced equation shows the molar amounts of Mg and S to undergo a balanced reaction where, if the reaction goes to completion, there is neither Mg or S left over. But what if the molar quantity of Mg is 1.5 rather than 1? Then you would expect:

$$1.5\ Mg + S \rightarrow MgS + 0.5\ Mg$$

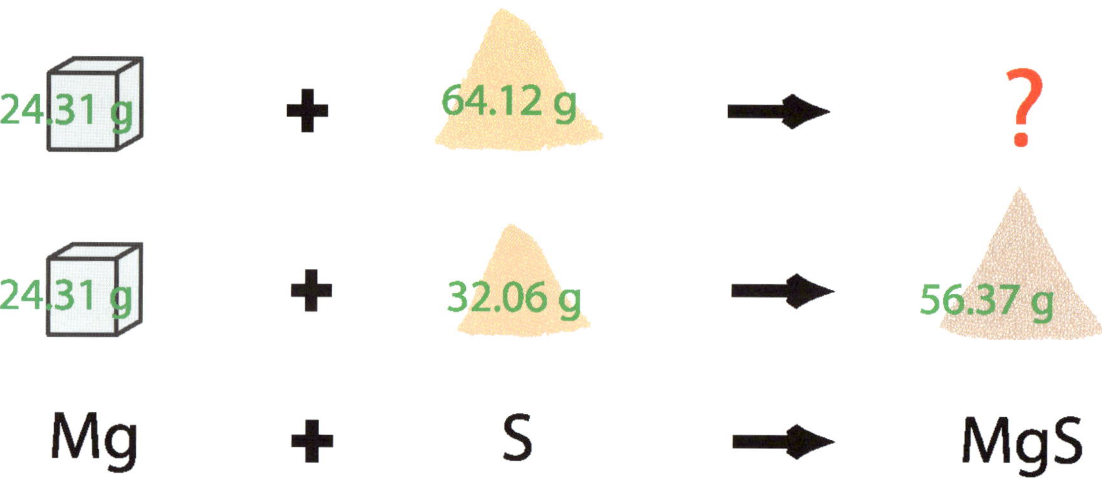


**Figure 1.** As you know, chemicals react in molar proportions. However, sometimes the chemicals are not present in their proper molar proportions. The set of chemicals just above the equation represent the proper molar proportions of magnesium and sulfur. When they react they produce the stated amount of MgS. What would happen to the amount of MgS you could get if the amount of sulfur were doubled (as it is in the set of chemicals on top)? Because there is yet only one mole of magnesium present, the amount of product you can get is absolutely prevented from being more than one mole! Magnesium is the *limiting reagent.*

Of course, in practice we often know the gram quantities of chemicals and need to calculate the molar quantities to see which substance might be lacking. We know that 24.31 g (1 mol) of Mg would react with 32.06 g (1 mol) of S to form 56.37 g (1 mol) of MgS. How much MgS can be produced by this reaction if you start with 24.31 g of Mg and only 20.00 g of S?

First, let's calculate how many moles of each element we have:

$$? \text{ mol Mg} = \frac{24.31 \text{ g Mg}}{} \left| \frac{1 \text{ mol Mg}}{24.31 \text{ g Mg}} \right. = 1.000 \text{ mol Mg}$$

$$? \text{ mol S} = \frac{20.00 \text{ g S}}{} \left| \frac{1 \text{ mol S}}{32.07 \text{ g S}} \right. = 0.6236 \text{ mol S}$$

Again, we know from the balanced equation that 1 mole of Mg reacts with 1 mole of S to produce 1 mole of MgS. Therefore, it is also true that if 0.6236 mol of S react, that means that 0.6236 mol of Mg have also reacted, and that 0.6236 mol of MgS have been produced. And once 0.6236 mol of S have reacted, no more MgS can be produced because, at this point, all the sulfur will have been used up.

In this kind of situation we refer to sulfur as the **limiting reactant** or **limiting reagent**, and we say that the Mg is present "in excess." In fact, we have an excess of 0.3764 mol of Mg (the original 1.000 mol minus the 0.6236 mol that reacted). And this was a pretty easy one to figure out because we had a simple 1:1 molar ratio between the reactants and the product.

Here is an example that's a little more involved. We have seen previously that nitrogen and hydrogen can react to form ammonia. How many moles of ammonia can be prepared from the reaction of 64.7 g $N_2$ and 12.1 g $H_2$? As always, the first step is to write out the balanced equation:

$$N_2 + 3\,H_2 \rightarrow 2\,NH_3$$

If you consider carefully, you will realize that there is more than one way to approach this problem. One way we could do it is to convert all the gram quantities to molar quantities. A balanced reaction would require the molar quantity of hydrogen to be three times that of nitrogen. We would be able to tell whether the moles of $H_2$ available were more than three times the moles of $N_2$.

Instead, we'll determine how much $NH_3$ could be formed if all of the available $N_2$ reacted. Then we'll do the same for $H_2$. The limiting reagent will be the one that forms the least ammonia. If neither is limiting they should form precisely the same amount of ammonia. The calculation to determine the amount of ammonia that can form from 64.7 g $N_2$ would look like this:

$$?\ \text{mol NH}_3 = \frac{64.7\ \text{g N}_2}{} \left| \frac{\text{mol N}_2}{28.02\ \text{g N}_2} \right| \frac{2\ \text{mol NH}_3}{\text{mol N}_2} = 4.62\ \text{mol NH}_3$$

Then we determine how much $NH_3$ can form from 12.1 g $H_2$:

$$?\ \text{mol NH}_3 = \frac{12.1\ \text{g N}_2}{} \left| \frac{\text{mol H}_2}{2.016\ \text{g H}_2} \right| \frac{2\ \text{mol NH}_3}{3\ \text{mol H}_2} = 4.00\ \text{mol NH}_3$$

In this case, once the reaction has proceeded to the point where 4.00 mol of $NH_3$ have been formed, the supply of $H_2$ will be exhausted. $H_2$ is the limiting reactant. While we decided to express the answer in moles in this calculation, you can calculate the limiting reagent using either moles or grams. Figuring it in grams would require an additional unit conversion in our calculation.

Let's do this one more time. How many moles of $Fe_3O_4$ (the form of iron in magnetite, called *ferrosoferric oxide*) can be obtained by reacting 12.3 g of Fe metal with 8.0 g of water, according to the equation below?

$$3\,Fe + 4\,H_2O \rightarrow Fe_3O_4 + 4\,H_2$$

Treat each reactant separately:

$$?\ \text{mol Fe}_3\text{O}_4 = \frac{12.3\ \text{g Fe}}{} \left| \frac{\text{mol Fe}}{55.85\ \text{g Fe}} \right| \frac{\text{mol Fe}_3\text{O}_4}{3\ \text{mol Fe}} = 0.0734\ \text{mol Fe}_3\text{O}_4$$

$$?\ \text{mol Fe}_3\text{O}_4 = \frac{8.0\ \text{g H}_2\text{O}}{} \left| \frac{\text{mol H}_2\text{O}}{18.02\ \text{g H}_2\text{O}} \right| \frac{\text{mol Fe}_3\text{O}_4}{4\ \text{mol H}_2\text{O}} = 0.11\ \text{mol Fe}_3\text{O}_4$$

Which is the limiting reactant? Less $Fe_3O_4$ can be formed from the amount of Fe present than from the amount of water we started with. The amount of Fe present can produce only 0.0734 mol of $Fe_3O_4$.

Having completed that calculation, do you realize that we might have arrived at the same answer to the question by doing the calculations around $H_2$ instead of $Fe_3O_4$? The reagent that will form the lesser amount of one product will also form the lesser amount of the other. Doing so would have changed only the last unit factor in each of our equations. Once again, we could obtain the answer in either moles or grams of $Fe_3O_4$.

### Percent Yield

Again, the problems you just completed were done in recognition of the fact that chemicals are not always present in nice stoichiometric quantities in real reactions. Another assumption we have been making all along is that chemical reactions always go to completion. Actually, for a variety of reasons which we will discuss throughout the remainder of this course, many chemical reactions do not continue until all of the limiting reactant is used up. Therefore, the quantities of products that we have been calculating in our reactions up to this point represent the **theoretical yield** of those products. Theoretical yield is the expression of the maximum amount of product that is theoretically possible given the amounts of reactants present. (See Figure 2.)

Scientists calculate theoretical values all the time. For example, we might want to calculate the rate of fall of a wad of paper dropped to the floor from a height of 10 m. In physics we learn to calculate the *theoretical maximum* rate of fall in a vacuum where there is no resistance from air. We may not be able to calculate the *actual* rate of fall precisely, but at least we'll be able to learn the limit on how fast it *can* fall by applying the theory of falling objects in a vacuum. We can also predict that the air will have less effect on a brick than it will on paper. So the brick should come very close to the theoretical maximum rate of fall, while paper will be, to some extent, held up by the air. The only way we can know the actual rate is to do an experiment to find out. We might learn that the actual rate of fall is some percentage (perhaps 98%) of the theoretical rate.

So it is with chemical reactions. No real reaction ever goes to *absolute* completion. There are always unreacted molecules present after any reaction. Some reactions go very nearly to completion—that is to say—their actual yield is very close to their theoretical yield. For practical purposes they may be considered equal after some reactions. However, if a reaction *does not*

**Figure 2.** The quantities of products shown in chemical equations are theoretical values. The assumption is that absolutely *all* of the mass of reactants ended up as products. Although this may be practically true, it is never true in the absolute sense. There is always some (albeit, sometimes extremely small) amount of unreacted product remaining after any reaction.

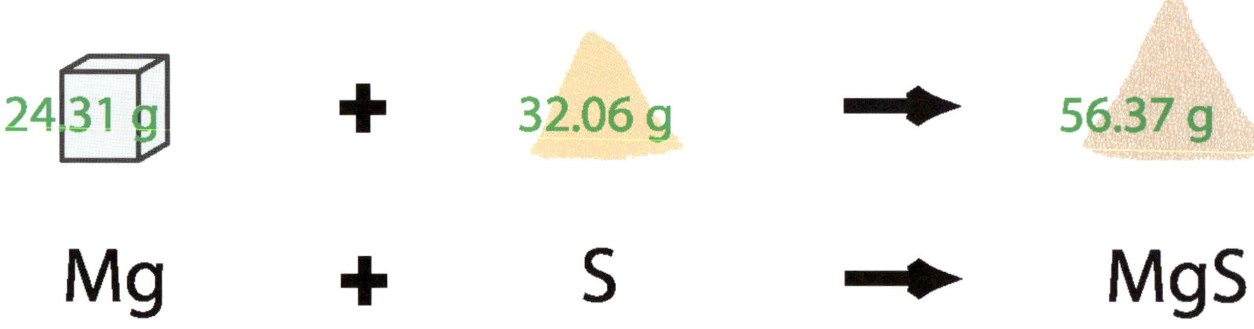

go to completion and produces *less* than the theoretical yield, then a **percent yield** of that product can be calculated. That quantity that we call the *percent yield* answers the question, "What percent of the theoretical yield of products did my reaction actually make?" (See Figure 3.)

The percent yield calculation, then, is very simple. It is the amount of product actually produced by a reaction (called the **actual yield**, or simply the **yield**) divided by the maximum possible yield according to the balanced equation (called the *theoretical yield*). This gives the fractional yield. And of course, multiplying any fraction by 100 converts it to a percent. Here is that simple equation:

$$\frac{\text{Actual Yield}}{\text{Theoretical Yield}} \times 100 = \%\ \text{Yield}$$

I already showed you how to calculate a theoretical yield. That's the amount of product that will be produced by the complete reaction of the limiting reagent according to the balanced equation. Since we can't ask you to do any experiments during your lesson time, we'll have to give you the actual yields if we wish for you to calculate % yields.

Let's try one together. Pick up that pencil and write along with me. Be sure you understand what's going on. Calcium carbonate will react with hydrochloric acid according to the following equation:

$$CaCO_3 + 2\ HCl \rightarrow CaCl_2 + CO_2 + H_2O$$

Let's calculate the percent yield if 146.0 g of $CaCO_3$ are treated with 236.0 g of HCl, and 125.0 g of $CaCl_2$ are formed.

Since we have no idea whether the quantities of reactants are present in their proper, stoichiometric ratio, we can either calculate to determine this, or we can just assume that they're not, and go on. Let's just bypass that calculation and assume that they are not. So, we'll first calculate the theoretical yield of $CaCl_2$, based on the amounts of each reactant. Notice that, since the

**Figure 3.** Only through experimentation can we learn the actual yield of a reaction. We express the actual yield as a percent of the theoretical yield. What is the % yield of this reaction?

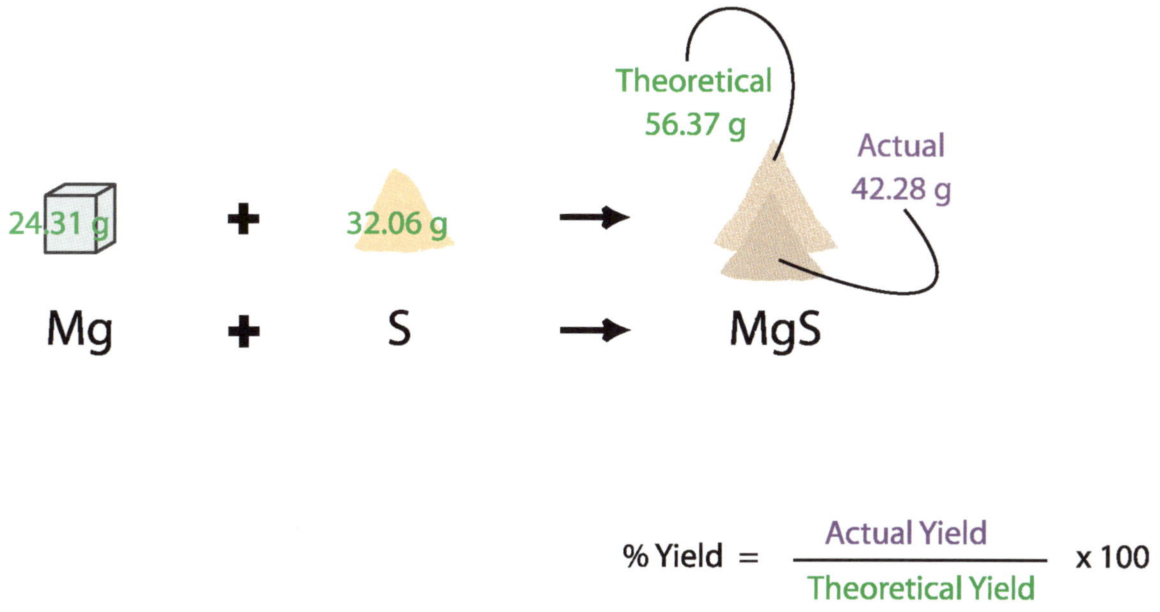

actual yield of $CaCl_2$ is given in grams, we'll calculate the theoretical yields (based on complete use of the available $CaCO_3$ and that based on the complete use of HCl) in grams as well:

$$? \text{ g } CaCl_2 = \frac{146.0 \text{ g } CaCO_3}{} \left| \frac{\text{mol } CaCO_3}{100.09 \text{ g } CaCO_3} \right| \frac{\text{mol } CaCl_2}{\text{mol } CaCO_3} \left| \frac{110.98 \text{ g } CaCl_2}{\text{mol } CaCl_2} \right. = 161.9 \text{ g } CaCl_2$$

$$? \text{ g } CaCl_2 = \frac{236.0 \text{ g HCl}}{} \left| \frac{\text{mol HCl}}{36.46 \text{ g HCl}} \right| \frac{\text{mol } CaCl_2}{2 \text{ mol HCl}} \left| \frac{110.98 \text{ g } CaCl_2}{\text{mol } CaCl_2} \right. = 359.2 \text{ g } CaCl_2$$

As before, the smaller of the two results, 161.9 g $CaCl_2$, represents the limiting reagent. The solution to the limiting reagent problem also provides the theoretical yield. Why? We said that the theoretical yield is the maximum possible yield. You can't possibly get a greater yield from a reaction than that provided by the limiting reagent!

As we said, the percent yield is the actual yield, divided by the theoretical yield, times 100:

$$\frac{\text{Actual Yield}}{\text{Theoretical Yield}} \times 100 = \frac{125.0 \text{ g}}{161.9 \text{ g}} \times 100 = 77.21\%$$

Now, please do the following problems. As before, see if you can figure out how to set them up without looking back at the example in the text. You don't really understand unless you can do it on your own.

**Exercises**

1) The reaction for the complete combustion of butane is:

$$2\ C_4H_{10} + 13\ O_2 \rightarrow 8\ CO_2 + 10\ H_2O$$

If 34.0 g of $C_4H_{10}$ and 60.0 g of $O_2$ are used, how many *moles* of water can be formed?

2) Sodium metal will react with gaseous ammonia to produce solid sodium amide, $NaNH_2$. The *unbalanced* equation is shown below:

$$Na\ (s) + NH_3\ (g) \rightarrow NaNH_2\ (s) + H_2\ (g)$$

If 60.0 g of sodium are mixed with 48.0 g of ammonia, how many grams of $NaNH_2$ can be formed? Which substance is the limiting reactant?

3) Silicon dioxide and graphite can react to form silicon carbide according to the following equation:

$$SiO_2 + 3\ C \rightarrow SiC + 2\ CO$$

What is the theoretical yield of SiC if 36.2 g of $SiO_2$ and 6.00 g of carbon are used? If 6.15 g of SiC are actually obtained in the reaction, what was the percent yield?

4) When 3.00 g of calcium chloride were reacted in an excess of silver nitrate, 6.4 g of AgCl were produced. What was the percent yield in this reaction?

## 40: Quiz 6

1) Given an equation, be able to calculate moles of any reactant or product given the amount (in either moles or grams) of any other reactant or product.
2) Be able to determine the limiting reagent when given the amounts of reactants in either moles or grams.
3) Be able to calculate the percent yield of a reaction when given the actual yield in either moles or grams.

Maybe it's just me, but I love chemistry!

State

## 41: The Big Squeeze

What are the properties of gases that make them different from liquids and solids? We have spoken of their loose arrangement of particles that causes them to move freely about their container. Implied in this loose arrangement is a lot of space among molecules. Because of this "intermolecular space" gases are *compressible*—that is, they can be forced together to occupy less space.

Because gases have low density, and are usually invisible, we might mistakenly think that gases have no mass—simply untrue. To prove this you could just do the following:

1) Place a lid on a container that is open to the air.
2) Weigh the container, including the air inside.
3) Use a vacuum pump to remove all the air from it.
4) Weigh it again.

You'll see that the container weighs less after the air is pumped out than it did before (see Figure 1). This is a clear demonstration that air has mass. Its density is approximately 1.293 g/L at 0°C and 1 atm pressure. These conditions are commonly referred to as "**standard temperature and pressure**," or "STP."

If you accept that gases have mass, then you will be struck by the fact that you have tens of miles of atmospheric gases above you being pulled toward the earth by gravity. That thickness of gas is compressing every square centimeter of your body's surface. The column of air extending directly above you applies the same amount of pressure as a column of water 10.3 m high or a column of liquid mercury 760 mm high.

Because mercury is such a dense liquid, it makes a good tool for measuring and comparing pressures. You will often hear of a **mercury manometer**, also called a **U-tube manometer**, that is used to measure gaseous pressures. Check out the illustration in Figure 2. If a piece of glass tubing were bent in the shape of a U and filled with mercury, the level of the mercury on both ends of the U would experience the same amount of pressure from the atmosphere above it, so they would be at the same height. However, if we were to suck the air out one end with a vacuum pump so that there were no air pressure on that end of the tube, the mercury would rise in that side of the tube. The height of the rise would balance against the pressure exerted by the air on the open side of the tube. At the same time, the mercury level on the other side of the tube would fall.

The difference between the two mercury levels in opposite sides of the tube represents a mass of mercury. That mass of mercury is equal to the mass that can be pushed upward by that amount of atmospheric pressure. The place where the two levels of mercury come to rest would be at balance against the pressure of the atmosphere on the open side. This would demonstrate that the pressure of the atmosphere is equal to that of a column of mercury approximately 760 mm high.

**Figure 1.** With air in it, this small desiccator weighs 150 g; with the air pumped out it weighs about 149 g. Air has mass, but because your body relies on its pressure all around you, you hardly notice it. If it suddenly went away, you'd notice (about the time your blood started boiling)!

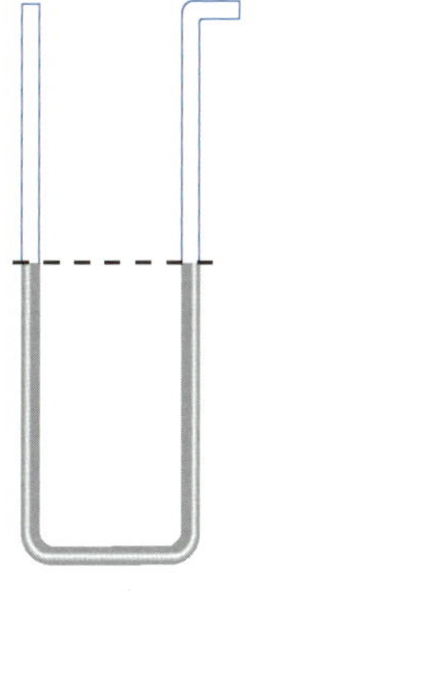

**Figure 2.** (Left, above) If mercury is placed in an open U-tube its level will be the same on both sides. (Left, below) If a vacuum is drawn on one side, the mercury will rise to the level where the mass of the liquid is balanced by the pressure of the atmosphere. That rise is 760 mm under the Earth's atmosphere. Of course, the atmospheric pressure varies with your altitude and changes slightly over time. 760 torr is "normal" atmospheric pressure at sea level.

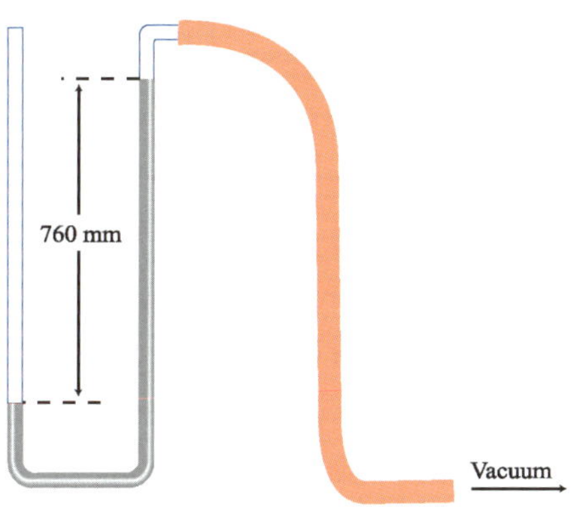


Because millimeters of mercury—mmHg—is such a common expression of pressure, it has come to be called by its own name. That name is **torr**. From now on, when you see 760 torr, you may interpret this to mean the pressure applied by a column of mercury 760 mm high.

Well, what if we changed the diameter of the glass tube? Would that change the height of the mercury column? No, because the diameter of the atmospheric column above the mercury will also have changed. (See Figure 3.) No matter how big the tubing is, the height will still be 760 mmHg. The *mass* of mercury will be greater in a larger-diameter piece of tubing than in a smaller one. But the *height* of the mercury will be the same. Furthermore the pressure applied by that amount of mercury per amount of area will be the same.

**Figure 3.** The diameter of the U-tube is of no consequence in the measurement of atmospheric pressure. A wider U-tube means a greater capacity for mercury, but it also means a greater area of atmosphere above it.

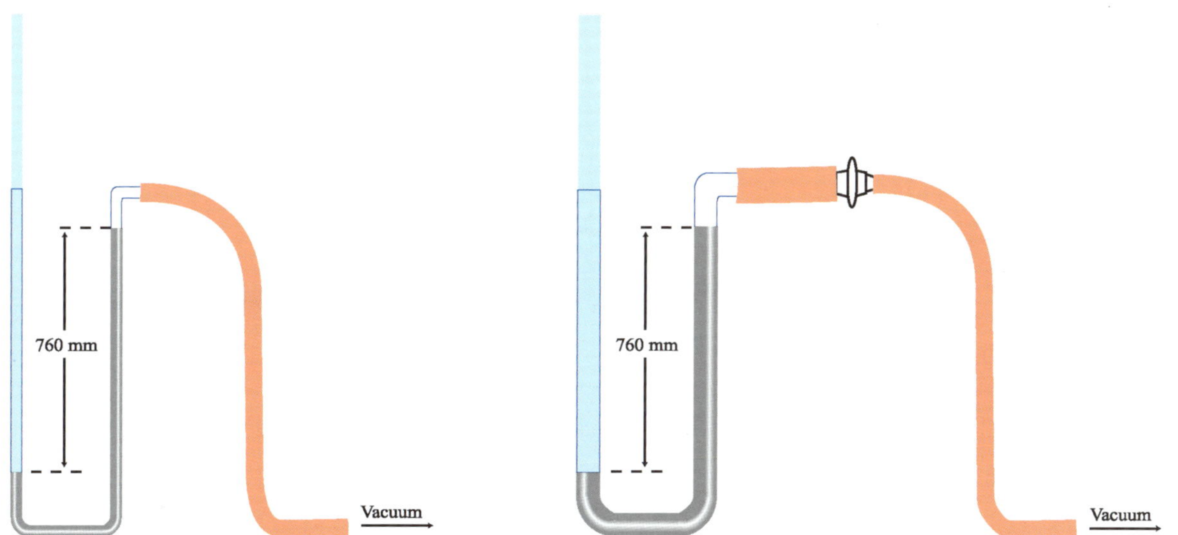

So then, the pressure applied by a column of fluid (liquid or gas) may be expressed in several ways. It may be expressed as:

- the mass of fluid divided by the surface area over which it applies pressure (any units of mass per unit surface area could be used.)
- some conversion of those values to a standard unit of pressure, such as lb/in$^2$ (PSI)
- the height of a column of a standard fluid, such as mmHg (Torr) or meters of water

The Table shows several standard units for pressure. Conversion factors to change from one set of units to another are provided inside the back cover of your text.

**Table.** Expressions for pressure and their equivalents in torr.

| Unit | Abbr. | Torr |
|---|---|---|
| atmosphere | atm | 760 |
| bar | NA | 750 |
| torr (millimeter of mercury) | mm Hg | 1 |
| meter water | m $H_2O$ | 73.6 |
| pascal | pa | 7.50 E(-3)* |
| pound per square inch | PSI, PPSI | 51.7 |

* The expression 7.50 E(-3) is shorthand for $7.50 \times 10^{-3}$. This form of shorthand is often used in tabular data. The E stands for "exponent" of ten.

## Boyle's Law

In the mid-1600's, Robert Boyle of Great Britain was one of the first people who began to understand matter as it is understood today. He performed a series of experiments concerning the behavior of gases under different conditions. What he did was simple, but what he learned was profound. Boyle placed different amounts of pressure on a column of air trapped in a U-tube. Figure 4 is an apparatus that we might make today that is similar to what he made. He then measured the air volume under compression. He found that the *pressure and the volume of the air were inversely related by a constant value.* In other words, as he applied greater pressure, the volume of the trapped air decreased predictably. Mathematically written, the volume of the air could be calculated as:

$$V = k/P$$

where k is a constant for a specific amount of any gas at a given temperature.

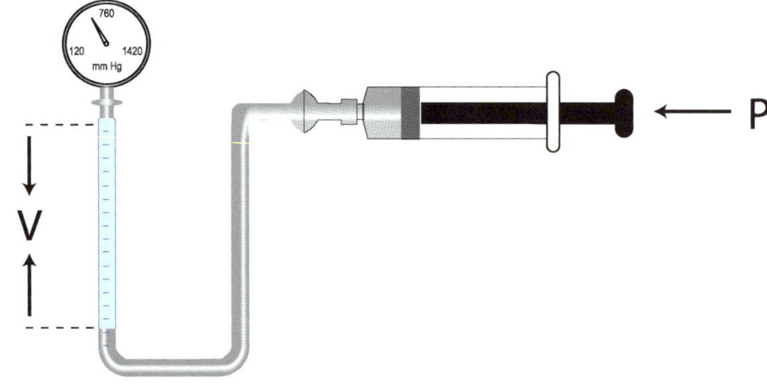


**Figure 4.** This apparatus demonstrates the principle of Boyle's law. As the pressure on the inside of the tube increases, the volume of gas trapped at the opposite end of the tube decreases in proportion. The volume is inversely proportional to the pressure.

This equation states that the volume and the pressure are related by a factor k. As pressure on the gas increases, the value of k/P decreases, so the volume decreases proportionally. By rearranging this equation, it's clear that the volume of gas, times the pressure applied, would be equal to that constant value:

$$VP = k$$

Since k is constant for all values of pressure, if the volume ($V_1$) were measured at any specific pressure ($P_1$), the product of P and V would be equal to their product at any other volume ($V_2$) and pressure ($P_2$). That is:

$$P_1 V_1 = P_2 V_2$$

This mathematical relationship, in whichever form you prefer, is now called **Boyle's Law**. This was the first of several historical steps that would lead to a much broader understanding of gases. Soon, you'll see the big picture, but for now...

**Exercises**

1) What volume will 3.89 L of a gas occupy if the pressure is reduced from 921 mmHg to standard atmospheric pressure?

2) A given mass of Ar gas has a volume of 25.0 L at 600 torr. What volume will the same quantity of gas occupy at 6.00 atm?

3) A nitrogen sample occupies a volume of 7.43 L at 1,100.0 torr. At what pressure would the volume be 5.00 L?

4) In an experiment, the pressure on a 7.0-L sample of $H_2$ is increased from standard atmospheric pressure to 1,750 mmHg. At the final pressure, assuming the temperature doesn't change, what is the volume?

# 42: Turn Up the Heat, Turn Up the Volume

## Charles' Law

It was not until about 150 years after Boyle that the significance of another simple observation was learned. I'm sure you've heard since you were a child that gases expand when they are warmed and contract when they are cooled. A French physicist named J.A.C. Charles determined that *a gas's volume can be predicted by its temperature*. (See Figure 1.) Some 100 years later this would eventuate into what is now called **Charles' Law**, which is written as follows:

$$V = kT$$

where k is a constant for a specific amount of any gas at constant pressure, and T is the absolute temperature (K).

Just as we did to obtain Boyle's equation, when we rearrange this equation we get:

$$V/T = k$$

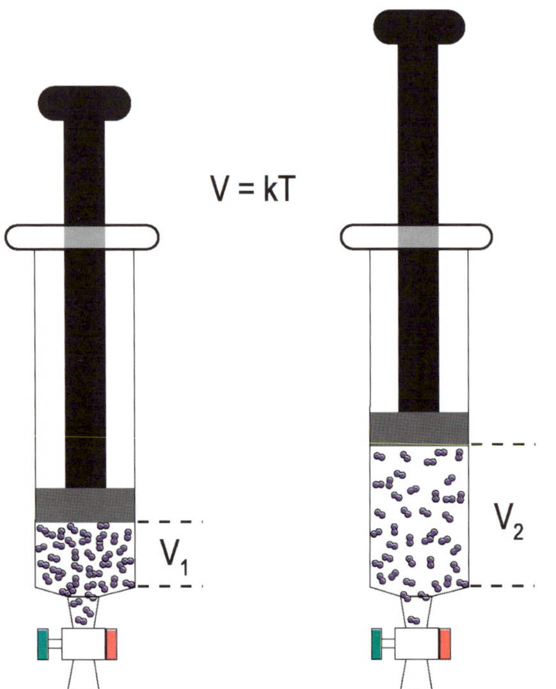


The volume of a gas divided by its temperature is a constant. That being true, any volume at its temperature would be equal to any other volume at *its* temperature. So, we can write:

$$\frac{V_1}{T_1} = \frac{V_2}{T_2}$$

## Gay-Lussac's Law

So far we have learned that the volume and pressure of a specific amount of gas at a constant temperature are related by a constant:

$$PV = k$$

And we learned that the volume and temperature of a specific amount of gas at a constant pressure are related by a constant:

$$V/T = k$$

We know that pressure and volume are related by a constant and that volume and temperature are related by a constant. By now, *you* can probably predict the next law that was discovered by a French chemist named J.L. Gay-Lussac, around 1800 (a contemporary of Charles). That's right—at a constant volume, the temperature and pressure of a set amount of gas are also related by a constant:

$$P/T = k$$

**Figure 1.** The volume of a gas is directly proportional to its temperature. This is known as Charles' law. Imagine that these two gas-tight syringes have frictionless plungers. The syringe on the left represents the volume ($V_1$) of a gas at a cooler temperature ($T_1$). When the syringe is warmed to $T_2$ the increased energy of the gas molecules will cause them to expand to $V_2$. The plunger, without friction to deter it, will be forced upward.

So then, at a fixed volume, any pressure divided by the temperature will equal any other pressure divided by *its* temperature:

$$\frac{P_1}{T_1} = \frac{P_2}{T_2}$$

All of these equations are expressions of what has come to be known as **Gay-Lussac's Law**. (See Figure 2.)

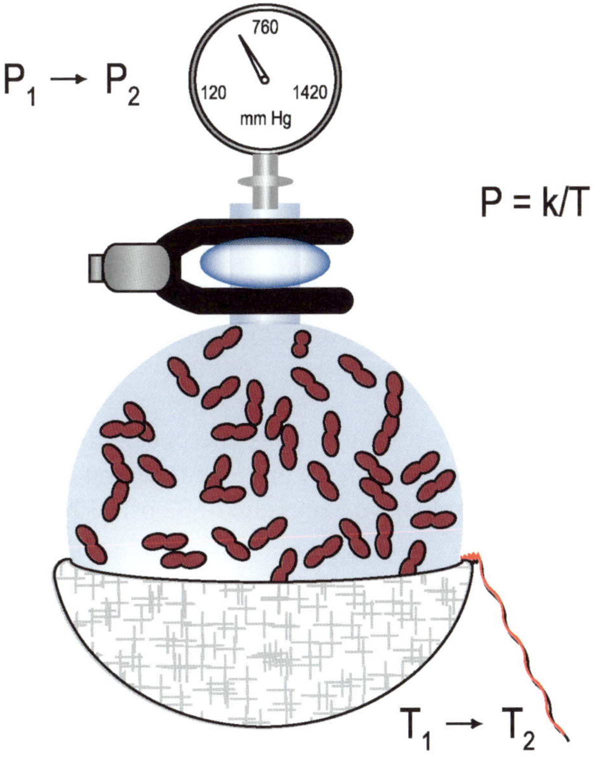


**Figure 2.** This is an illustration of Gay-Lussac's law. In this experimental system, the volume of gas is fixed. When the heating mantle heats up, the gas in the flask heats up. As the molecules absorb energy they have nowhere to go. They will collide with greater force against one another and against the container walls. This increases the pressure inside the container. The increase in temperature from $T_1$ to $T_2$ will result in an increase in pressure from $P_1$ to $P_2$. The increase in pressure is proportional to the increase in temperature.

## The Combined Gas Law

Gases may be easily controlled in laboratory situations. We may hold one factor constant while others are changed. However, in the real world, we can't expect things to happen so tidily. When filling up a hot air balloon, the temperature, the volume and the gas pressure inside the balloon change continuously. Surely with all of these factors related to one another by constant values there must be a simplified, combined form of the three laws that will allow us to predict one of these values as the other two change. The **Combined Gas Law** does just that. For a specific amount of any gas:

$$PV=kT$$

That is to say, the product of the pressure and volume is related to temperature by a constant value. It's easy to see how pressure, volume and temperature are related to this constant by rearranging this equation as we did each of the others:

$$PV/T = k$$

This equation says that the ratio of pressure times the volume (PV) to the temperature (T) is constant. Since PV/T is constant:

$$\frac{P_1 V_1}{T_1} = \frac{P_2 V_2}{T_2}$$

It's exercise time.

**Exercises**

Charles' Law

1) 2,500 mL of oxygen are heated from room temperature (27°C) to 45.0°C. Assuming the pressure remains constant, what is the new volume of the heated gas?

2) Suppose you have a 300 mL sample of He at 32°C. Assuming the volume of the container can vary so that the gas pressure is held constant, to what temperature would you have to heat the gas to increase its volume to 475 mL?

3) A given quantity of a gas occupies a volume of 3.80 L at 0°C. If the pressure is held constant, at what temperature will it occupy 2,750 mL.

4) If 3.6 L of nitrogen are cooled from 75°C to –15°C, what is the new volume?

Gay-Lussac's Law

1) A sealed container contains sufficient hydrogen to result in a pressure of 1.1 atm at 25°C. If the container is cooled to 0°C, what would be the new pressure in torr?

2) A given quantity of neon gas is confined to a glass cylinder. The pressure gauge on the cylinder reads 225 atm when the temperature is 22°C. What will the pressure be if the cylinder is warmed to 27°C?

3) In the early morning, before a man leaves for a long trip in his automobile, he checks the air pressure in his tires. The right front tire is found to have a pressure of $3\underline{0}$ psi. The temperature is 22°C. When he arrives at his destination the same tire is found to have a pressure of 32 psi. What is the new temperature of the gas in the tire?

4) A sealed container can hold gas safely up to a pressure of 40.0 atm. When a quantity of $SO_2$ is placed within the container at 27°C, it exerts a pressure of 21.9 atm. What is the highest temperature to which the container could be heated before exceeding its safe pressure limit?

The Combined Gas Law

1) If a quantity of helium occupies 16.5 L at 22.0°C and $76\underline{0}$ torr, what would be its volume at $6\underline{0}$°C and $95\underline{0}$ torr?

2) A given mass of $Cl_2$ occupies 12.5 L at 32°C and 760 torr. What would be the pressure of this gas if the volume were compressed to 8.5 L and if the temperature were raised by 15°C?

3) If a gas-filled balloon has a volume of 21.5 L at 27°C and 1.3 atm, what will be the volume if the temperature and pressure are raised to $3\underline{0}$°C and 1,200 torr, respectively?

## 43: Pulling It All Together

### A Unifying Theory

Every scientist in every discipline appreciates that all of nature operates around a series of simplifying principles. The scientists who first uncovered the unifying principles of nature were almost exclusively believers in God. Out of a culture that recognized a single, supreme mind behind the creation came a search for its supreme logic—natural laws constructed by a superior intelligence for His beings. It is impeccably reasonable to think that if The Creator fashioned the mind of humankind after His own, that His creations could understand His reasoning, provided He made it sufficiently simple.

The laws we have discussed in these lessons were confirmed through experimentation by several prominent names over a two-hundred-year span from the time of Robert Boyle in the mid-1600's to Gay-Lussac in the mid-1800's. We have just seen how those laws came full circle by recognizing that they were taking different points of view on the same natural principles and how those principles came together into a single combined law. However, the combined gas law is yet another limited view of an even greater unity in nature. This unity can be best observed if we try to understand gases as tiny particles.

The great experimenters of the 17$^{th}$ to 19$^{th}$ centuries tested the reactions of gases to every sort of physical stress and, by 1850, had summarized these reactions in the form of the equations revealed in the last few Lessons. All of the pieces of the puzzle were then in place for the theoreticians of the mid-19$^{th}$ century to develop a working model that explains why these factors participate as they do in gaseous behavior. The principles they came to rely upon for decision-making have been knit together and modified by more recent knowledge and have come to be known as the **kinetic-molecular theory** or **KMT**.

The KMT is an attempt to construct a framework for understanding gases. It begins with what those early experimenters had concluded about matter—that it consists of individual particles which had come to be called molecules. They were able to explain the properties of the three states of matter, by considering the attractions among their particles and the amount of intervening space. For example, if liquid water were heated to above its boiling point, it entered the gas phase. As it did, the amount of space it filled was considerably greater than the amount of space filled by the water itself. This implied that the particles making up water must have greater intermolecular distances (distances among molecules) than those making up liquid water. This leads to the first postulate of the theory:

Gases consist of individual particles with no attraction for each other. The particles are small in comparison with the distances among them, so a container of gas is primarily empty space.

Now consider the air inside a balloon. If the gas molecules were motionless, the elasticity (stretchiness) of the balloon would collapse the gas molecules in on themselves, and they would then presumably form a liquid because of their new closeness. But the gas molecules are not that easy to push around. It will be noticed upon blowing up a balloon that the inward pressure of the balloon wall is met with equally strong outward pressure from the gas inside. (See Figure 1.) The wall of the balloon may be forced outward by the gas despite this inward pressure. How can this be if gas particles are far apart? It can only be if those gas particles are in constant, rapid motion and if they collide against one another and with their container wall. This, restated, is a second postulate of the theory:

Gas particles are in rapid, random motion. Their particles will collide with one another and with the walls of any container in which they are held. Because there are no attractions among particles, their motions and their collisions will be random. The internal pressure on the walls of a vessel containing a gas results from collisions of gas particles with the walls of the container.

Suppose the gas molecules lost kinetic energy through friction every time they collided with one an-

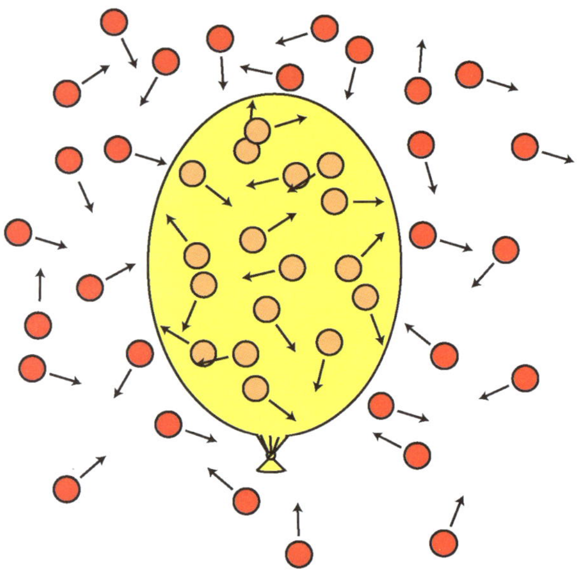

**Figure 1.** In order to increase in size, the gas pressure inside a balloon must overcome (1) the pressure of the atmosphere that surrounds it and (2) the inward force of the rubber wall. With each puff of air blown into the balloon, the resistance of the rubber increases, but the atmospheric pressure stays constant. The pressure inside the balloon will be equal and opposite relative to these two forces. (Yes, this is Newton's 3rd Law.)

other. The gases would give off heat while held at constant volume and pressure. Eventually, the particles would run out of energy, so their volume would decrease with time and they would end up in a puddle. In reality, the collisions among gas molecules are completely "**elastic**." That is, when they collide they spring back without the slightest loss of energy. Thus, we have our third postulate:

Particles do not lose heat through friction when they collide, because their collisions are elastic. Their kinetic energy is determined only by their absolute temperature.

Although there was no drum roll during that last sentence, you should stand up and place your hand over your heart when you read it. It really is quite a fascination. If the kinetic energy of a particle is determined only by its absolute temperature, then it shouldn't matter what *kind* of gas molecules you study; they should all have precisely the same kinetic energy at the same temperature. (See Figures 2 and 3.) It's that kinetic energy that will determine how much pressure will be placed on a container wall by a gas when held at a constant volume and temperature. We can assume, then, that *the pressure will be the same for the same number of particles of any gas at that volume and temperature!*

The kinetic-molecular theory of gases has greatly simplified all that we know about gases and has opened some new doors of understanding. We will peer through that first open door as we continue.

**Figure 2.** The kinetic energy is the same for all gases at a given temperature. That energy is directly proportional to the Kelvin temperature.

kinetic energy

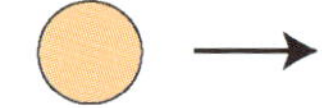

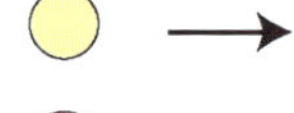

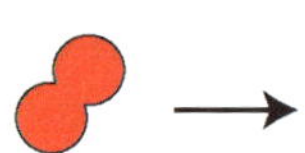

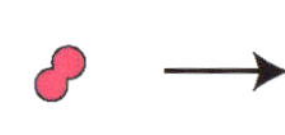

## Dalton's Law

If we hold the kinetic-molecular theory of gases to be true, it requires relatively little insight to see what Dalton saw. In any container of gases, the total pressure on the container walls is proportional to the *number of molecules* of all gases. Furthermore, the pressure from gas A plus the pressure from gas B is equal

**Figure 3.** The kinetic energy of a particle is $mv^2/2$. Since all gases have the same kinetic energy, but all have different masses (m), smaller particles must move at slightly greater velocities (v).

$$\text{kinetic energy} = \frac{mv^2}{2}$$

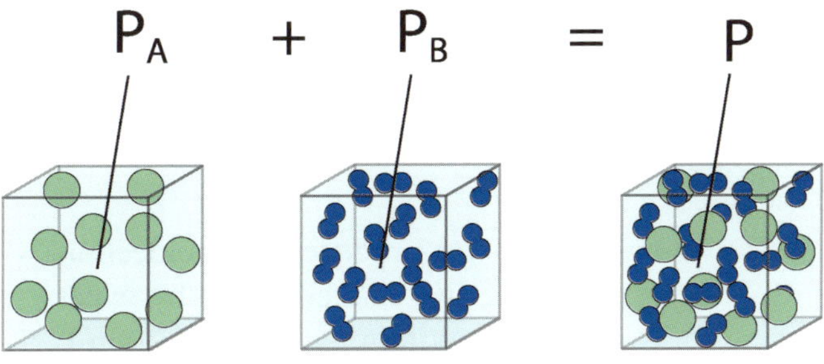


Figure 4. The total pressure of a mixture of gases is the sum of the partial pressures of the individual gases making up the mixture.

to the combined pressures of the two gases. Stated mathematically:

$$P = P_A + P_B$$

where $P_A$ and $P_B$ are called the **partial pressures** applied by gases A and B, respectively, and P is the total, combined pressure exerted by both gases. This is referred to as Dalton's law of partial pressures, and it holds true for any number of gases in combination. (See Figure 4.)

One common laboratory application of Dalton's law is in the collection of gases over water. An apparatus like the one illustrated in Figure 5 is used to collect the gaseous products of a reaction. As the gas collects it displaces water from the inverted test tube. Since the gas is in direct contact with water, the collected gas includes some portion of the gaseous product and some portion of water vapor. The experimenter needs to know how much of that gas is water. As we will learn in future lessons, the amount of water in the vapor is related to the amount of pressure that water vapor places on its container. So we refer to the *vapor pressure* of water in the headspace (gas phase) in the container.

The vapor pressure of water has been determined experimentally at various temperatures. We can simply look up the vapor pressure corresponding to the temperature at which our experiment was carried out. The *partial pressure* of water in the vapor above the liquid water in a closed container will be the same as its vapor pressure. As we learned, the total pressure of gas in the container is equal to the sum of the two partial pressures.

## Avogadro's Law

Like Dalton's law of partial pressures, Avogadro's law is also evident if the kinetic-molecular theory is held to be true. Avogadro determined that, at constant temperature and pressure, the same molar quantity of two different pure gases will occupy the same amount of space. In other words, one mole of oxygen occupies the same volume as one mole of hydrogen, one mole of argon or one mole of nitrogen. Experimentally, this has been verified. The volume is 22.4 L at STP. This volume is approximately 6 gallons, or the volume of a five-gallon bucket and a milk jug.

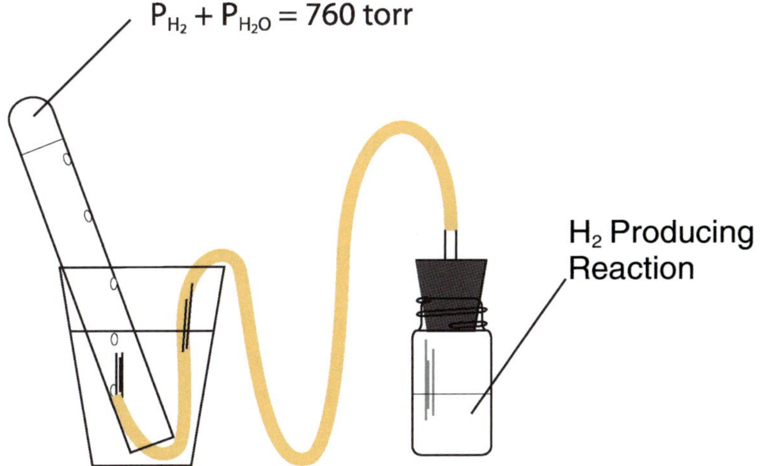


Figure 5. If any gas is held above water, the vapor in the trapped gas will consist of some portion of gaseous water molecules and some portion of molecules of the gas. The pressure on the walls of the container will be the sum of the pressures exerted by both. If the vapor pressure of water at a given temperature is known, then the proportion of water can be calculated.

**Figure 6.** On average, a gas particle has a space of 37 $Å^3$ to move about in at 0°C and one atmosphere of pressure. If the volume of a gas molecule is $1Å^3$, it has, on average, a sphere with a radius more than twice its own in which to move about without entering the sphere of another atom. Oxygen, shown here, has a slightly larger-than-average atomic radius. The length of the bond (from nucleus to nucleus) is 1.11Å.

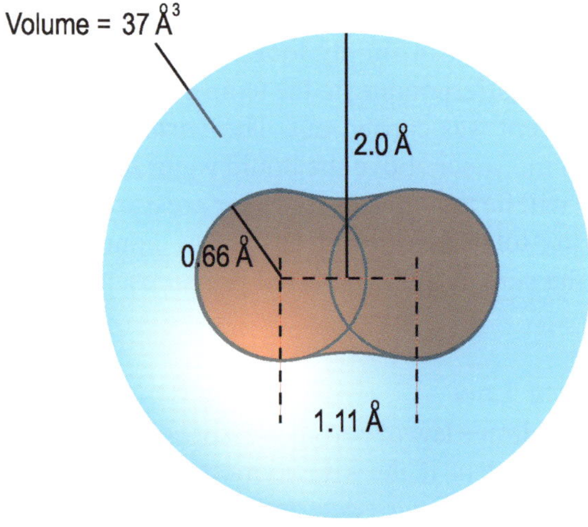


If equal volumes of different gases at the same temperature and pressure contain the same number of molecules, then, for gases, the volumes are equal to the molar quantities in a balanced chemical equation. For example, in the equation:

$$2\ H_2 + O_2 \rightarrow 2\ H_2O$$

each 2 L of hydrogen gas should combine with precisely one liter of oxygen gas to form water if the two gases are held under the same physical conditions.

Knowing that oxygen, hydrogen, argon and nitrogen have different gram molecular weights, and yet an equimolar amount of each occupies the same volume, it's easy to see that the densities of these gases must be different. Unlike solids and liquids, the densities of gases are expressed in g/L rather than $g/cm^3$. Otherwise the values of density are absurdly small.

**Exercises**

Dalton's Law

1) Oxygen gas was produced by the decomposition of $KClO_3$ and collected by water displacement. Within two hours, 350 mL of gas were collected at 24°C and 760 torr. What is the partial pressure of the $O_2$? (Vapor press. of $H_2O$ is 22.4 torr at 24°C.)

2) What is the partial pressure of $CH_4$ (g) collected over water at 301 K and 715 torr? (Vapor press. of $H_2O$ is 28.3 torr at 28°C.)

3) When acetic acid and sodium bicarbonate are mixed, one of the products is $CO_2$. A volume of 725.0 mL of $CO_2$ was prepared in this fashion and collected by displacement of $H_2O$ at 25°C and 75$\underline{0}$ torr. What volume would the dry $CO_2$ occupy at STP? (Vapor press. of $H_2O$ is 23.8 torr at 25°C.)

Avogadro's Law

1) If 3.00 L of a gas measured at STP have a mass of 5.30 g, what is the molar mass of the gas? By looking at your periodic table, what gas would you suspect?

2) 54.2 g of a given gas occupies a volume of 5.10 L at 30.0°C and 6.0 atm pressure. What is the molar mass of the gas?

**Hint: To solve this problem you need to know the volume of this same quantity of gas at STP—then you can convert the volume to moles of gas.**

3) Calculate the density of $NH_3$ (g) at STP.

4) How many liters of $H_2$, when reacted with $N_2$, would produce 4.5 L of $NH_3$ (g) at constant temperature and pressure?

## 44: An Ideal Gas Obeys the Law

Up to this point we have written lots of conclusions about the nature and behavior of gases. Let's summarize the relationships we have confirmed thus far:

| | |
|---|---|
| PV=kT | For any gas, the pressure and volume are proportional to the temperature. |
| PV = k | For any gas, the product of pressure and volume is a constant. |
| V = kT | For any gas, the volume is proportional to the temperature. |

To these relationships we add one more, namely:

$$PV = kn$$

where n is the molar quantity (the number of moles) of any gas. This equation says that the product of the pressure and volume is proportional to the molar quantity of gas. If either P or V is held constant, the other will increase or decrease with increases and decreases in n.

We now have four different equations containing four different variables. We will have to admit that, for all the years that these bright scientists were learning about the relationships among these variables, they were only poking around at the surface, for we now see the deeper relationship that ties them all together. These four variables can all be combined into a single expression called the **ideal gas law**, which is stated as follows:

$$PV = nRT$$

where R = the **ideal gas law constant**.

In every previous equation where k was used to show the relationships among variables, k contained and concealed R. Think about it for a moment. When we said:

$$PV=kT$$

k contained both R and n.

When we said:

$$PV = k$$

k contained n, R and T.

When we said

$$V = kT$$

k was concealing the product of n, R and 1/P. (See the Figure.) Now we have one simple relationship that includes and implies all the others. Furthermore, we have

$$PV = nRT$$

| Law | Stated | At Constant | Hidden in k | Alternate Statement |
|---|---|---|---|---|
| Boyle | k = PV | n & T | nRT | $P_1V_1 = P_2V_2$ |
| Charles | k = V/T | n & P | nR/P | $V_1/T_1 = V_2/T_2$ |
| Gay & Lussac | k = P/T | n & V | nR/V | $P_1/T_1 = P_2/T_2$ |
| "Combined" | k = PV/T | n | nR/V | $P_1V_1/T_1 = P_2V_2/T_2$ |

**Figure.** Each of the gas laws discovered over the period of two hundred years was a partial statement of the ideal gas law, PV = nRT. Each one of them contained a constant value (shown here as k). And, in each case, the value of the ideal gas constant (R) was hidden in the value of k. When all of these relationships had been recorded, it became obvious that they were all different viewpoints on the same law--the "ideal gas law."

a known value, which we call the ideal gas law constant, that will allow us to calculate any of these variables if we but know the conditions under which the gas exists.

If you can understand this relationship and can learn higher mathematics, you can be an engineer. That's because the ideal gas law is one of the most complex of the fundamental laws of nature known to humankind. Most of the basic laws of nature involve two variables. A few involve three. This one is unusually complex in that it involves four variables.

Notice that if we solve the ideal gas law for R, we get:

$$R = \frac{P\,V}{n\,T}$$

This means that R has units of pressure times volume, over moles times temperature. If we measure pressure in atmospheres, volume in liters and temperature in Kelvin, R is:

$$0.0821 = \frac{\text{atm L}}{\text{mol K}}$$

Of course, we can express any of these variables (except moles) in a variety of different units, and the numeric value of R will change as the units change. Until you have had plenty of practice, just memorize this one expression for R and convert the units of all the variables to these units.

**Exercises**

1) What pressure will be exerted by 0.75 mol of $CO_2$ in a 2,500-mL container if the temperature is 20°C?

2) How many moles of gas are contained in a 41.0-L tank if the pressure gauge reads 3.10 atm and the temperature of the tank is 17°C?

3) 7.76 g of an unknown gas occupies a volume of 4.15 L at a pressure of 1,140 torr and a temperature of 310 K. What is the molar mass of the unknown gas?

4) A gaseous mixture of inert gases contains 8.05 g of Ne and 4.42 g of He. What volume would this mixture occupy at STP?

5) If 6.6 g $O_2$ and 3.8 g $N_2$ are placed in a 3.5-L container at 33°C, what is the partial pressure of the oxygen? What is the total pressure?

6) Write as many equations as you can think of from PV = nRT, by substituting k for two or more of the variables. (R is a constant; all the other factors are variables.)

## 45: Grandson of Stoichiometry

You now have all the information that you will need to solve the following problems. You will need to remember three things:

1) You must remember the grand relationship, PV = nRT.
2) You must remember to always balance your chemical equations.
3) You must remember that if you have a balanced equation, you know not only the *molar* proportions of gases, but, according to Avogadro's Law, you also know the *volume* proportions of the gases. Don't make the mistake of thinking that one liter is equal to one mole. That's not true. But if one mole of reactant makes one mole of product, then one liter of reactant will also make one liter of product. Feel free to use this knowledge in coming up with unit factors that will help you in solving the following problems.

**Exercises**

1) Suppose you want to burn a piece of magnesium that has a mass of 7.27 g. What volume of $O_2$ would be needed at 23°C and 1.02 atm?

2) How many g of potassium must react with $H_2O$ to produce 1.60 L of $H_2$ at STP?

3) Consider the following reaction:

$$C_2H_2\ (g) + 2\ HF\ (g) \rightarrow C_2H_4F_2\ (g)$$

If 630 mL $C_2H_2$ are to be completely reacted, and if the pressure and temperature are held constant, what volume of HF would be required?

4) Magnesium carbonate can be thermally decomposed, producing magnesium oxide and carbon dioxide. What volume of $CO_2$ would be produced at 35$\underline{0}$°C and 0.95 atm if 7.3 g of $MgCO_3$ were completely decomposed?

5) The lead sulfide in galena (lead ore) can be leached (dissolved by reaction) with hydrochloric acid according to the following stoichiometry:

$$PbS\ (s) + 2\ HCl\ (aq) \rightarrow PbCl_2 + H_2S\ (g)$$

If 9.812 g of PbS are completely reacted, what volume of $H_2S$ gas would be obtained if it were measured at 3$\underline{0}$°C and 1 atm?

## 46: Quiz 7

1) Know all the gas laws (or be able to come up with them from the ideal gas law, PV = nRT). Be able to use each in calculations. In addition to the ideal gas law you will need Dalton's law and Avogadro's law. Their equations will not be provided for you.
2) Know the conditions that are called "STP," and that 1 atmosphere = 760 torr.
3) Know that the molar volume of any gas is 22.4 L at STP.

Yes, it's blue. But *why* is it blue?

## 47: Liquids, and Changes From Liquids to Gases

A liquid is a substance having particles that are attracted to one another strongly enough to eliminate nearly all of the compressible space among them, but weakly enough to allow those particles freedom of movement. Let's take a moment to realize what this means. We have introduced gases and have studied the Kinetic-Molecular Theory of gases. According to this widely accepted theory, gas particles have no attraction to one another, so they exist independently. The particles themselves are dwarfed by the space that exists among them. This gives plenty of room to squeeze gas particles together. When gases are squeezed together tightly enough to eliminate this space, or are cooled down so that their kinetic energy is reduced enough, they become liquid. (Figures 1 and 2.)

This unique balance of *intermolecular forces* gives liquids the unique properties that distinguish them from solids and gases. In the following paragraphs we will explore some of those properties.

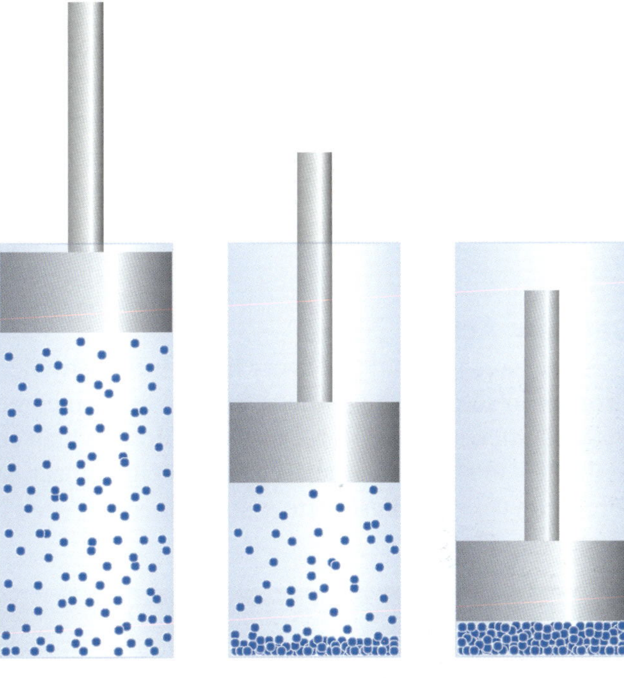

**Figure 1.** A liquid may be formed by compression of a gas to force it into less space. As the piston applies greater and greater pressure, the molecules of gas take up less and less space.

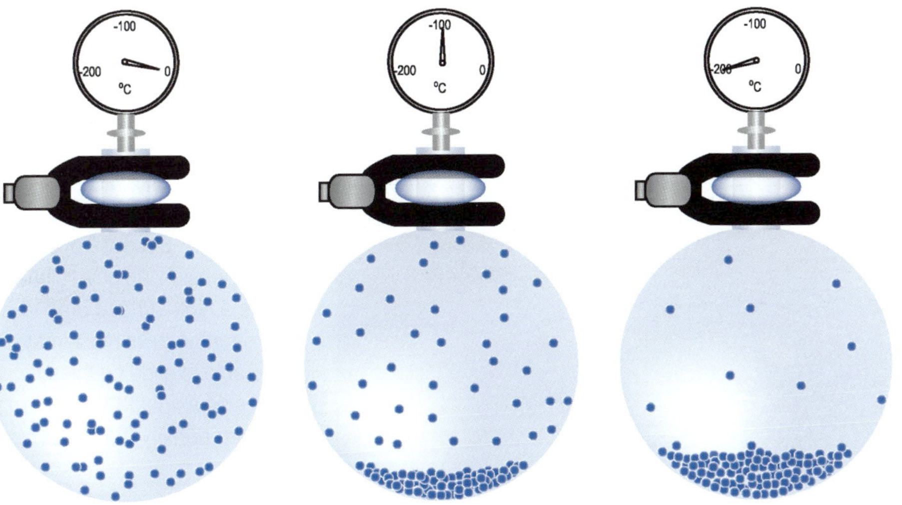


**Figure 2.** A liquid may also be formed by cooling of a gas to reduce the kinetic energy of the particles. As the individual particles lose energy, they enter the liquid phase. The lower the temperature gets (approaching absolute zero) the fewer the number of particles that will remain in the gas phase.

## Surface Tension and Evaporation

Because the particles (molecules or atoms) making up a liquid have some degree of attraction to one another, and because they lack any great attraction to the particles making up air, there is a kind of closeness among liquid molecules at the surface of a container where it meets with the air. This tightly-knit boundary forms a kind of molecular wall through which molecules move with relative difficulty. This wall makes it difficult for a molecule to break into a solution from the air. The resistance of the liquid surface to penetration is called *surface tension*.

Surface tension explains certain properties we observe in liquids, such as the ability of a needle, or a piece of wire mesh to float if it is laid on the water gently. (See Figure 3.) Once the surface tension is broken by such an object, it will fall rapidly through the liquid. Surface tension also explains the tendency of a liquid to bead up on a foreign surface and to occupy as little space in air as possible. (Although raindrops are usually drawn in the familiar pointy-headed configuration, in reality, they are nearly spherical.)

Surface tension is measured as the amount of energy required to break through the surface of a liquid. It is expressed in units of energy per square centimeter of surface area (e.g., dynes/$cm^2$). Water has such a high surface tension because of its uniquely high polarity—the positive pole of one water molecule attracts the negative pole of a neighbor. (See Figure 4.)

## Capillary Action

We have just said that every liquid has intermolecular forces that tend to hold it together, although they are stronger in some than in others. Surface tension comes about from these so-called **cohesive** [together + stick] **forces** (“stick together” forces). These forces which cause molecules to stick to one another may also cause the molecules to stick to the surface of a foreign object. When a liquid is attracted to a solid surface, the attractive forces are called **adhesive** [to + stick] **forces** (“stick to” forces). For example, water molecules have *cohesive*

**Figure 3.** Because of surface tension, an object that has a high surface area in proportion to its mass may be able to glide on the surface of a liquid that has a lesser density than itself.

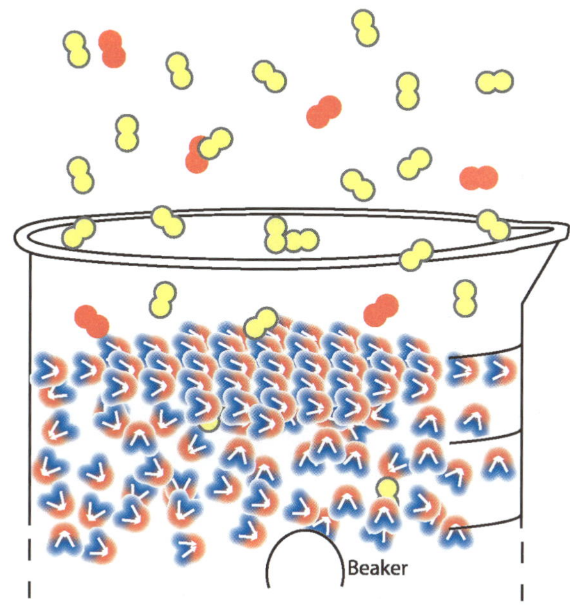


**Figure 4.** Because they lack any attraction to the molecules in the air, molecules near the surface of a liquid are more organized than those in the bulk liquid. This organization makes the surface of a liquid harder to penetrate. The more attraction there is among the liquid molecules the harder the surface will be to penetrate. This difficulty of penetration is called *surface tension*. Not only does it prevent air molecules from entering as readily. It also makes evaporation by liquid molecules more difficult.

forces, and they share *adhesive* forces with glass. When the surface of water is pierced by a glass capillary (a thin tube), the water level in the tube rises above the water outside the tube. This is called **capillary action** or **capillarity**. (See Figure 5.)

Capillary action has many practical implications. In the laboratory, glass measuring devices have to be corrected for the tendency of water to rise up the sides. This prevents the water from giving a level surface for reading volume. The curved surface is called a **meniscus**. (See Figure 6.) The graduations on the glass and other **hydrophilic** [water + loving] measuring devices have to be manufactured to account for the meniscus effect in order for the glassware to measure accurately. Measuring devices with plastic or other **hydrophobic** [water + fearing] surfaces may not form a visible meniscus because there are no mutual adhesive forces between them and the water.

Laboratory concerns aside, the capillary effect is important in many important natural processes. Among the most important is the tendency for water to rise in the small capillaries of the xylem in plants. This capillary rise (along with pull from evaporation of water at

**Figure 5.** *Capillary action* is the name given to the rise of a liquid inside a capillary tube. The capillary action increases as the diameter of the tube gets smaller. (As the tube gets smaller, there is greater surface area in contact with the water relative to the amount of water that has to be pulled up the tube.) Here a few drops of dye placed on a glass slide climb up between two other glass slides by capillary action.

**Figure 6.** A meniscus is the curve in the surface of the water where it meets with the air. It is caused by the water clinging to the surface of the glass. Volumes are read from glassware at the lowest point on the meniscus.

**Figure 7.** It's the same capillary action shown in Figure 5 along with suction that is created by the evaporation of water at the leaves that forces water from the roots of the tree to the top where it is needed to keep those leaves alive.

the leaves of a plant) brings water to the dizzying heights of tree tops from the soil below. (See Figure 7.)

## Heat of Vaporization

The molecules making up a liquid may act on their own to some extent. For example, one molecule may have more kinetic energy than another because it absorbed some energy from its surroundings. When energy is absorbed by a molecule, its kinetic energy increases. A particular molecule occasionally gets up enough energy to break through the surface boundary of the liquid into the neighboring air. We would say that that molecule *vaporizes* or *evaporates*. When it leaves, it takes that extra energy with it, leaving the liquid phase somewhat less energetic than it was before. As millions of such particles evaporate, they have a cooling effect on the entire liquid phase.

As you would expect, the amount of heat required for a specific kind of particle to break through the surface tension and pass into the air depends on the strength of the attractions among the molecules in the liquid. The stronger those attractions are, the less able a particle will be to leave. Since these intermolecular forces are different for every substance, the heat energy required to overcome them is also unique to each liquid. The result is that every liquid has its own unique energy requirement for vaporization (evaporation).

Since the amount of energy required to vaporize a liquid is dependent on the amount of liquid present to absorb that energy, energies of vaporization must be expressed in units of energy per amount of liquid. Common units of energy of vaporization are called the **molar heat of vaporization** (heat/mole) and the **heat of vaporization** (heat/gram). You could just as easily express it in heat per milliliter or heat per ton, but all the other scientists at the Annual Heat of Vaporization Convention would give you dirty looks.

Please be aware that condensation is the opposite change from that of vaporization. So it only stands to reason that the heat given up by liquid as it condenses is the negative of the heat of vaporization.

We have described two different properties of liquids (surface tension and heat of vaporization), each of which exists because of the intermolecular attractions among liquid molecules. If you were to guess that these properties change in proportion to one another, you would be correct. The higher the surface tension of a liquid is, the higher will be its heat of vaporization as well.

The **heat of vaporization** will be needed to calculate temperature changes in substances that involve a change of phase from liquid to gas or from gas to liquid. Suppose we wanted to convert water from some current temperature ($T_1$) to steam. How much energy would it take? If you'll recall, we saw a similar problem back in Lesson 7. Way back then we were able to calculate the amount of heat required to change a substance from one temperature to another. However, this problem is a little different. We are now involving a change of phase from liquid to gas. As I said, that change requires additional energy. The heat of vaporization is defined as the minimal amount of heat required to change one gram of water at 100°C to steam without changing its temperature. For water, that amount of heat is 2,259 J/g.

The energy requirement to bring about the temperature change is given by our old friend George:

$$\Delta\varepsilon_t = mh_s\Delta t$$

That is to say, the change in energy for a given change in temperature is equal to the mass of the water, times the specific heat of water, times the change in temperature. To place units on these values, the energy change in J is given by:

$$?\,(\text{J}) = \text{mass H}_2\text{O (g)} \times \text{specific heat}\left(\frac{\text{J}}{\text{g°C}}\right) \times \Delta t\,(°\text{C})$$

Now that we have water at the boiling point, it takes additional energy to convert it to steam. That's where the heat of vaporization comes in. The energy change required to convert water at 100°C to steam at that same temperature is its heat of vaporization:

$$\Delta\varepsilon_v = mh_v$$

where $\Delta\varepsilon_v$ is the energy of vaporization for the mass (m) of water and $h_f$ is its heat of fusion. In its units:

$$?\,(\text{J}) = \text{mass H}_2\text{O (g)} \times \text{heat of vaporization}\left(\frac{\text{J}}{\text{g}}\right)$$

Then we add the first energy ($\varepsilon_t$) to the second ($\varepsilon_v$) to get the total energy requirement to bring the liquid up to boiling and to make the transition to steam. Got it? I hope so—you'll need it!

The reverse process to bring steam from 100°C back down to some temperature would work exactly the same way using exactly the same value for the heat of vaporization. However, the energy change will be a negative value rather than a positive one. If we wanted to heat up the steam (under pressure) after the phase change, it would require additional energy using the specific heat of steam. (But that's a matter for a different day.)

**Exercises**

1) The heat of vaporization of water is 2,259 J/g. Suppose we now want to take 18.5 g of water which is at 32.3°C and convert it to steam at 100.0°C. How many kJ of energy would be required?

2) How many joules of energy will be given up by steam if it is dropped from 100.0°C to become liquid water at 0°C.

3) The normal boiling point of acetonitrile, $C_2H_3N$, is 80°C. How many calories of energy would be required to change 10.0 g of acetonitrile from 65°C to vapor at 80.0°C? (The specific heat of acetonitrile is 0.541 cal/g°C, and its heat of vaporization is 173.7 cal/g.)

## 48: Liquid-Vapor Equilibrium

The molecules of a liquid, like those of gases, are in constant motion (although it's not as rapid). Because some particles have more kinetic energy than others, molecules regularly leave the liquid phase to become gas (vapor). In a sealed container of a liquid, these particles are limited in how far they can go. They have only two possible fates—they can either remain in the headspace of the container or reenter the liquid. In fact, both of these are taking place constantly. That is to say, some of the particles will be present in the **headspace** while others are reentering the liquid.

In an open container, however, once a liquid particle vaporizes it becomes part of the atmosphere. As long as there is some energy entering the liquid, molecules will be gaining heat and vaporizing. How frequently molecules leave the solution, then, will depend on how fast they absorb energy from their surroundings. In an energy-intensive environment they will absorb heat and leave the liquid phase relatively quickly. Molecules leaving solution can be likened to popcorn popping. All of the kernels are heated, but when one gains enough energy to pop, it leaves the pan. As the oil in the pan heats up, the rate at which the kernels pop increases until they have all popped. (See Figure 1.)

**Figure 1.** Vaporization of a liquid is somewhat like the popping of popcorn. The entire mass of liquid heats up, but the individual particles are located in different positions relative to the heat source and don't reach that critical temperature at the same time. Once a particle has heated up enough to break through the surface, and is positioned close enough to the surface to leave without its energy being transferred to another particle, it pops out of solution.

As you can imagine, when a container that has been sitting open is first sealed, there will be relatively few vaporized molecules of that liquid in its headspace. As time goes on, more and more vapor particles will accumulate there. As the concentration of these particles increases, they will increasingly return to the liquid. This process will continue until there is a balance between the number of particles in the liquid phase and those in the gas phase. This is the point of **equilibrium**. We'll have much more to say about equilibria as time goes on. For now, notice that the word equilibrium doesn't mean that nothing is happening in the flask. Evaporation continues to take place as new molecules build up energy and leave solution. But, at equilibrium, every molecule that leaves the liquid is met by another molecule that has lost enough energy to reenter. We say that equilibrium is not **static** (unchanging), but **dynamic** (undergoing constant change). (See Figure 2.)

Of course, in discussing the properties of gases, we have learned that any gaseous molecule exerts pressure on its container. Molecules of a vapor above its liquid are no exception. They, too, exert pressure on the walls of their container. How much pressure they exert depends on the quantity of those molecules and upon their temperature. (This is an application of PV = nRT where the volume is constant. If the volume is constant and R is constant then the only variables left are n and T—the molar quantity and the temperature. Isn't it neat how these things always make sense?)

The vapor pressure is another measure of intermolecular attractions among liquid particles. The more a liquid tends to hang together, the fewer the molecules that will escape to the vapor phase. The fewer there are in the vapor phase, the lower the vapor pressure will be. Liquids that enter the gas phase readily (because they are less attracted to one another) are referred to by the term **volatile**. (See Figure 3.)

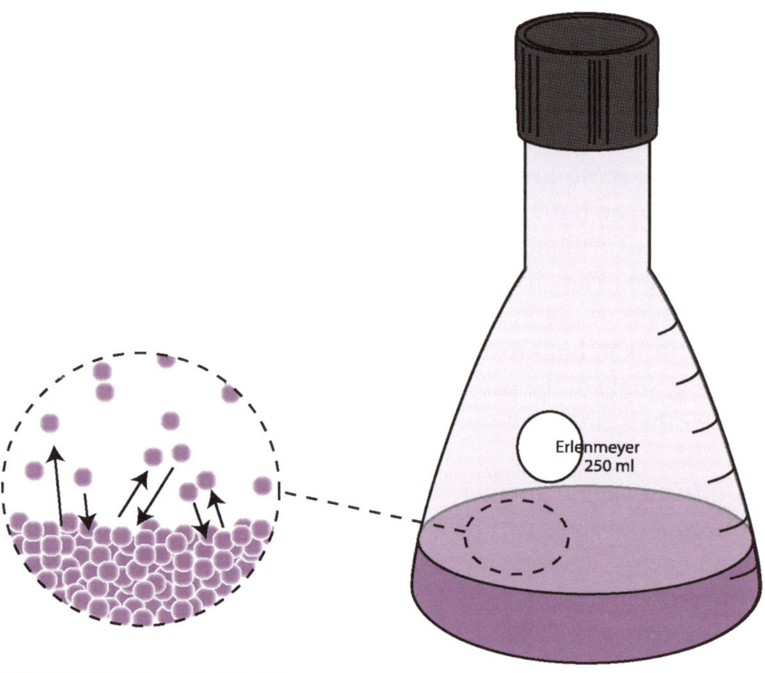


**Figure 2.** A liquid in equilibrium with its vapor may look boring enough on the surface, but if you could see what was actually going on at the particle level, you would see constant exchange of particles across the interface. On average, one particle of liquid enters the headspace for every particle of vapor that reenters the liquid phase.

As long as there is liquid in the bottom of a sealed container, that liquid will seek its unique dynamic equilibrium with its headspace gas. *The concentration of those particles in the gas phase will be the same at equilibrium regardless of the volume of liquid in the container*. (See Figure 4.) However, if there is so little liquid in the container that it all enters the gas phase, the usual equilibrium will not be reached. The concentration of its molecules in the gas phase will be less than the equilibrium quantity.

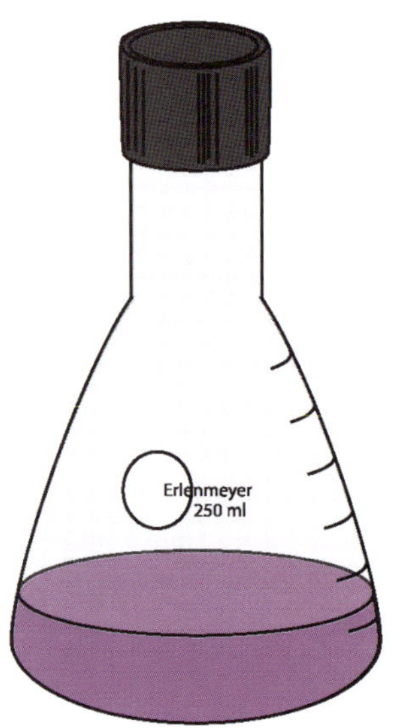


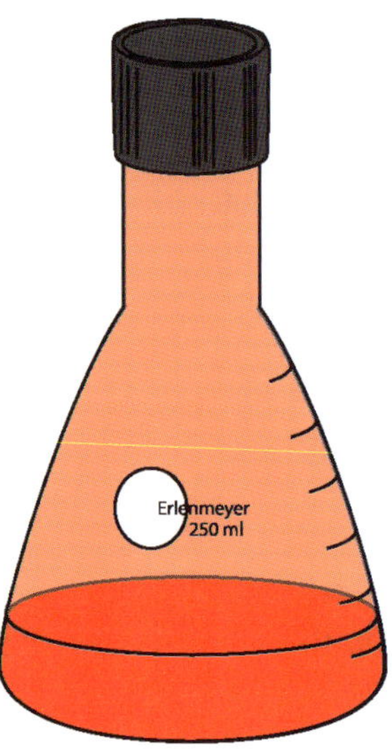


Now imagine that you place a liquid in a sealed container and allow time for the liquid to come to equilibrium with its vapor. Afterward you place the entire container in a warm water bath. What will happen? The warmth of the water bath will increase the temperature of the container's contents. The liquid and the gas

**Figure 3.** Volatile substances are those which have high equilibrium vapor pressures at room temperature. This means that the concentration of vapor phase molecules above the liquid will be high. Of these two substances, the one on the right is the more volatile. Its vapor-phase concentration is higher.

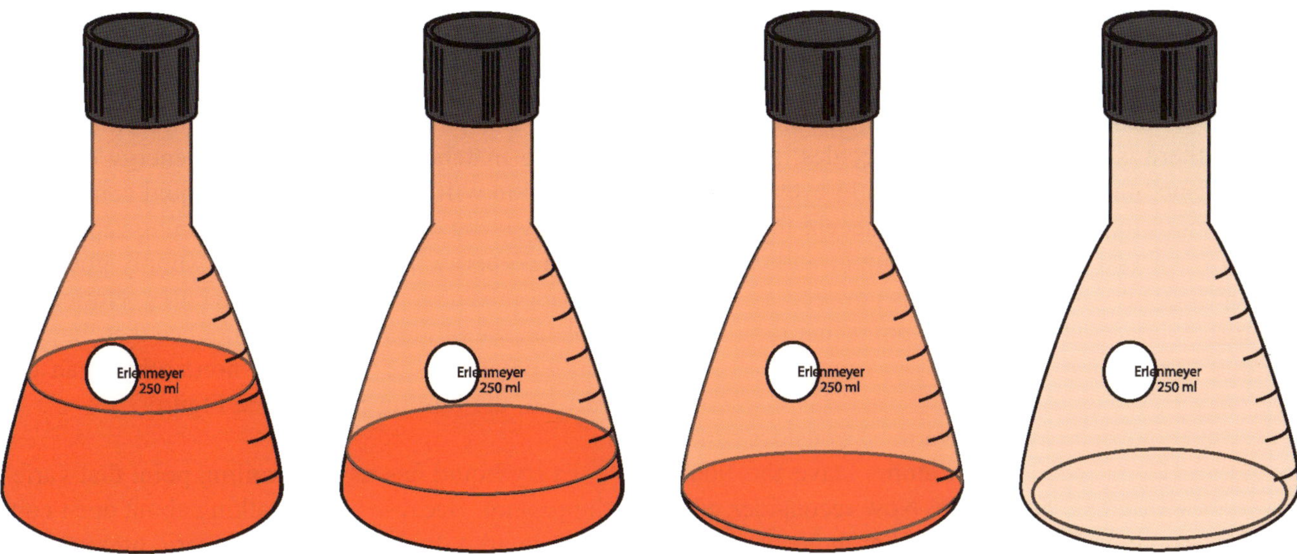


**Figure 4.** Because there is liquid in each of the first three containers, the vapor pressure in each of them will be the same, even though there are different amounts of liquid in each. The fourth container has no liquid in it, so its vapor pressure will depend on the amount of gas it contains and on the temperature.

phases will warm and more molecules will have sufficient energy to leave the liquid phase. A new equilibrium will be established that results in less liquid and a higher concentration of gas molecules above the liquid. Molecules will enter and exit the liquid phase at a greater rate than before, but, on balance, more molecules will exist in the vapor phase. The pressure placed on the container walls by the vapor will increase as well. (See Figures 5 and 6.)

## Boiling Point

Once you understand this basic principle of vapor pressure, understanding the boiling

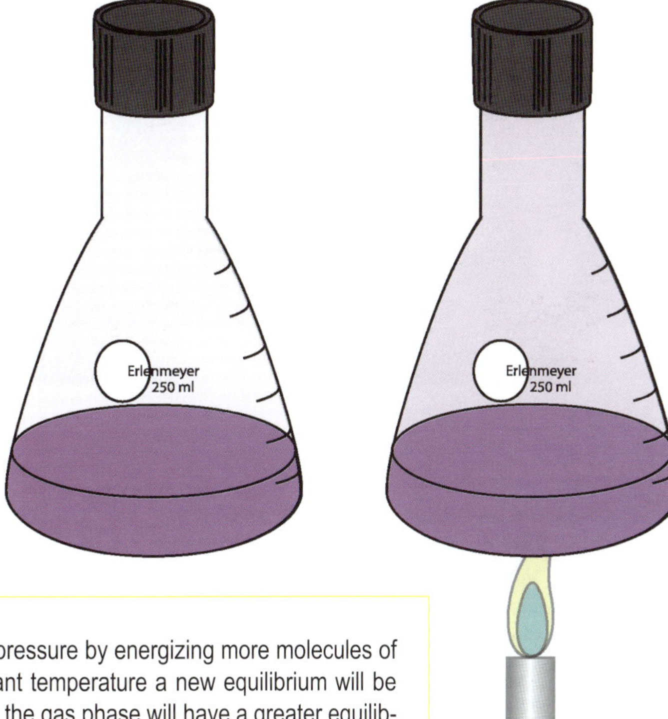


**Figure 5.** Warming a sealed container increases the vapor pressure by energizing more molecules of liquid, sending them into the gas phase. At a higher, constant temperature a new equilibrium will be established between the liquid phase and the gas phase, but the gas phase will have a greater equilibrium concentration. What will happen if we continue to heat this glass container? (Never do it!)

of liquids is pretty straightforward. We've been talking about the equilibrium that exists between the liquid and vapor phases in a sealed container. But what happens if the container is open to the atmosphere. Clearly there is no reason why molecules, once they have escaped their liquid phase, should ever return to it. The room for the atmosphere to accept those molecules is practically endless. Any liquid will continue to evaporate into the atmosphere as new energy comes along to excite molecules to the point that they leave the liquid.

But in an open system we must consider a new situation. If you recall, when atoms leave the liquid phase they take heat with them. Now reconsider what happens as the liquid is heated. Other molecules more rapidly reach their leaving energy and escape into the atmosphere. As they leave they take their energy with them. As the temperature rises it does so with greater difficulty because of most energetic molecules are leaving faster and faster at the surface of the liquid. Finally, the two stresses reach an equilibrium. The rate of heating matches the rate at which energy escapes. The solution will get no hotter. The liquid contains all the energy it can contain without becoming vapor. Any additional energy put into the liquid simply goes into the evaporation of additional molecules. Notice the complete lack of fanfare as we announce that the condition we have just described is none other than our old friend, the **boiling point**. (See Figure 7.)

Every liquid has its own boiling point that varies with pressure. Because we are talking about open containers we are talking about atmospheric pressure. So it's most common to think of *the* boiling point as *the one* that occurs under exactly one atmosphere of pressure. This we refer to as the **normal boiling point**. (See Figure 8.)

**Figure 6.** A pressure cooker is an example of a heated, sealed container. The temperature of the water inside can get higher than the normal boiling point of water because the vapor pressure goes higher than atmospheric pressure. The higher the temperature gets, the higher the internal pressure gets. When the pressure reaches a certain point (about 15 PSI), a weight on the steam outlet lets out any additional pressure. This prevents the cooker from building up too much pressure. A "pressure relief" will pop open if the steam outlet should get plugged. By using Gay-Lussac's Law, and by converting 15 PSI to atmospheres, you could calculate the temperature of the steam inside this vessel.

We have said that vapor pressure is generated by the tendency of molecules to leave the liquid phase until equilibrium is reached between liquid molecules exiting and vapor molecules entering. The vapor phase is more favored at higher temperatures, so the vapor pressure of a liquid increases as the temperature of the liquid increases. When this temperature rises to a certain point, the vapor pressure of the liquid equals the atmospheric pressure above the liquid. At that point, the liquid boils. So we can define the *boiling point* as *the temperature at which the vapor pressure of the liquid is equal to the pressure of the atmosphere above the liquid*. If the atmospheric pressure is 760 torr, then the boiling point of a liquid is the point at which the vapor pressure of the liquid is 760 torr. (See Figure 9.)

If the vapor pressure is plotted against temperature, its appearance

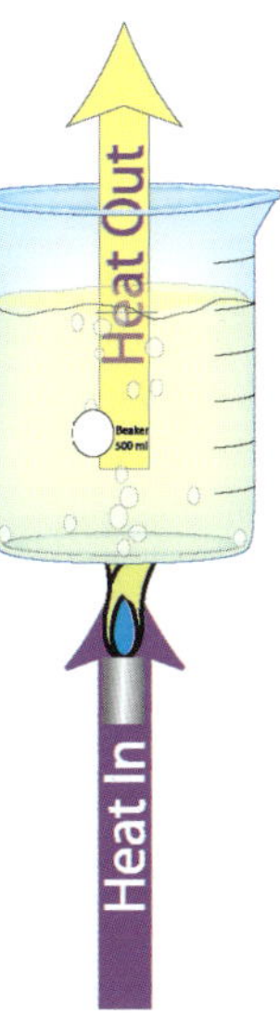


**Figure 7.** There are a couple of different ways of defining or describing the boiling point of a liquid. One way is to think of it as the temperature at which (under intensive heating) the rate of heating by the heat source is equal to the rate of cooling by evaporation at the surface.

**Figure 8.** The atmospheric pressure is about 760 torr at sea level. If you live quite a bit above sea level, your atmospheric pressure will typically be lower, so water will boil at a lower temperature at your house.

**Figure 9.** A second way of defining the boiling point of a liquid is in terms of vapor pressure. The boiling point of a liquid is the temperature at which its vapor pressure is equal to the atmospheric pressure. At absolute zero (A) the vapor pressure is theoretically zero. Above absolute zero, but below the boiling point (B), some evaporation occurs, but the vapor pressure is below atmospheric pressure. At the boiling point, the vapor pressure of the liquid is precisely equal to the atmospheric pressure above the liquid. At that point, any additional energy input will send the liquid entirely into the vapor phase.

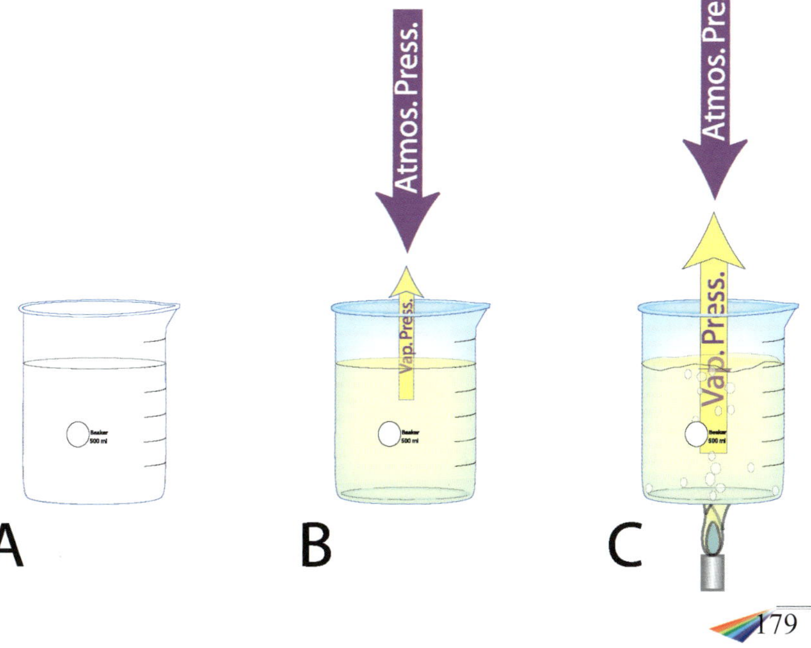

will be like that of the **vapor pressure curve** for water shown in Figure 10. The normal boiling point of water is shown where the vapor pressure curve hits 760 torr. (Of course, you know that the normal boiling point of water is 100°C by definition.) The boiling point at any other pressure can be read off the chart as the temperature that corresponds to that vapor pressure.

For example, looking at Figure 10, we can determine that isopropyl alcohol has a vapor pressure of 60 torr at 30.5°C. If you place a container of isopropyl alcohol under a vacuum so that the pressure in the headspace is 60 torr, it will boil at 30.5°. Its normal boiling point is 82.5°C. You can get vapor pressure curves, or data from which you may draw your own curve, from almost any chemistry or chemical engineering handbook. Any chemist or chemical engineer will have access to one or own one. I have several. You can access one through your local library.

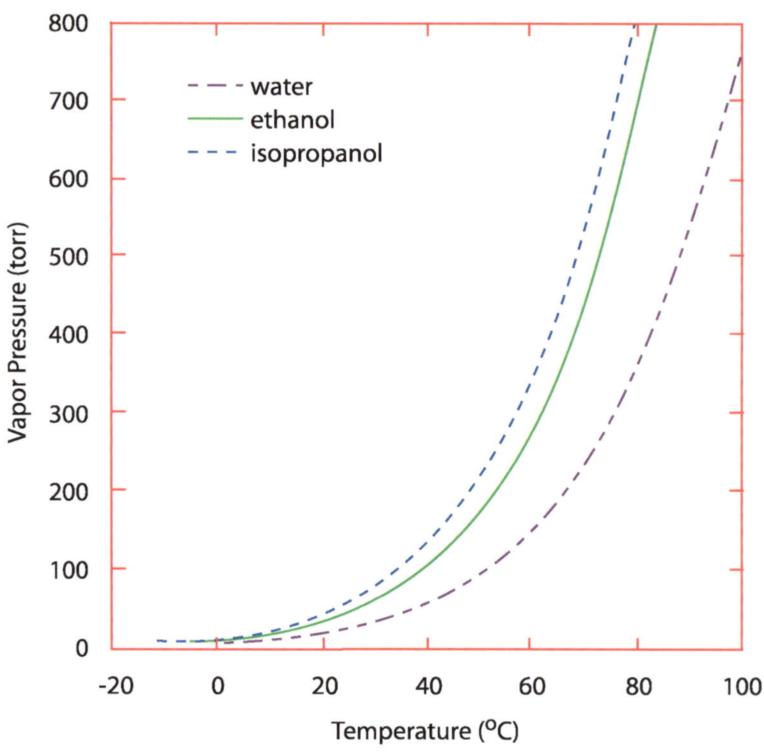


**Figure 10.** These are vapor pressure curves (vapor pressure as a function of temperature) for three common liquids. The boiling point of a liquid is the point at which the vapor pressure equals the atmospheric pressure at any given temperature of the liquid.

**Exercises**

1) Given the vapor pressure curves (Figure 10) for ethanol, water and isopropyl alcohol, answer the following questions:

   a) What is the boiling point of water at 400 torr?
   b) What is the vapor pressure of isopropyl alcohol at 25°C?
   c) At what temperature will ethanol boil if the pressure over the liquid is 500 mm Hg?
   d) What is the normal boiling point of isopropyl alcohol?

2) Given the data in the table on the right, plot a vapor pressure curve for bromobenzene. From your curve, what would be the boiling point of bromobenzene at a pressure of 0.85 atm?

| T (°C) | $P_{vap}$ (torr) |
|---|---|
| 28 | 5 |
| 40 | 10 |
| 54 | 20 |
| 69 | 40 |
| 78 | 60 |
| 91 | 100 |
| 110 | 200 |
| 132 | 400 |
| 156 | 760 |

# 49: Liquid-Solid Equilibrium, Etc.

Just as the change from liquid to vapor is a dynamic equilibrium, so is the change from liquid to solid. After all, both are characteristics of molecules which have either gained too much energy to stay in the liquid phase or have lost so much energy that they become sluggish and adhere to their neighbors. However, the change from liquid to gas involves molecules spreading out, whereas the change from liquid to solid requires very little compression. So while we might expect the atmospheric pressure to have a considerable effect on the boiling point (the spreading out of the liquid to form gas), we would not expect it to have as much effect on either solidification ("freezing") or melting. Indeed it doesn't. In the case of water, pressure has almost no effect on the **freezing point/melting point**—the temperature at which a substance's liquid state solidifies or its solid state liquefies. (These temperatures are one and the same.)

Like the change from liquid to vapor, the change from liquid to solid or from solid to liquid requires a change in energy. A block of ice in liquid water, heated slowly and uniformly, will remain at a constant temperature until the ice is gone. The added heat energy is used for the sole purpose of changing the phase of the water from its solid to its liquid without any accompanying increase in water temperature. Once the ice is completely converted to water, the temperature of the water will begin to increase.

Although we have been speaking primarily about water, you should realize that the principles we have covered apply to other liquid/solid systems as well. Just as each substance has its own unique temperature of boiling and solidification, each has its own characteristic amount of heat that is required to change it from the solid state to the liquid state. As we said earlier, the amount of heat required to change one gram of a *liquid* to its *vapor* at its *boiling point* is called the *heat of vaporization*. Well, the amount of heat required to change one gram of a *solid* to its *liquid* at its *freezing point* is called its **heat of fusion**. (The heat required for the molecules to "fuse" into solid form or to "unfuse" the molecules of a solid.) And, just as the vapor at boiling point could be changed back to liquid by removing the heat of vaporization, so the liquid at freezing point can be changed back to solid by removing the heat of fusion.

In just a few minutes you'll be calculating energies required for all phase changes and temperature changes among solids, liquids and gases. Calculations involving conversions from liquid to solid or from solid to liquid work exactly the same way as our previous problems (Lesson 47) of calculating the conversion of liquid to gas. The general equation is George, for a change in temperature:

$$\Delta\varepsilon_t = mh_s\Delta t$$

where $\varepsilon_t$ is the energy required for a change in temperature, m is the mass of the substance, $h_s$ is the specific heat of the substance, and $\Delta t$ is the change in temperature. If we add units to this equation, it becomes:

$$?\,(\text{J}) = \text{mass}\,H_2O\,(\text{g}) \times \text{specific heat}\left(\frac{\text{J}}{\text{g}°\text{C}}\right) \times \Delta t\,(°\text{C})$$

For a phase change from liquid to gas, or the reverse, use the equation from Lesson 47. For a change from solid to liquid, or the reverse:

$$\Delta\varepsilon_f = mh_f$$

where $\varepsilon_f$ is the total energy of fusion for the mass (m) of the substance, and $h_f$ is the heat of fusion. Here is the same equation with units:

$$?\,(\text{J}) = \text{mass}\,H_2O\,(\text{g}) \times \text{heat of fusion}\left(\frac{\text{J}}{\text{g}}\right)$$

Just remember that anytime you add heat to the system, the energy change is positive, and if you remove heat from the system the energy change is negative. Also notice that the heat of fusion for water (335 J/g) is considerably lower than its heat of vaporization (2,259 J/g). It takes much less heat to change ice to water than it does to change water to steam.

## Solid-Gas Equilibrium?

A liquid can usually be thought of as an intermediate state between the solid and gaseous states. Solids melt to become liquids, then liquids evaporate to become gases. Gases condense to become liquids and liquids freeze to become solids. *Usually!*

Have you ever seen what happens to an ice cube that is left in the freezer for a long time? You know, the one that gets caught down behind the ice maker and is left there for years? When you find it, it's a lot smaller than a normal ice cube. Being in the freezer, why does it get smaller? It's because, in a dry environment, some of the ice bypasses the liquid phase and goes straight from the solid phase to the vapor phase. This is especially likely to happen at low atmospheric pressure.

While living in Minnesota, we noticed that snow can disappear over time even though the temperature is –20°F. Sunlight strikes the ice directly at the surface and causes molecules to gain enough energy to leave the frozen state. They pass directly into the vapor phase. (You'll never see any liquid water at –20°.) This process that occurs in a period of hours in sunlight might take months in the freezer where energy is in short supply. The process of direct vaporization of a solid at low pressure is referred to as **sublimation**. Sublimation is a neat process to watch when it happens quickly because the solid literally vanishes into the air. (Don't go sit in the freezer to watch ice cubes disappear.)

Just as distillation can be used as a method for purifying liquids according to their boiling points, sublimation can be used as a method for purifying solids that have high vapor pressures. For example, when iodine is heated at normal, atmospheric pressure, it doesn't go through a liquid phase, but rather, is converted from solid directly to vapor. If the vapor passes over a cold surface, it will form solid crystals right on that surface. Any solid impurities in the iodine will be left sitting at the bottom of its original container. Many substances that ordinarily pass through a liquid phase can readily be made to sublime if warmed under low pressure.

## Phase Equilibrium Diagrams

If you make a y/x chart of pressure versus temperature, you can see how all the phases of a given substance interact. We have provided two such diagrams in Figures 2 and 3. First, consider that water freezes/melts at 0°C at 760 torr (1 atm). Since we know that pressure has very little effect on freezing points, Figure 1 shows a line that is practically vertical drawn for all values of pressure at 0°C. (Pressure has so little effect on temperature that, even under strong vacuum at 5 torr, the freezing point is still – 0.01°.) Anytime water exists at a temperature to the left of that line segment, it will be in the form of ice.

**Figure 1.** Anytime you can smell a solid you can be sure that it has a high vapor pressure or it would never have reached your nose in sufficient concentration to be detected. These mothballs are made of naphthalene. It sublimes at room temperature. It's in the class of chemical compounds called "aromatic hydrocarbons." Benzene is in the same class.

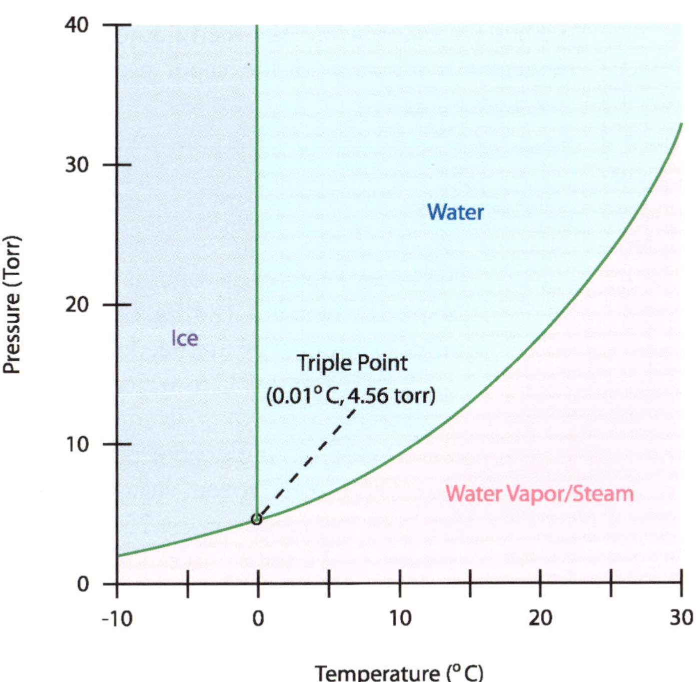


**Figure 2.** This is a phase diagram for water. The *y* axis displays pressures that are well below atmospheric pressure so that you can see the way the curve behaves as it approaches the freezing point. Temperatures below 0°C are also shown. This allows us to see how ice fits in the picture. This figure shows the temperature and pressure ranges at which you might expect ice to pass directly from the solid to the gas phase (sublime). These are the conditions under which you might "freeze-dry" something. The lines represent equilibria between phases. For example, the line that separates the water phase from the vapor phase represents points at which water is turning into vapor at the same rate as vapor is turning into water. The point at which all three phases are at equilibrium is called the *triple point*.

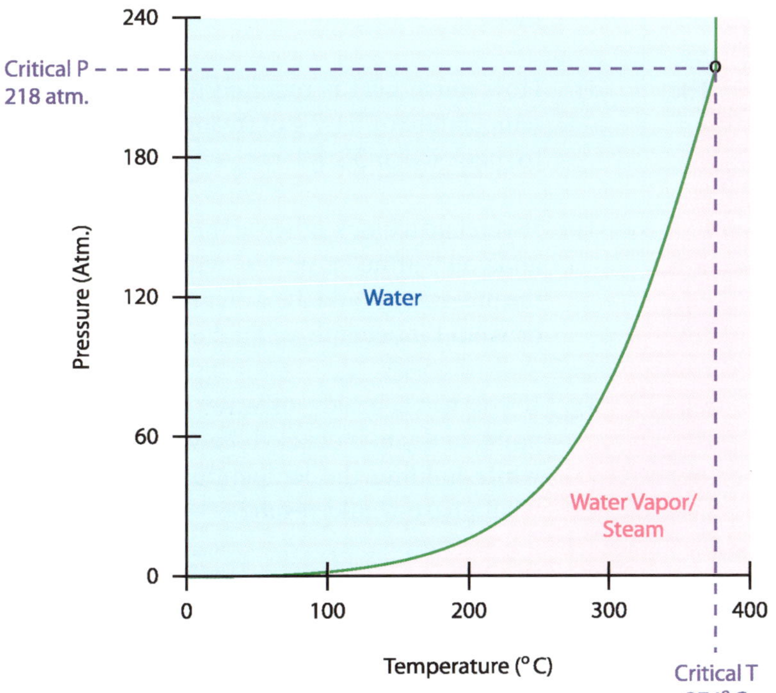


**Figure 3.** This, too, is a phase diagram for water. Notice the immense pressures on the *y* axis. Notice also that no liquid water occurs at temperatures greater than the *critical temperature*, even at higher pressures. This phase diagram is identical to the vapor pressure curve discussed in Lesson 47. Each point on the curve represents a boiling temperature for water at that pressure—the point at which liquid water and water vapor are at equilibrium.

At temperatures to the right of the line segment, water exists as a liquid. Right on the line both water and ice exist at equilibrium. By this, you may know that anytime you have ice and water existing together at equilibrium, the temperature of the slurry will be 0° regardless of the pressure.

As the pressure becomes quite low and/or the temperature becomes high, water enters the vapor phase. (If it's cool we call it *water vapor*, if it's above the normal boiling point we tend to call it *steam*.) Notice in particular that there is a place in the lower left corner of the graph in Figure 2 where water vapor and ice are directly in equilibrium. For example, if the temperature is below 0° and the pressure is less than 3 torr, the conditions are right for ice to sublime very rapidly.

Now don't be confused. Ice will sublime in an *open* container at *any* temperature and pressure (if it doesn't melt first). Similarly, water that is below its boiling point will evaporate from an open container. But, in a *sealed* container once an *equilibrium* is established among the phases, ice will not sublime, nor will water evaporate, unless their temperature and pressure are in the proper quadrant of the graph.

There is one final feature that you notice. Look at the single point at which all three phases are at equilibrium. This point is called the **triple point**. When a substance exists at its triple point, any change of temperature or pressure *will* result in a change of phase. The triple point of water is at 0.01°C at 4.56 torr.

Turn your attention now to Figure 3. There is a temperature above which it is impossible to liquefy a vapor regardless of how much pressure is applied. For example, steam will not be forced to become liquid water if its temperature is greater than 374°C. This temperature is called the **critical temperature**. The critical temperature may be defined as *the highest temperature at which molecules may be compressed into the liquid phase*. The amount of pressure that must be applied to accomplish this reversion to the liquid phase at critical temperature is called the **critical pressure**.

Both of the diagrams we just examined are called **phase equilibrium diagrams** (or just *phase diagrams*) for water. You can understand why a scientist or engineer would want to have such a diagram on hand for any substance he works with routinely that is subject to changes of phase. Using a well-done phase diagram you can accurately predict what phase a substance will be in at any combination of temperature and pressure. Also notice that the part of the phase diagram that describes the equilibrium between liquid water and water vapor is identical to the vapor pressure curve we saw in Lesson 47.

**Exercises**

1) The heat of fusion of water is 335 J/g and its specific heat is 4.184 J/g°C. How many joules (J) of energy are needed to change 18.5 g of ice at 0.0°C to water at 32.2°C?

2) If you were having a pool party, but the pool water was frozen, how much heat would be required to bring 25,000 gallons of water and 2,000 lb of ice to a comfortable 27°C? Give your answer in kilowatt hours using the conversion factors in the back of the book. At $0.07 per kilowatt hour, how much would the required energy cost? Is a good party worth the money?

3) How much heat is required to reduce the temperature of 21.2 g of steam from 212°F to make ice at 32°F?

4) If you had a block of ice at the triple point, and you warmed it up 1°C without raising the pressure, would the water remain solid, or would it become a liquid or a gas?

5) If you had a beaker of water at the triple point, and you increased the vacuum on it while holding the temperature constant, what phase would result?

6) If you had a pressure chamber containing water at 300°C and 60 atm, what phase would it be in?

7) If you placed a beaker of water in a vacuum chamber at 10°C, what phase would the water be in when the vacuum reached 5 torr?

## 50: Solutions

Perhaps the best way to start our discussion about solutions is to define the term *solution*. (Duh.) A solution, in its simplest form, consists of one substance completely dispersed within a second. When we say "completely dispersed," we mean at the *molecular* level—the molecules of one substance are completely dispersed among the molecules of another. Or, if one of the substances is ionic, the ions of the one are completely dispersed in the other.

Although a solution may consist of combinations of any of the three phases of matter, people most often use the word to refer to dispersions in liquids. Technically the word can refer to a complete dispersion of any substance in any other, as long as the dispersion is complete.

Complete dispersion also implies that the molecules or ions are dispersed *uniformly* among one another. This uniformity comes about by random motion and collisions among particles. The random dispersion of one substance within another is called **diffusion**. Because gases have the most active particles, diffusion among their particles is rapid. If you open a container of an odoriferous volatile liquid or gas on one side of the room, you may detect particles of that gas at the other side of the room in seconds. (See Figure 1.) However, if you release a small amount of liquid in a swimming pool the same size as that room (with no pumps operating to stir the water), it might take hours or even days for the released substance to travel the same distance.

Liquids and gases can diffuse readily through some solids that are chemically matched to their properties. For example, there are some substances that, when placed in the palm of your hand may be tasted in your mouth a few seconds later. That's because the substance diffuses through your skin and into your bloodstream, then circulates to your mouth and nose where taste sensations are registered.

Of course, diffusion of solids through other solids is typically *very* slow. The particles have such low kinetic energy, they are so tightly bound, and there is so little intermolecular space that diffusion is extremely difficult. Although solid particles may work their way through another solid, it might take years for it to diffuse one centimeter. If bars of gold and lead are clamped together, molecules of gold are detectable within the lead bar after a long period of time. The longer they remain together, the deeper that penetration will become. This illustrates that the particles of solids are in motion and that their intermolecular forces, although strong, are not entirely impenetrable by atoms.

Diffusion is the process by which a solute dissolved in a solvent maintains its uniform composition. The constant, random rearrangement of particles makes and keeps a solution homogeneous—uniform in composition throughout. Any sample taken from it will have the same composition as any other. Every solution is completely uniform by definition. If it's not uniform, it's not a true solution.

**Figure 1.** Diffusion is the process by which molecules, by random motion and interaction, mix themselves with other molecules over time. This figure illustrates the diffusion of a colored gas through air. Notice that the concentration of the gas is highest at the source. The farther from the source you look, the lower the concentration. This is a "concentration gradient" or "diffusion gradient." The gas will continue to diffuse until its concentration is uniform throughout the entire space it occupies.

## General Properties of Solutions

Not all substances will readily disperse in one another. Whether they are compatible or incompatible depends upon the forces that act among their molecules. Substances that have strong electromagnetic attractions among their particles will surely dissolve in one another. The less attractive those particles are to one another, the less interaction there will be among them.

Molecular size also affects solubility. Large, lumbering molecules intrude less readily into pools of molecules with which they are poorly compatible. The amount of kinetic energy in a particle is a second important factor. A small, fast-moving molecule will break the surface of an incompatible liquid more readily than a large, slow-moving one will.

Solids and liquids will usually increase in solubility when heated because faster-moving molecules are more interactive and dispersive by nature. (They collide more and move faster, so diffusion is favored.) However, gases become too energetic to remain in liquid solution when they are heated, and they will leave to enter the vapor phase.

Since vapor particles don't interact, they are largely unaffected by the nature of their neighbors. Any gas will occupy the same space as any other and the molecules of the two will intermingle.

While the word **soluble** may suggest that particles are interactive, the word **insoluble** is less well defined. That's because there is always some amount of interaction among particles of any two substances. The question is "*How much* interaction will there be?" If the two substances are compatible, any amount of substance 1 will interact with any amount of substance 2. If the substances are poorly compatible, there will be some, but relatively little, interaction.

As we said earlier, most of the practical applications of solutions in the laboratory involve a solid, liquid or gaseous **solute** dissolved in a liquid **solvent**. For example, we may speak of oxygen gas from the air dissolving in water. In this case, oxygen is the solute and water is the solvent. We may speak of benzene dissolving in water. Because benzene is not very soluble in water, benzene is the solute and water is the solvent. However, if we place a small amount of water in lots of benzene, water is the solute and benzene is the solvent. So then, the majority substance is the solvent, while the minority substance is the solute. These definitions become a little less clear in cases where two liquids or gases are completely soluble in one another. In a solution of one mole of water with one mole of acetone, which is the solvent and which is the solute? And, by the way, who cares?

## What Types of Particles Interact?

But just what are those forces that are responsible for compatibility or incompatibility between two substances? Let's consider the solution of an ionic solid in a molecular liquid. An ionic solid is solid because of the attraction of a positive ion, say sodium, with a negative ion, such as chlorine. This attraction will tend to prevent sodium chloride from separating into its ions. Any solvent that separates the sodium cations from the chloride anions must overcome the attractions between them. Will this be more readily done by a solvent that has no electrical character to it or by one that has positive and negative attractions among its molecules? Ions can be pulled apart only by greater electrical attractions than those which hold it together.

What kind of liquid has ionic character? Let's take a look at some of the more common liquids. Several molecules are shown in Figure 2. What decides ionic character? It's the distribution of electrical charges around the molecules. Benzene is a uniform molecule. There is no reason why the electrons should distribute themselves in any way other than uniformly. Carbon tetrachloride is the same in this respect. The charges are uniform about this molecule because it has one central carbon atom and four chlorine atoms that are distributed uniformly about its surface. Hexane, too, has a uniform distribution of charges.

Now look at hexanol. Here is the first hint of nonsymmetry. One end of this molecule is different from the others. Notice which molecule changes the look. It happens to be oxygen. Oxygen is known for its strong electron-pulling character. (Remember? This is what makes it a strong oxidizer.) It also has two non-bond-

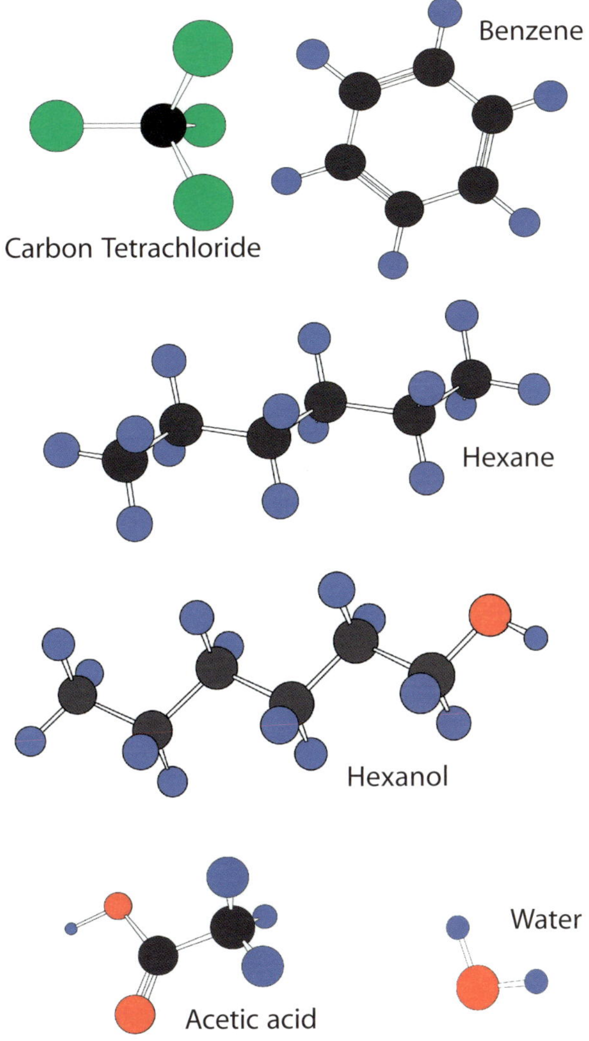


**Figure 2.** Which of the molecules are polar? They will be the ones that have electrons pulled in one direction or another that makes the molecule's electron cloud lopsided. This usually involves an electronegative atom on only one side of the molecule. Lots of nonpolar atoms tend to make up for the effects of one polar atom, so the most polar molecules will be small molecules with one, lopsided, highly electronegative atom. These molecules are arranged from least polar to most polar. Carbon tetrachloride and benzene are nonpolar. Their electron clouds are uniform. The length of the carbon chain in hexane allows electrons to run up and down, giving it only a minute polar character. Hexanol, on the other hand, has that piggish oxygen molecule on one end making it considerably more polar. Acetic acid is more polar yet, and water is the most polar of them all. There are few liquids that are more polar than water.

ing pairs of electrons on it. This will put a negative charge on this molecule in the vicinity of oxygen, causing it to have a somewhat positive charge on its neighboring atoms. Since there are many atoms downstream to absorb this effect, hexanol is not terribly polar, but it does have some polar character.

Next, look at acetic acid. The unique design of those oxygen atoms with their high concentration of negative charges makes this molecule like a little static-charge magnet. In other words, it has poles of greater and lesser negative charges. This **polar** character of acetic acid makes it easy for an ionic solute to dissolve in it. There are positive and negative locations among the solvent molecules that attract the positive and negative aspects of the acetic acid, causing it to break apart into its ions and dissolve.

### The Uniqueness of Water as a Polar Solvent

Although, at first glance, water appears to be a symmetrical molecule (the left side of H-O-H looks identical to the right side), first glances can be deceiving. Once again, oxygen is to blame. Look at the electron dot diagram of water in Figure 3. Oxygen has two unshared pairs of electrons. The negative charge from those electrons repels the negative charge on the electrons of hydrogen atoms so that the molecule bends like a boomerang. Now the electron-sucking nature of oxygen can work to make the oxygen side of the molecule more negative than the hydrogen side—it is among the most polar of the liquids. (See Figure 4.)

Water molecules rarely, but occasionally, separate into positive and negative ions. These ions hold a particularly important place in chemistry, so they are given their own names. The positive hydrogen ($H^+$) ions are called **hydronium ions**, and the negative ($OH^-$) ions are called **hydroxide ions**. Later we'll have much more to say about the polar character of water and this tendency it has to ionize. For now, let it suffice to say that water has a strong attraction to ionic substances.

There are other, nonionic (molecular) substances that have polar character. These too will dissolve in water. Sugars are an example of polar, molecular compounds of high solubility.

Figure 3. If the unshared pairs of electrons on oxygen are drawn on the same side of the water molecule, it makes it clear why water is a polar solvent. It has a partial negative charge on the oxygen side of the molecule and a partial positive charge on the hydrogen side.

## How Soluble is Insoluble?

A nonpolar molecule like benzene is not considered to be very soluble in water. You may even see it recorded as "insoluble" in some reference books. However, as we discussed, even incompatible molecules interact to some extent. If a layer of liquid benzene is floated on a layer of distilled water, an occasional benzene molecule will enter the water phase, and vice versa. As the benzene concentration in the water increases, the tendency of these molecules to leave the water and reenter the benzene layer increases until an equilibrium is reached between the two layers. At this point, as many benzene molecules are going in one direction as the other. The same is true for the water molecules entering and exiting the benzene layer. This is yet another example of dynamic equilibrium.

At equilibrium, if the water layer were analyzed for benzene, its concentration would be found to be on the order of 1 part benzene per 1,000 parts of water by weight. This is the same as 0.1% wt benzene. As nonpolar as benzene is, it is still a rather small molecule. A large, nonpolar molecule, such as tetrachlorodibenzodioxin, or decachlorobiphenyl (see Figure 5) is far less soluble. These compounds have equilibrium solubilities in the 1 part per 1,000,000,000,000 range. And *that* ain't much.

## Saturated, Unsaturated, Supersaturated

When a solute has reached its equilibrium solubility in a given solvent, the solution is said to be **satu-**

Figure 4. Water is one of the few solvents with enough polar character to wedge itself between two ions having opposite charges. That's what makes it such a good solvent for most chemicals used in the laboratory. It has been called "the universal solvent."

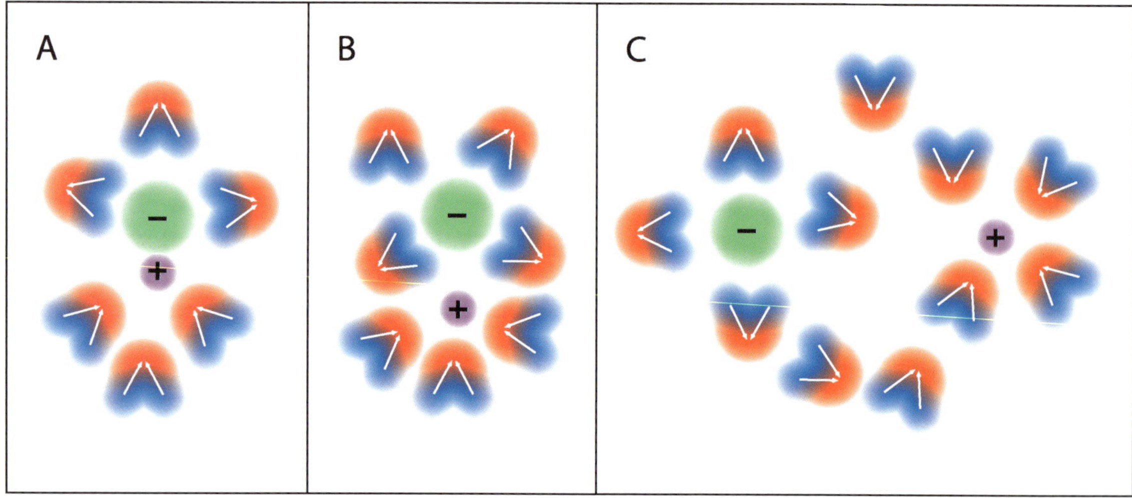

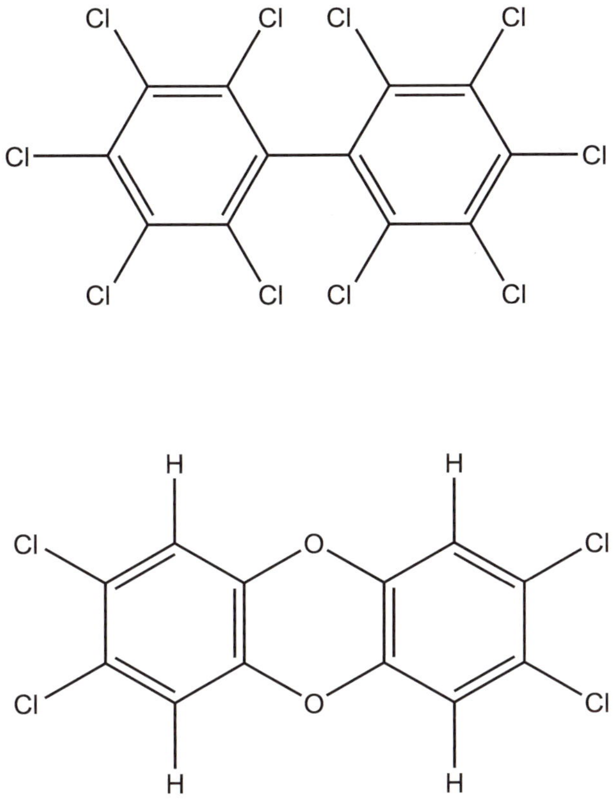

**Figure 5.** The structure on the top is decachlorobiphenyl. It is the most heavily chlorinated compound in a series of compounds called polychlorinated biphenyls (PCB's). These chemicals were manufactured for use as insulating and cooling liquids until they were banned by the US government in the 1970's. Because of their extreme nonpolar character accompanied by their large size, they have equilibrium solubilities in water on the order of a few parts per trillion.

Likewise, the chemical below is tetrachlorodibenzodioxin (TCDD). It was made famous as the chemical toxin in the "Love Canal" contamination. It and similar compounds also have equilibrium solubilities in the parts per trillion range. Based on toxicity tests with guinea pigs, this chemical was believed to be among the most toxic chemicals known to mankind. Untold millions of dollars have been spent cleaning up sites contaminated with it. It has recently been shown that guinea pigs are especially sensitive to TCDD, and (although somewhat toxic) it is not nearly as toxic as it was once believed. (How embarrassing!)

**rated**. Anything less than saturated is **unsaturated**, and anything more than saturated is **supersaturated**. Please spend a few moments with me understanding what is taking place at the molecular level when a solid solute is brought in contact with a liquid solvent.

The first thing that happens is that the solute molecules begin to disperse. Some solutes will disperse in a particular solvent quickly, while others will dissolve slowly in the same solvent. Of course, the rate at which a solute dissolves depends on how tightly the molecules of the solute itself cling to one another, but it also depends on how readily these foreign molecules are accepted by those of the solvent.

As the solute dissolves, its concentration in the solvent increases. As the concentration of solute molecules increases, the rate at which more molecules enter the solution decreases. If a beaker containing a solute is left static (unmixed), the concentration of dissolved solute will be highest near the solid. The concentration of solute will decrease with greater distance from this solid so that a "**concentration gradient**" will exist across the solution. Because the highest concentration is nearest the solid, the solid will be slow to enter the solution.

We can solve this concentration problem by shaking, stirring or otherwise agitating the solution. (Just don't agitate it until it gets hacked off at you.) Stirring a solution as the solute enters brings a lower concentration of solute molecules nearer to the surface of the solid. This allows the solute to enter the solution more quickly.

Solute continues to enter the solution and to disperse itself evenly by diffusion. Molecules of solute will continue to enter the solution until an equilibrium is reached between the solid and dissolved fractions. At that point the solution is said to be saturated. On average, for every molecule of solute that enters the solution at equilibrium, another exits to regain the equilibrium balance. The closer a solution is to saturation, the more slowly the solute dissolves. Figure 6 models the **rate of dissolution** for a solid in a liquid over time. As the concentration approaches the equilibrium concentration, the rate of dissolution decreases toward zero.

The rate of dissolution is highest when the solution concentration is lowest.

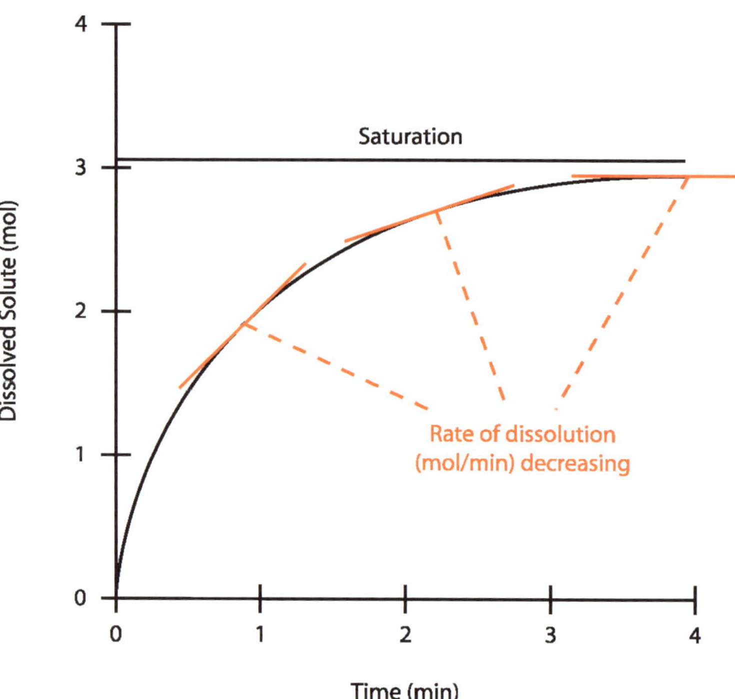


**Figure 6.** When a pure solvent first contacts a solute, the rate of solution is at its highest. As the solvent comes closer and closer to its saturation point, the rate of dissolution decreases. At saturation, the rate of dissolution is equal to zero. The rate of dissolution is the slope of the curve at any point. When you take calculus, you'll learn that the slope of the curve at a given point in time is the "derivative" of the equation of the curve at that point.

Increasing the temperature of a solution increases the kinetic energy of the particles. This increase in kinetic energy favors more interaction among molecules. This increase in interaction *usually* results in an increase in solubility, especially if heat absorption is required to break the stable crystalline structure of an ionic compound. In most cases, the concentration of the solute will continue to increase as the liquid approaches its boiling point.

If given a solubility curve as shown in Figure 7, or the data from which it is drawn, you can determine the solubility of a compound at any temperature in the temperature range shown. By *solubility* I mean the mass of solid dissolved in the liquid when the liquid is *saturated*. Or, I might have said *the mass of solid in the liquid when the dissolved and undissolved solute fractions are at equilibrium*. If an amount of solute is added that exceeds the amount shown on the graph at that temperature, some undissolved solute will remain and the solution will be saturated. If less solute is added than that shown on the curve at that temperature, then the solution will be *unsaturated*. Remember this. You'll see it again in the Exercises.

When a warm solution containing a relatively high concentration of solute is cooled, the solubility of the solute decreases. When this happens the concentration of solute may actually be higher than its equilibrium solubility. Such a solution is said to be *supersaturated*. The concentration of solute may even be considerably above its equilibrium concentration before crystals begin to re-form. Once crystallization has begun, it will continue until the equilibrium is reestablished.

**Solutions of Liquids in Liquids.** While liquids are more likely to enter a liquid solution upon heating, an increase in temperature also favors a higher vapor pressure. Heating to increase the rate of dissolution will cause the solute to enter the solvent, but it will also cause the dissolved solute to vaporize.

**Solutions of Gases in Liquids.** Although we just said that increasing the temperature of a solvent makes a solid or a liquid enter it more rapidly, gaseous solutes are different in this respect. Gases are made up of small, non-interactive, high-energy particles. For this reason, they tend to be poor solutes in most liquids. They don't like interacting with liquid particles; they enter and escape quickly because of their small size; and once in there they move too quickly to stay. With notable exceptions, they usually have solubilities in the range of a few or a few tens of parts gas per million parts liquid.

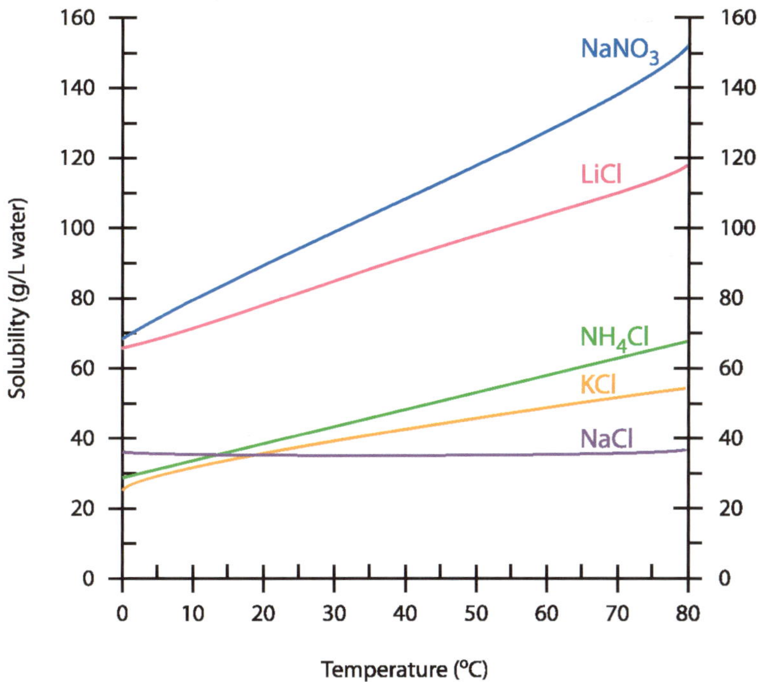


**Figure 7.** These solubility curves show the solubilities of different salts over a range of temperatures. Notice that most (but not all) of them increase insolubility with an increase in temperature. The one that increases the least with increasing temperature is our old friend sodium chloride.

While pressure has little effect on the solubilities of solids, it tends to force gas molecules into solution. Perhaps the best-known example of this principle is seen in carbonated beverages. In the bottling of soft drinks, carbon dioxide gas is forced into solution under pressure. When the bottle is opened the vapor pressure of $CO_2$ is reduced. Over time the dissolved carbon dioxide leaves the liquid phase. That's why it "goes flat" when you leave it sitting open too long. When the cap is placed on a half-used bottle of pop, there is no longer $CO_2$ in the headspace. Carbon dioxide will leave solution until a new equilibrium is reached between the dissolved $CO_2$ and the $CO_2$ in the headspace. So a recapped bottle of pop will get flatter and flatter each time it is opened until all the carbonation is gone.

As an added note just for you popologists—because heat tends to drive gases out of solution, soft drinks will hold their carbonation longer if they are cold when opened. They will also maintain their carbonation longer in the refrigerator than they will out in the warm air. If you are willing to work to preserve that last little bit of pop, transfer it to a smaller, sealable (preferably cold) bottle that will be full when the transfer is complete. Very little carbon dioxide will escape to a small headspace. But transfer it gently or you'll make all the fizz come out. (See Figure 8 on the following page.)

There are a variety of problems caused by the low solubility of gas in water. One of the more significant to wildlife is the return of heated water to receiving streams. Because hot water doesn't hold oxygen very well, it tends to starve fish and other oxygen-requiring creatures of the oxygen they need.

**Exercises**

1) Will a solution made by adding 8.50 g of LiCl to 10.0 g of water be saturated or unsaturated at 40°C? (Use Figure 7.)

2) Will a solution made by adding 11.0 g of KCl to 20.0 g of water be saturated or unsaturated if the temperature of the solution is 60°C?

3) A saturated solution is prepared by dissolving 140 g of $NaNO_3$ in 100 g of water at 70°C. If the solution is cooled to 20°C, approximately how many grams of $NaNO_3$ will fall out of solution? (Assume that no water is lost due to evaporation and that the cooled solution is not supersaturated.)

**Figure 8A.** When a soft drink bottle is filled at the bottler, it's filled under pressure with carbon dioxide. Some of the carbon dioxide dissolves, and some remains in the headspace.

**Figure 8B.** When the bottle is opened, the pressure is relieved. The carbon dioxide in the headspace is replaced by air and carbon dioxide bubbles form within the liquid. When the cap is replaced, a new equilibrium will have to be established between the carbon dioxide in solution and the air in the headspace. Some air will dissolve and some carbon dioxide will vaporize. The larger that headspace is, the more carbon dioxide will have to escape solution in order to equilibrate.

**Figure 8C.** To prevent the loss you have to do two things. First, you have to eliminate the headspace. Second, you have to pressurize the bottle again, so that more of the carbon dioxide is forced to remain in solution. (An increase in pressure forces more gas into solution.) This C-clamp works nicely. (Although it is a little cumbersome in the refrigerator.) Oh, yes. One more thing. Gases are more soluble in low-temperature solutions than in high-temperature solutions. Keep it cool!

## 51: Concentrations I—Express Yourself

When dealing with pure substances, an amount of that substance is easy to express. A solid is usually expressed as a mass (mg, g or kg). The quantity of a liquid or a gas is often expressed as either a mass or a volume (mL or L). When two substances are combined, the situation is no different except that you now need an expression for each of those substances. There are many ways of expressing how much of one substance is contained in another. The types of expressions used will be different depending on the types of materials under consideration. There are even differences in the ways concentrations are expressed from one scientific discipline to another. The scientists in some disciplines even adopt their own units for specific purposes. Examples include the "dietary calorie," "activity units" of an enzyme and "international units" of vitamins.

The types of expressions explained in the remaining paragraphs of this lesson are summarized in the Table on the following page. Mainly, you should observe that it's a monstrous table. Am I right? By the time you have finished this section of chemistry, you'll be able to use any of these forms of expression for concentration and convert from one to the other. Are you impressed? You should be.

### Weight Per Weight

As I said, the most convenient expression of a solid is the measure of its mass. When one solid is contained within another we usually express the combination as a mass within a mass. Take the case of a gold necklace made of only gold (2.9 g) and nickel (2.1 g). There are several ways this could be expressed based on what we wish to communicate. We could say on every occasion that the necklace is 2.1 g Ni per 2.9 g Au. However, if the necklace were broken into two separate pieces (a tragedy, you will agree), each piece will have an unknown quantity of gold and nickel. We can weigh the two lengths of chain and determine the total mass of each, but we have no basis for determining how much of that weight is nickel and how much of it is gold.

However, if we assume that the mixture of nickel and gold is a true solution, so that it is completely homogeneous, *then* the proportion of nickel to gold (or gold to nickel if you prefer) will be the same in each piece of the chain. If we know these proportions in the entire chain, we know their proportions in a segment of any size.

How many ways are there to express the weight proportion of these two elements? There are several possibilities. It may be expressed as the proportion of:

1) weight of nickel to weight of gold (wt Ni/wt Au)
2) weight of gold to weight of nickel (wt Au/wt Ni)
3) weight of nickel to the weight of the entire necklace or segment of necklace [wt Ni/ (wt Ni + wt Au)]
4) weight of gold to the weight of the entire necklace or segment of necklace [wt Au/ (wt Ni + wt Au)]

These are all different expressions of the amounts of these two substances in the homogeneous mixture.

If we determine the weights of the segments, and if we know any one of the proportions in 1-4, the weights of gold and nickel in any segment of the necklace can be calculated. Notice those proportions are of two types. The first two expressions involve comparison of the mass of one element to the mass of another. The last two expressions involve a comparison of the mass of one element to the total mass of the two elements combined. Which should we use?

There are two reasons that the second pair of expressions are better for most purposes. The first reason is easy to see. What if we have more than two components in our solution? Should we express all of them in terms of one other component? It begins to make more and more sense to express them in terms of a total mass rather than the mass of an arbitrarily chosen substance. The other reason is a little less straightforward, but it's a good exercise to go through in case you ever see a concentration expressed in a mass proportion like this.

**Table.** Common expressions of concentration used in the practice of chemistry.

| Name | Numerator | Denominator | Typical Expression | Other Units | Typical Use |
|---|---|---|---|---|---|
| Mass per unit mass | | | | | |
| Mass Ratio or Mass Proportion | mass of substance 1 | mass of substance 2 | 1:1 (wt) | | practical lab work at the bench; standard procedures |
| Mass Fraction | mass of substance 1 | total mass of all substances | g/g; % (wt) | mg/kg | solid in a solid; dense liquid in a solid |
| Engineering Units | mass of substance 1 | total mass of all substances | parts per million (ppm) | parts/billion by mass (ppb) | common for expressing small mass proportions in water solutions, comparable to mg/L or ug/L (density of dilute water solutions is very close to 1, so 1 L is very close to 1 Kg) |
| Mass per unit volume | | | | | |
| Mass/Volume Fraction | mass of substance 1 | total volume of all substances | g/L | mg/L | solid in a liquid; dense liquid in a liquid; general laboratory purposes; one of the most common expressions of concentration |
| Expressions involving moles | | | | | |
| Mole Fraction | mole of substance 1 | total mole of all substances | dimensionless | mol/mol | reaction chemistry, esp. where the solvent participates in the reaction |
| Mole Percent | mole of substance 1 | total mole of all substances | % | | reaction chemistry, esp. where the solvent participates in the reaction |
| Molarity | mole of substance 1 | liter of total solution | M | mol/L | most chemistry where water is used as the solvent; reaction chemistry, esp. where the solvent does not participate in the reaction |
| Molality | mol solute | Kg solvent | m | mol/Kg | calculations involving colligative properties |
| Normality | mol $H^+$ or $OH^-$ ions | liter water | N | mol $H^+$/L mol $OH^-$/L | acid/base chemistry |
| Volume per unit volume | | | | | |
| Volume Ratio or Volume Proportion | volume of substance 1 | volume of substance 2 | 1:1 (v) | | practical lab work at the bench; standard procedures |
| Volume Fraction | volume of substance 1 | total volume of all substances | L/L; ml/L; % (v) | parts/million by vol. (ppmv) | practical work with gaseous or liquid/liquid systems |

Let's use the first expression to calculate the weight of Ni, given a segment weighing 3.2 g. Here is what we have to do:

Let x = wt Au (g) in the segment
Let y = wt Ni (g) in the segment.

Since the weight of the segment of the necklace is 3.2 g, then:

$$3.2 = x + y \qquad \text{eq. A}$$

The proportion of the wt Ni to wt Au (y/x) is:

$$\frac{y}{x} = \frac{2.1}{2.9} = 0.724$$

or:

$$y = 0.724\,x \qquad \text{eq. B}$$

So we now have two equations in two unknowns which can be solved by substitution as follows. First, by substituting the expression for y found in equation B into equation A, we get:

$$3.2 = x + 0.724\,x$$

or:

$$3.2 = 1.724\,x$$

or:

$$x = 1.856$$

Now substitute x into equation B to get:

$$y = 0.274 \times 1.856$$

or:

$$y = 1.343$$

So then, to the correct number of significant digits:

wt Au = 1.9 g
wt Ni = 1.3 g

and the original equation is true:

$$3.2\,\text{g} = x + y = 1.9\,\text{g} + 1.3\,\text{g}$$

This is all very fun and interesting (right?), but it's also a lot of work to find a simple answer. It seems like there should be an easier way. Well, there is. Look back at the last two expressions of the concentration of gold and nickel (expressions 3 and 4). These expressions make this calculation considerably easier. There'll be no multivariable equations.

Let's use expression 3 to describe the *fraction* of Ni ($F_{Ni}$) in the chain:

$$F_{Ni} = \frac{2.1\,\text{g Ni}}{2.1\,\text{g Ni} + 2.9\,\text{g Au}} = 0.42\,\text{(dimensionless)}$$

(*Dimensionless* means no units.) Now we can use $F_{Ni}$ to calculate the mass of nickel in any portion of the chain that has the same fractional composition of Ni. In our 3.2 g segment, the amounts of Ni and Au are:

$$?\,\text{g Ni} = 3.2\,\text{g segment} \times F_{Ni} = 3.2 \times 0.42 = 1.34\,\text{g}$$

$$?\,\text{g Au} = 3.2\text{g total} - 1.34\text{g Ni} = 1.86\,\text{g Au}$$

To the correct number of significant digits, those values are 1.3 g Ni and 1.9 g Au.

Expressions 3 and 4 are so much easier to use than the others that they are by far the most commonly used. In a solution containing more than two substances, every substance present must have its own fraction if we are to describe its concentration. About the only time you'll see mass ratios is in standard procedures where a 1:1 mass ratio might be specified so you can easily "double the recipe" by making the mass ratio 2:2, for example.

While we have used the fraction, F, to describe the concentration of Ni, it would be just as easy to express it as a % by weight. You know well that any fraction multiplied by 100 is a percent:

$$\text{Ni}\,\%\,(\text{w}) = F_{Ni} \times 100 = 42\%$$

Although it may be bad grammar, the term **"weight percent"** or **"mass percent"** is used to refer to the percent by weight of one substance in a mixture. For simple, whole proportions of one solid within another, these problems can be worked out in your head. For example, if 25 g of sulfur are melted into 75 g of paraffin, the product (100 g total) is 25% wt sulfur.

### Weight Per Volume

Because it is so convenient to measure liquids in volumes it is convenient to express concentrations of solutions in some weight of solute per volume of liquid. For example, if 15 g of a solid are dissolved into 1 L of water, how much more simple can the expression be than 15 g/L? Afterward, you can take out a portion of that solution and, because solutions are homogeneous, the concentration of that small portion will still be 15 g/L. Even though the amount is less than a liter, the amount of solute carried is less than 15 g and the amount of solute is proportional to the amount of solution.

### Engineering Units

Occasionally, especially in the engineering world, concentrations of a solid in a liquid (usually water) will be measured in weight proportions. Rather than percents (parts per hundred), units of parts per thousand, parts per million ("ppm"), parts per billion ("ppb") or parts per trillion of solute in total weight of solution are common. Perhaps the most common of these measurements is parts per million since this is the concentration range of solutes in which many biological systems operate. One part per million is equivalent to 1 mg solute per 1 L (1,000,000 mg) water.

**Exercises**

Weight Per Weight

1) Give the weight percent of gold in 18 karat. (1 karat is 1/24 by weight.)

2) If you were making a purchase strictly for the amount of gold you were getting, would you buy 25 kg of 18 karat or 45 kg of 14 karat?

3) The mass of a human being is approximately 70% water, with 80% of the remaining 30% being carbon. What is the mass of non-carbon, non-water matter in an average, full-grown, 70-kg person?

4) What is the mass percent of KOH in a solution that is made by dissolving 15.5 g of KOH in 63.0 g of water?

5) A concentrated nitric acid solution (70.0% $HNO_3$ by mass) has a density of 1.424 g/mL at 60°C. At that temperature how many grams of $HNO_3$ are contained in 800.0 mL?

6) A $3\underline{0}$-g, dry soil sample is placed in a muffle furnace at 500°C to oxidize all the carbon to carbon dioxide. When the oxidation is complete, the sample weighs 22 g. Assuming all of the weight loss was due to carbon oxidation, what was the weight percent of carbon in the original sample?

Weight Per Volume

1) If 25 g of NaOH are dissolved in $50\underline{0}$ mL of water, what is the concentration in g/L?

2) If $10\underline{0}$ mL of that solution were separated into a 200 mL Erlenmeyer flask, what would its concentration be?

3) How much sodium hydroxide (g) are in that $10\underline{0}$ mL of solution?

4) If 4.5 g of ammonium chloride were dissolved in 200 L of water what would be the ammonium chloride concentration in mg/L?

## 52: Quiz 8

1) When given the specific heat, heat of vaporization and heat of fusion of a substance, be able to calculate the energy requirements for increasing its temperature and for changing its phase.
2) Be able to determine boiling points at different pressures using a vapor pressure curve.
3) Be able to calculate mass/mass and mass/volume concentrations given the amounts of substances involved.
4) Know how to use solubility curves to determine whether or not a solution is saturated at a given temperature and concentration of solute.

Phase is affected by atmospheric pressure which is, in turn, affected by altitude. Temperature is also influential. Condensed water droplets in the atmosphere (fog) are more dense than air and tend to fill low-lying areas, especially at high altitudes where temperatures are cooler.

# 53: Concentrations II—Our Furry Friend Gone Fractional

## Moles Per Mole

As we discovered early in our study of chemistry, the concept of a mole is very important in determining how substances react with one another. They react in correct molar proportions. For this reason, reaction chemistry is usually expressed in terms of moles of solute per mole of solvent, especially when the molar quantities of the two are nearly the same, or when both solvent and solute participate in the reaction. For example, one mole of ethanol might be added to one mole of water. The mole fraction of ethanol ($X_e$) in water is then expressed as:

$$X_e = \frac{mol_e}{mol_e + mol_w}$$

where:

$$mol_e = \text{moles of ethanol}$$
$$mol_w = \text{moles of water}$$

Since we said earlier that $mol_e = 1$ and $mol_w = 1$, then:

$$X_e = \frac{1\,\text{mol}}{1\,\text{mol} + 1\,\text{mol}} = 0.5\,(\text{dimensionless})$$

Notice that in both the case of the mole fraction (moles/mole) and the weight fraction (g/g) the units in the numerator and denominator cancel, leaving these expressions without units, or "**dimensionless**." Either of these fractions can be multiplied by 100 if you prefer your answer to be expressed as a percent.

## Moles Per Volume

In many kinds of chemical and biochemical reactions, the solvent, often water, is simply a carrier for the solute. Water itself doesn't participate in the reaction. In that case, we don't care about the molar quantity of water. For example, we might have a solution that contains one mole of calcium chloride in one liter of water. We call this a one **molar** solution of calcium chloride; we say that the concentration is a **molar** concentration or that it is expressed in **molarity**. The abbreviation for molar is **M**. (I used to have a chemistry teacher whose last name was Moeller. I've never fully recovered.)

Now suppose we have a 1 M solution of sodium carbonate and a 1 M solution of calcium chloride. Note that the weights of the calcium chloride and sodium carbonate would be different in their solutions because one mole of calcium chloride (110 g) is slightly heavier than one mole of sodium carbonate (106 g). The reaction between the two salts is written as follows:

$$CaCl_2 + Na_2CO_3 \rightarrow 2\ NaCl + CaCO_3$$

Notice that the equation is balanced when the two reactants are present in the same molar amounts. If you use 100 mL of your 1 M calcium chloride, how much of the 1 M sodium carbonate solution will be required to have the two reactants represented in balanced proportions? 100 mL. See? When solutions are prepared according to their molar concentrations, stoichiometry is easy. If you want to react calcium chloride with a two-time molar excess of sodium carbonate, you add the same amount of a 2 M solution.

## Moles Per Mass of Solvent

Just to confuse new chemistry students, someone came up with the bright idea of inventing a new unit to express moles of solute per kilograms of *solvent* (mol/kg). (*Warning:* It's not moles of solute per kg of *solution*!) This kind of unit is useful in the chemistry of colligative properties (which we will discuss shortly). It wouldn't be so bad had they not chosen to name the expression **molality**. So now we have it, and we can't get rid of it. The abbreviation for *molality* is m, and its units are mol/kg.

Please don't confuse your molarity (M) with your molality (m). If I ask you to calculate the molarity of a solution made by dissolving 1 mol of sucrose in water and bringing the volume to 1 L, the answer will be 1 M (one molar). But if I ask you to calculate the molality of a solution made by *adding* 1 mol of sucrose *to* 1 kg of water, the answer will be 1 m (one molal). *Notice that the final volume of the 1 M solution will be 1 L, but the volume of the 1 m solution will not.* It will be the volume of 1 kg of water (1 L) plus the volume of one mole of sucrose (unknown). Molality is not a volume-based measurement! If you want my advice, you'll set up every molarity problem by showing the units "mol solute/L sol'n" in place of "M," and you'll set up every molality problem by showing the units "mol solute/kg solvent" in place of m. Once you have completed the problem, you can replace these units with "M" and "m" in your answers.

So then, let's calculate the molarity of a solution made by dissolving 3.13 g of lithium chlorate ($LiClO_3$) in enough water to make 200 mL of solution. First you have to calculate the molar mass of $LiClO_3$ (90.39 g/mol), then set up the problem this way:

$$\frac{?\ \text{mol}}{\text{L sol'n}} = \frac{3.13\ \text{g LiClO}_3}{200\ \text{mL sol'n}} \left| \frac{\text{mol LiClO}_3}{90.39\ \text{g LiClO}_3} \right| \frac{1{,}000\ \text{mL}}{\text{L}} = 0.173\ \text{M}$$

On the other hand if you were asked to calculate the molality of a solution prepared by dissolving 173.0 g of ethanol ($C_3H_2OH$) in 750.0 g of water, once again you would have to determine the molar mass of ethanol (46.07 g/mol) then set up the problem this way:

$$\frac{?\ \text{mol}}{\text{kg slvnt}} = \frac{173\ \text{g EtOH}}{750.0\ \text{g water}} \left| \frac{\text{mol EtOH}}{46.07\ \text{g EtOH}} \right| \frac{1{,}000\ \text{g}}{\text{kg}} = 5.007\ \text{m}$$

(Here we have used a "trivial" abbreviation for ethanol as EtOH. Methanol is sometimes abbreviated as MeOH.)

### Volume Per Volume

The only occasions for expressing concentrations in units of volume per volume is when both the solvent and the solute are liquids or when both substances are gases. For example, a solution of benzene in hexane may be expressed in L/L or mL/mL. Likewise, nitrogen and argon gases may be allowed to combine in a proportion of $m^3/m^3$. These concentrations are referred to as **volume fractions**. Like weight fractions, volume fractions may also be expressed in **volume percents**. Unlike expressions involving moles, volume fractions don't have any special significance for the understanding the of chemistry of substances, but are simply used as a matter of laboratory convenience. When we write a report for presentation, we usually convert those values to more meaningful molar equivalents.

In cooking, all measurements are made in volumes. This creates some amount of imprecision. For example, when you measure flour you don't weigh it. You fill a cup with it and scrape the top of the cup to level the flour. How tightly the flour is packed can make a significant difference in the way your brownies turn out. Likewise, if there are air pockets in the shortening, you will be a little shortening short. Weight measurements of solids are really much better for consistency. Rest assured that food scientists don't measure their flour by volume until they wish to translate their results to home cooks.

The price of an electronic analytical balance is now becoming pretty affordable. Balances have become so convenient to use that measuring cups are comparatively difficult for some applications. How long do you suppose it will take people to convert to the more precise, more convenient method of measurement in their kitchens?

### Unit Conversions Among Concentrations

The major take-away point from this discussion is that there are many different forms of expression for concentrations. You'll have to be able to convert one type of expression to another with ease if you wish to become a chemist, an engineer, a biologist or any kind of investigative scientist.

**Exercises**

Mole Fraction and Mole Percent

1) Calculate the mole fraction of 2.4 mol acetone in 9.6 mol water.

2) Calculate the mole fraction of 10$\underline{0}$ g methanol ($CH_4O$) in 10$\underline{0}$ g of chloroform ($CHCl_3$).

3) Calculate the mole percent of acetone in a 1:1 (wt) solution of acetone ($C_3H_6O$) in water.

Molarity and Molality

1) What is the molarity of a solution containing 0.85 moles of $NH_4CHO_2$ in 40$\underline{0}$ mL of solution?

2) How many grams of $CuSO_4$ are required to prepare 700.0 mL of 1.25 M $CuSO_4$ solution?

3) What is the molarity of a solution made by dissolving 12.2 g of sodium sulfide, $Na_2S$, in enough water to make 40.0 mL of solution?

4) How many mL of 0.880 M solution can be prepared from 23.0 g of potassium chlorate ($KClO_3$)?

5) How much water should be added to 5$\underline{0}$ g of sucrose, $C_{12}H_{22}O_{11}$, to prepare a 0.20 m solution?

**Figure.** In cooking, all measurements are of volume, including the solid measurements. You can blame some of the problems of cooking on the imprecision in measuring flour, shortening and salt this way.

# 54: Changing the Concentration of a Solution

Often, it will be necessary to change a solution of one concentration to a different concentration by either diluting or concentrating it. It is important for the scientist to come to a good understanding of dilution and concentration processes. It is also important to understand how to calculate the concentration after such a change is made.

## Dilution

There are endless examples of real-life situations where the ability to calculate the effect of dilution on a solution's concentration is important. Laboratories are unique in that dilution calculations are not just part of life, they are a way of life. Figures 1 and 2 illustrate some of the best-known and most heavily used tools in the laboratory. These are tools used to carry out precise and accurate dilutions.

Not only are glassware tools needed to carry out a dilution properly, you need the appropriate mathematical tools as well. A useful equation for calculating the change of concentrations due to dilution is:

$$V_1C_1 = V_2C_2$$

That is, for a single solution that is diluted by adding additional solvent, the original volume ($V_1$) times the original concentration ($C_1$) is equal to the final volume ($V_2$) times the final concentration ($C_2$). For example, if 2 mL of a solution containing 250 mg/L of manganese sulfate were diluted to 1 L, what would be the concentration of the dilute solution?

$V_1 = 2 \text{ mL} = 0.002 \text{ L}$
$C_1 = 250 \text{ mg/L}$
$V_2 = 1 \text{ L}$
$C_2 = \text{unknown}$

Note that the units of volume and concentration must be the same on both sides of the equation to form an equality. If you are given $V_1$ in mL, be sure $V_2$ isn't in L, or your dilution will be off by a factor of 1,000! We calculate:

$$(0.002 \text{ L})\left(\frac{250 \text{ mg}}{\text{L}}\right) = (1\text{L})\left(\frac{C_2 \text{ mg}}{\text{L}}\right)$$

Now, since we assured ourselves that all of the units matched before we started, we can drop them out to make it simple to do the calculation:

$$\frac{0.002 \cdot 250}{1} = C_2$$

or:

$$C_2 \frac{\text{mg}}{\text{L}} = 0.5 \frac{\text{mg}}{\text{L}}$$

**Figure 1.** The volumetric flask is the main tool used in laboratories for diluting solutions with great precision. The red line etched on the neck of the flask marks the precise volume that the flask contains. The narrow neck assures that there is little error in the dilution since even small errors in volume will be visible. A precise amount of solid or liquid containing the solute is measured into the flask and the solvent is used to fill the flask to the appropriate volume. Afterward, the flask is stoppered and the solution is mixed well. A well-equipped laboratory will contain flasks ranging in volume from 10 mL up to perhaps 2 L.

More information on the subject of dilution is included in the Appendix, Method 4. There instruction is given on how to carry out and calculate serial 2-fold and 10-fold dilutions. Serial dilutions will not be covered in the next quiz. Your laboratory activities should give you plenty of experience.

**Figure 2.** If 1.00 mL of solution that is 2.50 M in methyl red is placed in the 500 mL volumetric flask and diluted to the mark, what is the resulting concentration of methyl red? In a real lab we might do this kind of calculation several times a day.

## Concentration

The formula above for calculating the concentration of a solution after diluting by adding more solvent also works for determining the new concentration after removing solvent by evaporation or distillation. In both cases the amount of solute stays the same while the amount of solvent changes. You might be asked to calculate the concentration of a solution after concentrating it from $V_1$ to $V_2$, where $V_2$ is a smaller volume than $V_1$. Use the same equation you just used. It doesn't matter that the volume is going down instead of up; the relationship between volume and concentration still holds.

Finally, you might need to know the amount of solute present in a solution of a given volume and concentration. The amount of solute present in any solution can be calculated at any point in time as follows:

$$S\ (\text{mg}) = C\ (\text{mg/L})\ V\ (\text{L})$$

If you know the volume and the concentration, the amount of solute can be easily calculated, regardless of what you have done to it beforehand. For example, let's say we have a solution of 6.00 M sodium hydroxide and we want to know what weight of sodium hydroxide is present in 10.0 mL of this solution. We just do the math:

$$?\ \text{mg NaOH} = \frac{6\ \text{mol}}{\text{L}} \left| \frac{10.0\ \text{mL}}{} \right| \frac{1\ \text{L}}{1{,}000\ \text{mL}} \left| \frac{50.0\ \text{g}}{1\ \text{mol}} \right| \frac{1000\ \text{mg}}{1\ \text{g}} = 3.00 \times 10^3\ \text{mg NaOH}$$

Now let's prove ourselves by solving some problems.

**Exercises**

1) Calculate the molarity of a solution prepared by diluting $10\underline{0}$ mL of 12.0 M HCl with $35\underline{0}$ mL of water. **Caution:** I said we *diluted it with* 350 mL of water. The final volume ($V_2$) is not 350 mL.

2) Calculate the molarity of a solution prepared by diluting 630 mL of 0.525 M $K_2Cr_2O_7$ with $1{,}00\underline{0}$ mL of water.

3) What volume of 16.0 M $H_2SO_4$ should be used to prepare $50\underline{0}$ mL of 0.50 M $H_2SO_4$ solution?

4) What is the concentration (mg/L) of 250. mL of a 525 mg/L solution of KCl that has evaporated to a volume of 150 mL?

5) What is the concentration of a solution that was 1.2 g/L in $Cu(NO_3)_2$ if 75 ml of it are diluted to a total volume of 1.00 L?

6) What is the mass of aluminum in 17.5 mL of a 30.0 M solution of aluminum hydroxide?

# 55: Collig-igative Properties-operties

Most properties of solutions, including surface tension, density and effects on light, to name a few, are different for different solutes. For example, if you were to shine a light on a copper sulfate solution, a different color of light will come out the other side than if you were to shine that same light on an iron chloride solution. However, there are a few characteristics of solutions that are the same regardless of the kind of solute under consideration. These characteristics are called **colligative properties**.

The focus of this discussion is on solutes and what they do to the properties of their solvents. There are two kinds of solutes to consider—molecular compounds and ionic compounds. For these first few sections that follow, it's important that we focus our attention only on the *molecular* ones. Afterwards we will expand our viewpoint to include ionic ones. Please keep in mind that, although we may spend more time talking about water, these principles apply to all solvents.

During the course of this lesson you will learn that there are four related properties of a solvent that change when a solute is added to it. The unique thing about these properties is that it doesn't matter *which* solute is added only *how much*. The four properties are: boiling point, freezing point, vapor pressure and osmotic pressure. In the following sections we will take a turn at examining each of these properties.

## A Solute Increases the Boiling Point of Its Solvent

In order to discuss the way solutes affect the boiling point of a solvent, we should talk only about nonvolatile solutes, or our conversation will evaporate. When a nonvolatile, molecular solute is added to a solvent, the solute raises the boiling point of the solvent. This is called **boiling point elevation**. For example, ethylene glycol ("antifreeze"), when added to water, increases the boiling point of the water. This allows the water in the radiator of an automobile to reach a higher temperature without boiling out. The more ethylene glycol we add to water, the higher the boiling point gets. The equation that describes the elevation of boiling point based on the amount of solute is as follows:

$$\Delta t_b = k_b m$$

where $\Delta t_b$ is the change in the boiling point, $k_b$ is a constant for a specific solute (for water it's 0.515 °C kg solvent/mol solute) and m is the molal concentration of the solute.

Why does the boiling point change? Because the interaction of water molecules with solute molecules holds them tighter in the liquid phase.

If sucrose is added to water, the boiling point of the water increases. Once again the amount of the increase is related to the amount of sucrose added. But do you think that the amount of the increase will be the same for one mole of sucrose as it is for one mole of ethylene glycol? Guess again. Boiling point elevation is a colligative property of a solution. It doesn't matter what molecular solute you pick, it will have the *same* effect on boiling point as any other. This suggests that boiling point elevation is related to the *concentration* of the solute rather than the *kind* of solute. (See Figure 1.)

## A Solute Decreases the Freezing Point of Its Solvent

With very little imagination you can anticipate what will be said in this paragraph. The freezing point of a solution decreases with an increase in the concentration of a molecular solvent. Just as sucrose causes an *increase* in the boiling point of water, it also brings about a *decrease* in the freezing point. This decrease, called **freezing point lowering** or **freezing point depression,** is also a colligative property of solutions. (See Figure 2.) Every molecular substance brings about the same amount of change in the freezing point with an equivalent molal amount of solute as follows:

$$\Delta t_f = k_f m$$

Figure 1. If every molecular solute has the same effect on boiling point, then why can't you use sugar in your car's radiator? You can—until it candies in your car's engine block! Ethylene glycol has several chemical properties that make it a good choice as an antifreeze/coolant. It's stable even at high temperatures. At those concentrations it's toxic to microbes that would otherwise grow in your radiator, yet it is relatively environmentally friendly. (It's biodegradable and not very toxic at lower concentrations).

What do you think $\Delta t_f$ stands for? You're right. It's the change in the freezing point. And m is still m. The value of this new constant, $k_f$, is 1.86 °C kg solvent/mol solute when water is the solvent. In a short while you'll use these equations to calculate some changes in boiling points and freezing points when we add different molal quantities of solute to different solvents.

Why does the freezing point of a solvent decrease as a solute is added? Because the interaction of water molecules with solute molecules makes it more difficult for the water molecules to crystallize. So the same stuff that keeps your car's radiator from getting too hot in the summer also keeps it from freezing in the winter.

Figure 2. An ice cream maker takes advantage of freezing point lowering. Salt and ice are layered around the canister. The ice cream is mechanically stirred while the salt drops the freezing point of water by 20°C (36°F). This is cold enough to freeze the ice cream in about 20 to 30 minutes.

### A Solute Decreases the Vapor Pressure of Its Solvent

As you'll recall, the vapor pressure of a solution is the amount of pressure exerted on the walls of a container by a vapor in equilibrium with its liquid. A solvent, having a relationship with a dissolved solute, will have less of a tendency to leave its container. This is called **vapor pressure lowering** or **vapor pressure depression**. Like boiling point elevation and freezing point lowering, vapor pressure lowering is a colligative property that depends on the molar amount of (not the type of) dissolved, molecular solute. Although this colligative property is given its own name, it is directly related to the other two in ways which we'll examine.

### A Merging of Two Concepts

A decrease in the vapor pressure implies a higher boiling point. Think about it: Water boils at 100°C. Add a molecular solute to this boiling water and the interaction of the solute molecules with the solvent molecules prevents water from evaporating so readily—the vapor pressure decreases. That means that those wildly energetic water molecules can't leave the liquid phase. The result is that the liquid phase gets hotter by keeping those overheated molecules. Eventually it will get hot enough to overcome the restriction of the higher vapor pressure and it will begin to boil, but at the time it boils its temperature will be higher. Thus, the addition of solute brings about an increase in boiling temperature, and at the same time, a decrease in the vapor pressure. You can't get one without the other.

As you'll recall, back in Lesson 48 we defined the boiling point in two different ways. First, we defined it as the temperature at which, under intensive heating, any energy put in is met by an equal amount of energy lost by evaporation. Second, we defined it as *the temperature at which the vapor pressure of the liquid equals the atmospheric pressure above it.*

### Freezing Point Lowering, Too

The freezing point of water is the temperature at which water and ice are at equilibrium—the rate at which water freezes is the same as the rate at which ice melts. Both water and ice have their own vapor pressures at different temperatures. *The freezing point is the only temperature at which the vapor pressure of water is equal to the vapor pressure of ice.* In this way the freezing point and the vapor pressure are related.

As I said before, when a solute is added to water, the interactions among the molecules prevent liquid water from leaving solution as readily—the vapor pressure is reduced. At the same time, and because of those same interactions, ice doesn't form as readily. So by the same interactions that lower the vapor pressure, the freezing point decreases.

We have now pointed out relationships among three of the four colligative properties of solutions. The uniting concept is that water molecules rather enjoy their relationship with solute molecules. This relationship tends to make it more difficult for water to freeze and for vapor to form, so it brings about proportional decreases in the freezing point and the vapor pressure, and a proportional increase in the temperature required for boiling.

### The Fourth Colligative Property

A fourth colligative property, which has a special significance in living systems, has to do with membranes—thin films or sheets. In order to understand this property, which we will come to call **osmotic pressure elevation**, considerable background information is required. We will reserve most of this discussion for our study of biology. For now let this brief description suffice:

If a sac prepared from a special plastic is filled with sugar water and dropped into a beaker of more dilute solution (it may be the same sugar or another substance altogether), the water will pass through the wall of the sac and into the more concentrated solution. This process, called **osmosis**, is defined as *diffusion of a substance across a membrane resulting in the decrease in concentration on the other side.* Osmosis is important in biology because it is a method by which living cells bring materials across their membranes (which serves the same purpose as our plastic sac).

When the water passes into the sac, the concentration of the material in the sac will decrease. But, because water has flowed into the sac, the pressure inside will increase. This increase in pressure will be related to the concentration of the solute. The greater the solute concentration is, the higher the osmotic pressure will go. (Cells placed in too high a concentration will explode.)

Interestingly enough, because the molecules of water are passing through the membrane one molecule at a time, it's as if the water were in its gas phase (and in fact, a portion of it is). The increase in the osmotic pressure is directly related to the vapor pressure above the water.

So then, there are four colligative properties: boiling point elevation, freezing point depression, vapor pressure depression and osmotic pressure elevation. All four of these properties are related to one another and are controlled by the amount of a solute that is dissolved in a solution.

**Figure 3.** Various salts are used to melt ice from roads and sidewalks. They do so by lowering the freezing point of the water making up the ice. If the freezing point can be lowered enough to make the water turn to liquid, it will run off the sidewalk and/or evaporate. If the temperature outside is lower than the new freezing point of the water, however, the salt will have no effect. The salt will simply lie on top of the ice.

## Ionic Solutes

If you'll recall, earlier we decided to restrict our discussion to molecular solutes because they represent the simple dispersion of a single type of particle in a solvent. When an *ionic* substance enters a solution it breaks into at least two different types of particles—an anion and a cation. These two types of particles put a double whammy on colligative properties. The effects are exaggerated in comparison with those of simple, molecular substances. While the effect of a molecular substance depends only on its molar concentration, the effect of an ionic substance depends also on the nature of the ion itself. In other words, the effect is slightly different from one ion to another. Because ionic substances have a dual effect on colligative properties, salts are more effective gram per gram than molecular substances in terms of their effect on colligative properties. (See Figure 3.) However, there are some down sides to the uses of salt. (See Figure 4.)

**Figure 4.** Salt helps keep cars traveling during the icy northern winters, but it also makes them corrode more quickly. The salt serves as an electrolyte to increase the oxidation rate of the metal. Car owners in the north can count on replacing the exhaust system every few years. The appearance of a car has little to do with how well it functions as transportation. But, when it looks this bad, I have trouble convincing my son that it's okay to drive it--not cool, but okay!

**Exercises**

1) What would be the boiling point and the freezing point of a solution of acetamide, $C_2H_5ON$, if 3.0 moles of acetamide were dissolved in 750 g of water?

2) A solution is made by dissolving 75.0 g of urea, $(NH_2)_2CO$, in 135 g of water. What would be the expected freezing point and the expected boiling point of this solution?

3) The main component of antifreeze is ethylene glycol, $C_2H_6O_2$. If a car radiator holds 5 kg of water, how much ethylene glycol would be needed to protect the radiator from temperatures as low as −15°C?

4) A 39.0-g sample of an unknown compound is dissolved in 120 g of ethanol. The resulting solution had a boiling point of 84.0°C. What is the molar mass of the unknown compound? (The boiling point of ethanol is 78.3°C, and its $k_b$ is 1.16°C kg solvent/mol solute.)

## 56: Quiz 9

1) Be able to calculate molarity.
2) Be able to calculate molality.
3) Know how to calculate the effects of dilution or concentration using $V_1C_1 = V_2C_2$.
4) Understand the colligative properties. Be able to calculate the freezing point depression and boiling point elevation of a liquid when given the boiling point constant and molal concentration of solute.

Natural waters are always complex solutions and suspensions. Even waters that are supposedly "pure" because they are remote from industrial and domestic activities may have considerable concentrations of dissolved ions. In fact, most people think that natural waters taste better than distilled water which has had all of its ions removed.

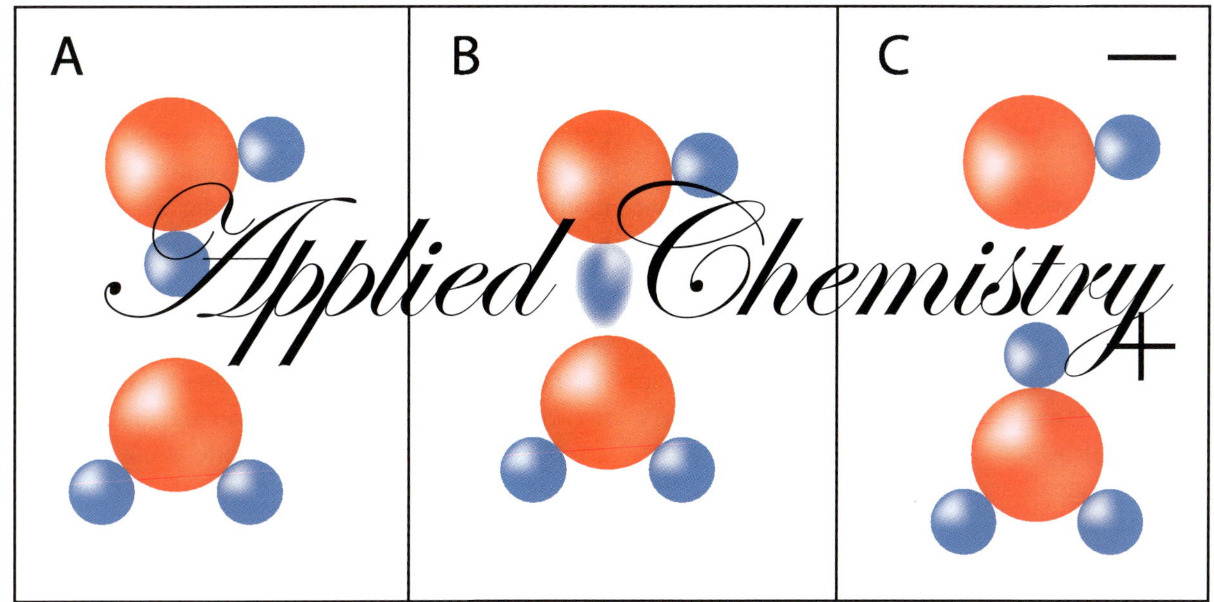
A
B
C
—
Applied Chemistry
+

## 57: Water

We have spent a lot of our time together talking about characteristics of liquids and solutions. While we have focused considerable attention on water, we haven't yet addressed water as a topic in and of itself. Because of its uniqueness as a solvent and as a foundation stone in the building and sustaining of life, it is certainly deserving of our attention. Because of its critical role as a reaction medium in the chemistry laboratory and in nature, entire courses are dedicated to the unique chemistry of this molecule.

There are a great many molecules and atoms that you can point the finger at and say, "If that molecule or that element didn't exist, neither would life." For example, carbon almost invariably forms four bonds that serve as the skeleton upon which living molecules are built. Phosphate has just the right oxidation energy to serve as a proper molecular energy currency for every living thing. Nitrogen has just the right electronegativity to give hydrogen-bonding ability to amino acids and nucleic acids that give shape to the molecules of life. Among all the unique particles whose properties come together to permit the activities that constitute living things, none affect our possibilities as greatly as simple $H_2O$.

Water is extreme in just about every respect. It is extremely polar and extremely dense for a liquid. Because of its unique crystal structure, it is relatively light when it's frozen and heaviest when it is at 4°C, making it unique in its temperature/density profile. It possesses a unique type of intermolecular bonding seen in few other molecules. It has an extremely low vapor pressure, an extremely high surface tension, extremely low freezing point and boiling point constants, and an extremely high specific heat. Werc it not for any of these properties our concept of life would be so radically modified one would have to wonder whether it would be feasible.

The properties of water come together with the properties of our solar system to provide a living environment suitable to life as no other known combination of atmosphere, planet, ocean and sun do. (Figure 1.) While we know relatively little about planets outside our own solar system, other life-sustaining systems become more and more difficult to envision as we learn more about the strict requirements of life. Those who spend their time gazing into space for other life forms no longer bother looking in places where water is known to be lacking.

**Figure 1.** Atmosphere and ocean work together to provide just the right living environment for us. Were the water molecule even slightly different, life would have to be different too—perhaps unimaginably so.

## Water as a Natural Resource

Water is among the most abundant natural resources in the Earth's crust. Approximately 71% of Earth's surface is covered with it at an average depth of 3.8 Km. While this may seem deep to you, it is but a thin skin on an Earth that is 13,000 Km thick. The distribution of water in its various reservoirs is shown in Table 1. A thorough study of water might include studies of many disciplines, some of which are summarized in Table 2.

New water molecules are constantly created during the oxidation of organic matter by living things (mostly bacteria and fungi). Water is broken down during photosynthesis. These are examples of several processes that operate to affect the water balance of the earth.

If you had to guess, what portion of the earth's water would you say is held in rivers? Guess again!

**Table 1.** Distribution of water in its global reservoirs.

| Reservoir | % |
|---|---|
| Oceans | 97.61 |
| Polar Ice and Glaciers | 2.08 |
| Ground Water | 0.29 |
| Freshwater Lakes | 0.009 |
| Saline Lakes | 0.008 |
| Soil and subsurface moisture | 0.005 |
| Atmospheric water vapor | 0.0009 |
| Rivers | 0.00009 |

Modified from Wetzel, 1982.

## Molecular Structure

We have already spoken in Lesson 47 of the unusually high surface tension of water that comes about from the two unshared pairs of electrons that belong to its oxygen atom. Let's review this feature of water and cover some details that we left out of our first discussion.

The simplest view of water is revealed in its molecular formula, $H_2O$. It's not until you examine the electronic structures of hydrogen and oxygen that you come to appreciate the detail of this molecule. Hydrogen atoms have a single, unshared valence electron; oxygen atoms have six valence electrons. Thus, when hydrogen and oxygen are covalently bound into a water molecule, there remain four unshared electrons. We also know that oxygen, like all the atoms in its group, has a tendency to be something of an electron pig, while hydrogen (like all IA elements) tends to be generous in sharing. The negative charge of the molecule gets directed internally to the oxygen atom. The unshared pairs of electrons cause the otherwise linear molecule to collapse into a tetrahedron with two hydrogen atoms and two unshared pairs of electrons at the vertices. (For review, See Lesson 19.)

**Table 2.** Scientific disciplines involving chemistry and natural waters.

| Discipline | Involves |
|---|---|
| Water chemistry | Lab and field studies of the unique chemical nature of water |
| Biology of waters | Includes subdisciplines of freshwater, marine and subsurface biology and ecology |
| Hydrogeology | Ground water |
| Atmospheric science | Atmosphere and evaporated water; often includes close contacts with oceanography |
| Limnology | Freshwater |
| Oceanography | The ocean and the water cycle |
| Wetlands ecology | Coastal swamp and marshlands and freshwater wetlands |

Were the water molecule linear, it would also be nonpolar. By its behavior we can see that this is not true. There are few liquid molecules in existence that are more strongly polar than water. Molecules that are more polar than water typically crystallize and become solid. The shape of lowest-potential-energy for water occurs if the bond angle collapses from 180° (linear) to 104.5°. The result is a molecule that has a net negative charge on one side and a net positive charge on the other side. This unusual molecular shape, and the position of oxygen, give rise to a second phenomenon that occurs only in a few other compounds. That phenomenon is **hydrogen bonding**.

## H-Bonding

Hydrogen bonding occurs only when hydrogen is bound directly to an atom that is small and highly electronegative. This reduces the entire periodic table to three atoms—fluorine, oxygen and nitrogen—in order of their H-bonding strength. These three atoms occupy a place in the periodic table that is at the extreme of small electron radius and high electronegativity. (See Lessons 16 and 17 for review.) From this set of restrictions, the molecules with hydrogen-bonding ability are evident. Ammonia ($NH_3$), alcohols (R–OH, where R is any organic group of atoms) and amino compounds (R–$NH_2$) are a few. Fluorine, the most strongly electronegative of the small atoms, forms the strongest hydrogen bonds. The resulting H–F molecule has such a terrific hydrogen-bonding strength that these molecules tend to cluster in solution because of their attractions to one another.

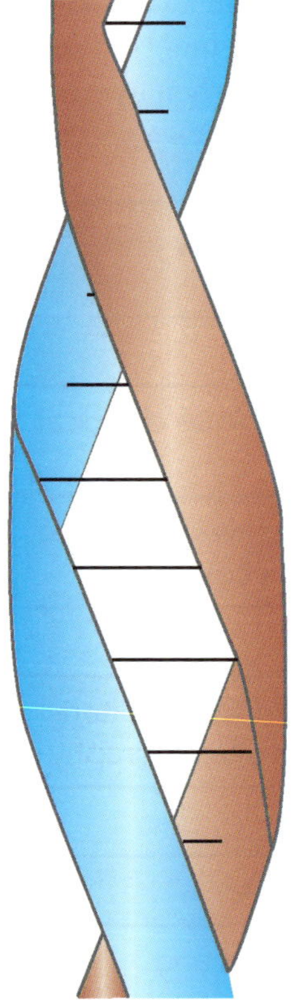

One hydrogen-bonding molecule may also attract a molecule of a different kind. It is this type of bond that holds large biomolecules, such as proteins and nucleic acids, in precise complex shapes. The two long strands that make up a DNA molecule (the central molecule of living things) are held together by hydrogen bonds (See Figure 2). Hydrogen bonds are weak in comparison with covalent and ionic bonds, but still strong enough to significantly change the properties of molecules that possess them. While covalent and ionic bonds affect atoms *within* molecules, hydrogen bonds are the strongest bonds affecting attractions *among* molecules.

## Chemical Properties

Water is extremely stable, so only radical substances will react with it. This makes it uniquely suited as a reaction medium for a large variety of reactions without the water molecule itself participating in the reaction. Nevertheless, there are several reactions that you can count on water to participate in. A few metals—namely sodium, potassium, and calcium—react with water to form the metal hydroxides and hydrogen gas. The reactions of water with sodium and potassium are violent ones. These metals are stored under oil to prevent them from reacting with the water vapor in the atmosphere.

Oxides of both metals and nonmetals react with water to form metal hydroxides. The metal oxides (e.g., CaO and $Na_2O$) are called **basic anhydrides**, because they form bases when they are added to water. Nonmetal oxides (e.g., $SO_2$, $CO_2$ and NO) that react with water are called **acid anhydrides** because they are acidic in water solution. These reactions are environmentally important. When $CO_2$ dissolves in water it becomes carbonate ions. These ions are received by algae for photosynthesis and by certain bacteria that use them the way people use molecular oxygen—for respiration. Because it can exist in three dissolved forms ($CO_3^{2-}$, $HCO_3^-$ and $H_2CO_3$), depending on how much free acid is present, it serves to "buffer" natural waters to prevent rapid changes in pH.

Sulfur dioxide is also environmentally important but for all the wrong reasons. It is produced by bacteria that use sulfate instead of oxygen for respira-

**Figure 2.** Every twist in a length of DNA has 10 pairs of hydrogen-bonding molecules connecting the two strands. In this diagram black lines represent bonding pairs. You can see several of them peeking out from between the strands. Each pair is connected by either two or three hydrogen bonds. A single DNA molecule may have tens of thousands of such bonds holding the two strands together.

tion. It is also produced during the combustion of fossil fuels that contain minor amounts of reduced sulfur compounds. Once in the air it dissolves in the falling rain where it forms sulfuric acid. This acid has been found to be in higher concentrations downwind of industrial and municipal plants that burn sulfur-containing fuels. It is held responsible for destruction of wildlife by decreasing the pH of freshwater habitats and by increasing the concentrations of unwanted minerals in those habitats. It is also believed to increase the rate of destruction of treasured landmarks made of vulnerable materials, such as marble (e.g., the Acropolis in Athens, Greece), iron and copper (e.g., the Statue of Liberty).

## Hydrates

Many chemicals are more structurally stable if they bind with water molecules to form a hydrate as they crystallize. They actually scavenge water out of the atmosphere to bring about this welcome stability. Such substances are called **hydrates**. Hydrates are ionic compounds, and their atoms display strong electrostatic attractions to the polar water molecule. These attractions are not as powerful as covalent bonds or ionic bonds.

The water that is bound to a hydrate may be driven off by heating. (See Figure 3.) The amount of heat required to drive off water molecules is different for every hydrate. Each hydrate has its own characteristic heat for driving off a specified number of water molecules from its crystalline structure. This means that the occurrence of water molecules in hydrates is not random. It is a property that is specific to the structure of the compound. Some chemicals form multiple hydrates that give up water at different temperatures.

In the absence of this **water of hydration** the would-be hydrate is said to be **anhydrous**. Among the strongest hydrate-forming molecules are the metal chlorides which may retain their water at extremely high temperatures. These most strongly water-loving hydrates will even draw enough water from the atmosphere so that they dissolve. On returning after some period of time to a container of one of these in the laboratory, you may find a jar full of solution where you thought you had left a jar of powder some months before. These water-pooling chemicals are referred to as **deliquescent**. (See Figure 4.)

As we said earlier, hydrates are more stable with water bound into their molecules. This implies, as we have said, that an input of energy will be required to drive off the water. It also implies that the formation of the hydrate from the anhydride is exothermic. That is, heat is actually given off when water is added to the anhydride. Often the water will boil and spatter from the rapid local heating that takes place.

## Water As a Solvent

The unusual polarity and hydrogen-bonding characteristics of water make it one of the few solvents that can push its way between two ions having opposite charges. It is one of the few solvents in which ionic compounds will completely dissociate into their ions. Acids and bases of all kinds (including organic ones) are ionic by definition. Water is also a good solvent for a few molecular compounds that have a polar quality, especially those with hydrogen-bonding capabilities.

**Figure 3.** Copper sulfate pentahydrate is blue. When the water is driven off, the anhydride is whitish. Hydrates of cobalt salts give similar color effects, but they tend to be red or purple.

These include sugars, alcohols, ammonia and amino ($-NH_2$) compounds.

**Electrolyte Solutions.** An electrolyte is a compound which, when added to water, promotes electrical conductivity in the solution. Pure water itself is a poor conductor, but even fractional molar amounts of ionic substances may increase its conductivity a millionfold.

Compounds are categorized based on the extent to which they ionize in solution. The categories are: nonelectrolytes, weak electrolytes and strong electrolytes. Nonelectrolytes are nonionic compounds and are therefore nonconductors in solution. Weak electrolytes are ionic compounds that partially dissociate into their ions when added to water. The portion of the compound that is dissociated at any given moment is usually a minority (e.g., trisodium phosphate and hydrogen fluoride). The portions of dissociated and non-dissociated molecules exist in equilibrium. Finally, as you might guess, strong electrolytes (such as chlorides, nitrates and sulfates) dissociate completely for practical purposes. These molecules also exist at equilibrium with their ions, but the equilibrium strongly favors the ions over the intact molecules.

The electrical conductivity of a solution increases with the solute concentration. However, as the concentration of solute increases, the interactions of the solute ions and water molecules prevent the ions from acting independently. At higher concentrations the conductivity may fall off from the expected.

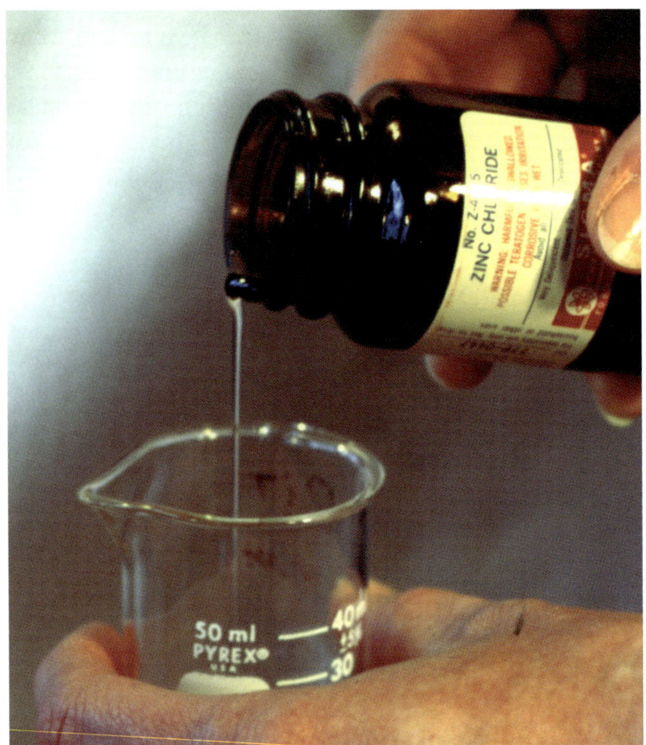


**Figure 4.** A deliquescent substance is one that collects water from the atmosphere. "Okay, who replaced my zinc chloride with this syrup?"

**Exercises**

1) For which of the following compounds would you expect hydrogen bonding to occur?

a) ethyl ether ($H_3C—O—CH_3$)
b) methyl amine ($H_3C—NH_2$)
c) phosphorus trihydride ($PH_3$)
d) benzene ($C_6H_6$)
e) nitromethane ($H_3C—NO_2$)

2) Which liquid from each pair of liquids would you expect to have the higher boiling point?

a) $H_2O$ or $H_2S$
b) $H_2O$ or $CH_3OH$
c) $CH_3CH_2NH_2$ or $CH_3CH_2OH$

# 58: Acids/Bases

## Ionization of Water

We have spoken of the extreme polarity of water due to the electronegativity of oxygen opposite the relative ease with which hydrogen (as a group IA element) gives up control of its electron. The water molecule has a somewhat negative charge on the oxygen side and a positive charge on the hydrogen side. The term applied to a molecule having opposite charges in different locations is a **dipole** (having "two poles").

Because of the dipolar nature of water, neighboring water molecules have an effect on one another. The negative, oxygen region of one molecule pulls on the positive, hydrogen proton regions of neighboring molecules. That pull is strong enough to occasionally remove a proton resulting in the formation of two ions. The following equation shows the ionization of water molecules:

$$H_2O \rightarrow H^+ + OH^-$$

Although chemists have long used this equation as the standard for showing the dissociation of water, and it is a convenient equation for certain calculations, the more accurate equation is:

$$2\,H_2O \rightarrow H_3O^+ + OH^-$$

In the presence of polar water molecules, a free proton ($H^+$) would be an extreme rarity. It would quickly add itself to the electronegative end of $H_2O$ to make $H_3O^+$. In either case, when you see $H^+$ or $H_3O^+$ they mean the same thing, and either may be referred to as a hydrogen ion, even though the proton itself is bound to another water molecule. This process of the ionization of water is illustrated in Figure 1.

## Hydrogen/Hydroxide Ion Imbalances

While the separation, or "**dissociation**," of hydrogen protons from the rest of the water molecule doesn't

**Figure 1.** Occasionally, when two water molecules meet (A), the electronegative portion of one molecule will attract a proton from a hydrogen atom on another molecule (B). When this happens water can ionize to form $H_3O^+$ and $OH^-$ ions in an equimolar proportion (C). The concentration of such ions in pure water at a given time is $10^{-7}$ M each.

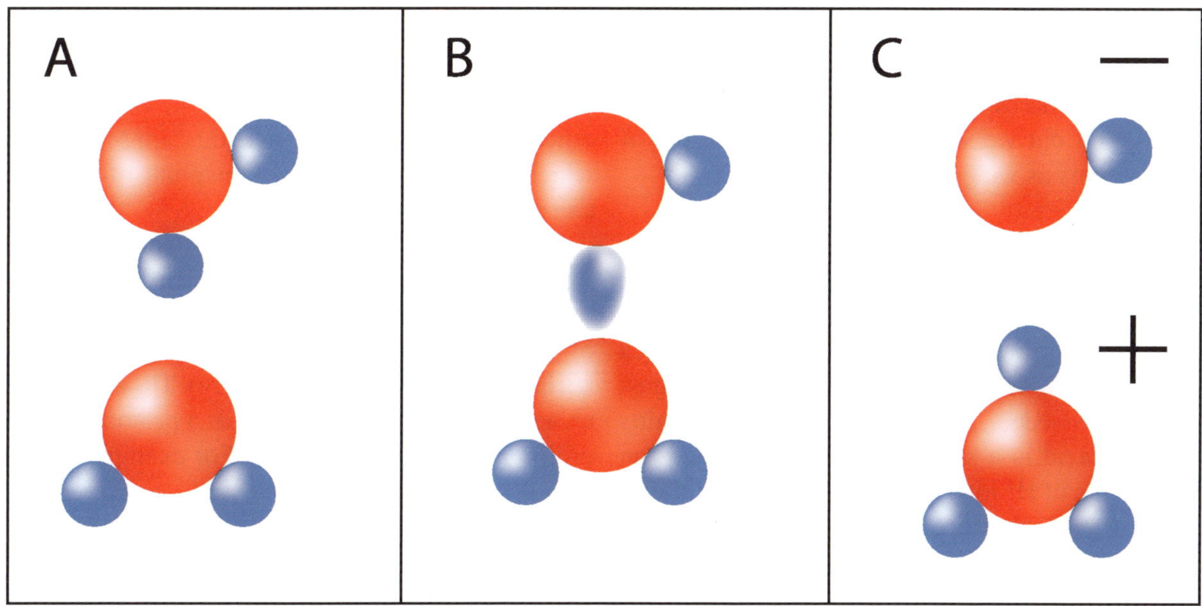

last long, some number of molecules are in this ionized state at all times. Notice that the molar concentrations of $H_3O^+$ and $OH^-$ are identical in pure water, because they are equally formed when water dissociates. Their concentrations are $1 \times 10^{-7}$ M. Although, in theory, the concentrations of these two ions are exactly the same, in practice they never are. A solution in which the hydrogen ion concentration is greater than it is in pure water is said to be **acidic**. Solutions in which the hydroxide ion concentration is greater than it is in pure water is said to be **basic**.

Solutions in which the concentrations of both hydrogen and hydroxide ions are greater than they are in pure water are said to be pure water ☺. (Free hydroxide ions react with free hydrogen ions to form water.) Because the concentrations of the two opposing ions are in balance, the solution is neither acidic nor basic. It is said to be **neutral**.

This brings us to the subject of the definition of an acid. According to the viewpoint we have taken about acids and bases, an **acid** is any substance that raises the *hydrogen* ion concentration of pure water. A **base**, then, is any substance that raises the *hydroxide* ion concentration of water. In fact, if you look at the list of acids in Table 1, you will see that they dissociate in water to form hydrogen ions. The bases in Table 2 dissociate in water to form hydroxide ions. Although our treatment of acids is a simple one that relies on the hydrogen ion, there are three ideas proposed by different chemists as to what should be called an acid. We will briefly summarize the three so that you will have heard of them, but we will rely primarily on one of those definitions as we go forward.

**Table 1.** Common strong acids and their formulas.

| Name | Formula |
|---|---|
| Hydrobromic | $HBr$ |
| Hydrochloric | $HCl$ |
| Hydroiodic | $HI$ |
| Nitric | $HNO_3$ |
| Perchloric | $HClO_4$ |
| Sulfuric | $H_2SO_4$ |

**Table 2.** Common strong bases and their formulas.

| Name | Formula |
|---|---|
| IA Metal Hydroxides | |
| Sodium Hydroxide | $NaOH$ |
| Potassium Hydroxide | $KOH$ |
| IIA Metal Hydroxides | |
| Magnesium Hydroxide | $Mg(OH)_2$ |
| Calcium Hydroxide | $Ca(OH)_2$ |
| Barium Hydroxide | $Ba(OH)_2$ |

## 1923—A Very Good Year (If You Like Acids)

So far, we have defined acids and bases in terms of the ions that form when water dissociates. This idea was first proposed in the late 1800's by Svante August Arrhenius of Sweden. According to the "**Arrhenius concept**" an acid is a substance which produces an overabundance of $H^+$ ions when added to water. Likewise, a base is a substance which ionizes to produce $OH^-$. While this definition is generally satisfactory, there are some ideas that it does not incorporate. Two other theories have modified this approach, both of which slightly broaden the viewpoint of what makes an acid.

The other two working models of acids and bases were advanced in 1923. The first of these was by a pair of scientists working independently in Denmark (Brönsted) and in England (Lowry). They proposed that an acid and a base are defined as a proton donor and a proton acceptor, respectively. For example, HCl is an acid because, in the reaction:

$$HCl + NaOH \rightarrow NaCl + H_2O$$

HCl has donated a proton to $OH^-$. $OH^-$ is a base because it has received that proton. Under this concept, acids and bases travel in pairs. For every acid, there is a "**conjugate base**." The conjugate base is what's left of the compound after the proton has been donated. In the reaction above, $H^+$ is the acid, so $Cl^-$ must be the conjugate base. Likewise, $Na^+$ is the "**conjugate acid**" of the $OH^-$ ion.

The way the **Brönsted-Lowry concept** broadened the definition of acids and bases was by suggesting that this process of donating a proton can take place outside of a water solution. This means that acid and base reactions are no longer uniquely aqueous reactions. In fact, any negative ion can act as a base if it forms a covalent compound by accepting a proton. Please just remember that the Brönsted-Lowry concept is known as the "**proton donor**" concept of an acid. It's the one we will refer to most often in this text. It is probably the most commonly used viewpoint.

The third idea that broadened our definition of acids was proposed by G.N. Lewis—in 1923. While the Brönsted-Lowry concept is known as the "proton donor" concept, the Lewis concept is known as the "electron pair" concept. Instead of thinking of proton donors and acceptors, Lewis thought of electron pair donors and acceptors. These two thoughts are not much different. Anytime a proton is donated by an acid, it is accepted by an unshared electron pair on a base. For example, when HCl donates its proton, it is received by an unshared pair of electrons on an $OH^-$ group as shown in Figure 2. This broadens the definition of an acid to any compound that forms a covalent molecule by tying up an unshared pair of electrons. Many cations that do not increase the $H^+$ concentration of water can do that. Please just remember that the Lewis concept is the **electron donor** concept.

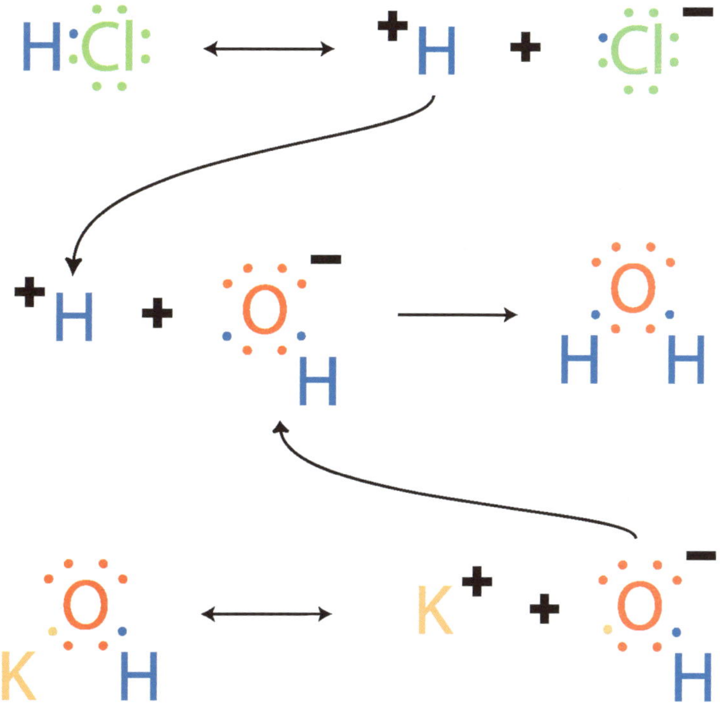


**Figure 2.** When HCl dissociates, a free proton is available. That makes HCl a proton donor and, therefore, an acid. When $OH^-$ (from KOH), with its three unshared pairs of electrons, encounters that proton, water is formed. You could say that KOH accepted the proton that was donated by HCl. That makes it a base by definition.

## Ionization of Acids/Bases

For our purposes, an **acid** is a substance which increases the hydrogen ion concentration of pure water or that "donates" a proton in a chemical reaction—a **proton donor**. Any ionic compound which dissociates in pure water to form a hydrogen ion qualifies. A few examples of acids are:

$$HCl \rightarrow H^+ + Cl^-$$
$$H_2SO_4 \rightarrow 2\,H^+ + SO_4^{2-}$$
$$HNO_3 \rightarrow H^+ + NO_3^-$$
$$HC_2H_3O_2 \rightarrow H^+ + C_2H_3O_2^-$$

You can see how each of these, when added to water, would bring about an imbalance between the natural concentrations of $H^+$ and $OH^-$ in favor of $H^+$.

As you might guess, then, a **base** is defined as a substance which increases the hydroxide ion concentration of pure water or that "accepts" a proton in a chemical reaction—a **proton acceptor**. Any ionic compound which dissociates in pure water to form a hydroxide anion fits this definition. While acids are proton ($H^+$) donors, bases are proton acceptors. A few examples of bases are:

$$NaOH \rightarrow Na^+ + OH^-$$
$$KOH \rightarrow K^+ + OH^-$$
$$Ca(OH)_2 \rightarrow Ca^{2+} + 2\ OH^-$$
$$Al(OH)_3 \rightarrow Al^{3+} + 3\ OH^-$$

When added to water, each of these would bring about an imbalance between the natural concentrations of $H^+$ and $OH^-$ in favor of $OH^-$.

## Acid/Base Nomenclature

The more common strong acids and the stronger of the weak acids will be recognized by their common names. Hydrogen chloride, for example, will be recognized as "hydrochloric" acid. The same is true with hydrobromic and hydroiodic (or hydriodic) acids.

Naming an acid that is formed from one of the "-ate" or "-ite" endings of polyatomic ions is also simple. If you have an ion that ends in "-ate", and it joins with a hydrogen ion to form an acid, you change its "-ate" to an "-ic." For example, nitr*ate* ($NO_3^-$) joins with a hydrogen ion to form nitr*ic* acid. If you have an ion that ends in "-ite," you change it to "-ous" (pronounced like the word *us*). Therefore, a nitr*ite* ion joins with a hydrogen ion to form nitr*ous* acid. Remember:

"Change '-ate' to '-ic' and '-ite' to '-ous'."

If you say that three times really fast, you'll never forget it (even though you might want to).

## Acid/Base Strength

Being ionic in nature, acids and bases are electrolytes in solution. Like other electrolytes they are described as **strong** or **weak** based on the extent to which they dissociate in water. An acid that completely ionizes in water is a strong acid. Naturally, then, a weak acid is one that does not completely dissociate to its ions in pure water. The strongest bases tend to be made from the IA metals, because they dissociate/dissolve so easily. Similarly, the strongest acids involve nitrate, sulfate and chloride anions—once again, the most readily soluble. Table 1 provides a list of the strong acids and Table 2 provides a list of the strong bases.

Within these Tables it's difficult to say which acid or base is the "strongest" since (for practical purposes) all of them dissociate completely in water, making them all ultimate proton donors/acceptors. However, as an example of the factors that influence acid strength, sulfuric acid can be manufactured at greater purity than hydrochloric acid can. So, in this respect, sulfuric acid can be purchased at greater acid "strength." Because of its high vapor pressure and its strong attraction to water, HCl is difficult to concentrate above 12 M. Sulfuric acid, on the other hand, can be manufactured as > 98% pure $H_2SO_4$. Also notice that it can donate two protons per mole rather than just one:

$$H_2SO_4 \rightarrow 2\ H^+ + SO_4^{2-}$$

When non-chemists refer to a strong acid, they may be referring to other chemical properties besides true acidity. For example, HCl is truly a strong acid, but it is also strongly corrosive. Sulfuric acid is a strong acid, but it is also an exceptionally strong oxidizer. In these chemical properties, other acids may be "stronger" than some of the strong acids, but they have nothing to do with acidity *per se*.

As with a strong acid, the strength of a strong base may also be limited by its solubility in water. The IA metal hydroxides are deliquescent, so they are infinitely soluble in water, whereas the IIA metal hydroxides are not very soluble. The latter are classified as strong bases because, to the extent that they dissolve, they ionize completely.

## Acid/Base Concentrations in Solution

The strength of an acid solution may be expressed in terms of its molar hydrogen ion ($H^+$ or $H_3O^+$) concentration. We have already pointed out that, in pure water, this concentration is $1 \times 10^{-7}$ M. Likewise, the strength of a base may be expressed in terms of the concentration of $OH^-$ ions in solution. Because of the equilibrium that exists between $H^+$ and $OH^-$ in solution, the concentration of one changes in inverse proportion to the concentration of the other. (That is, as the concentration of one goes up the concentration of the other goes down.) Imagine having a beaker of water having an $OH^-$ concentration that is described by the following:

$$H^+ + OH^- \rightarrow H_2O$$

What would the $OH^-$ concentration do if there were a truckload of $H^+$ ions in solution? All of those $OH^-$ ions would be bound up to $H^+$ ions, and there would still be $H^+$ ions to spare. For this reason, knowing that the concentration of one implies the concentration of the other, it has become our custom to express the strengths of acids and bases in terms of the hydrogen ion concentration alone.

Since the concentration of a strong base in solution would imply a hydrogen ion concentration of perhaps $1 \times 10^{-14}$ M, you'll agree that this expression is a little cumbersome. About a hundred years ago, a simpler expression came into use and is now more common in practice. In order to understand it, though, you need to be familiar with a family of mathematical functions called *common*, or *base 10, logarithms*. See you next time!

**Exercises**

1) Write equations to show the *complete* ionization of each of the following strong and weak acids:

   a) Hydrobromic acid (HBr)
   b) Nitrous acid ($HNO_2$)
   c) Iodic acid ($HIO_3$)
   d) Phosphoric acid ($H_3PO_4$)
   e) Boric acid ($H_3BO_3$)

2) Write equations to show the complete ionization of each of the following strong and weak bases:

   a) Potassium hydroxide (KOH)
   b) Calcium hydroxide ($Ca(OH)_2$)
   c) Ammonium hydroxide ($NH_4OH$)

3) Tell whether each of the following is an acid, a base or a salt:

   a) $K_2SO_4$
   b) HCN
   c) CsOH
   d) $Na_3PO_4$
   e) $H_2SeO_4$
   f) $Sr(OH)_2$

# 59: pH

## Logarithms

I'm not going to try to explain how exponential functions and logarithms come about from mathematics. These are exercises that any serious science student should go through at some point in his mathematics training. Instead, in this section I'll teach you some of the practical aspects of logarithms so we can understand expressions of acidity and basicity.

Figure 1A is a graph of $y = \log_{10} x$. Let's take a look. First, notice that there are no negative values of $x$, so there are no logs of negative numbers. Similarly, the log of 0 is undefined—the curve never crosses the $y$ axis. In other words, you can't "take the log" of zero or of a negative number.

Next, notice that the log of 1 is equal to zero. This means that the log of any number greater than 1 will be positive, and the log of any number smaller than 1 will be negative. For example, the $\log_{10}(10) = 1$, but $\log_{10}(0.1) = -1$.

The unique feature of this curve which makes it so useful in the sciences is that a 10-fold change in $x$ results in a change in $y$ of 1 (See Figures 1A through 1C). Table 1 shows the usefulness of this relationship as a means of writing large numbers in shorthand. Soon, you'll come to appreciate this feature for abbreviating large, cumbersome numbers. Instead of saying $x = 1 \times 10^{23}$ we can simply say $y = 23$ or $\log x = 23$. The size of the log will tell you the size of the $x$.

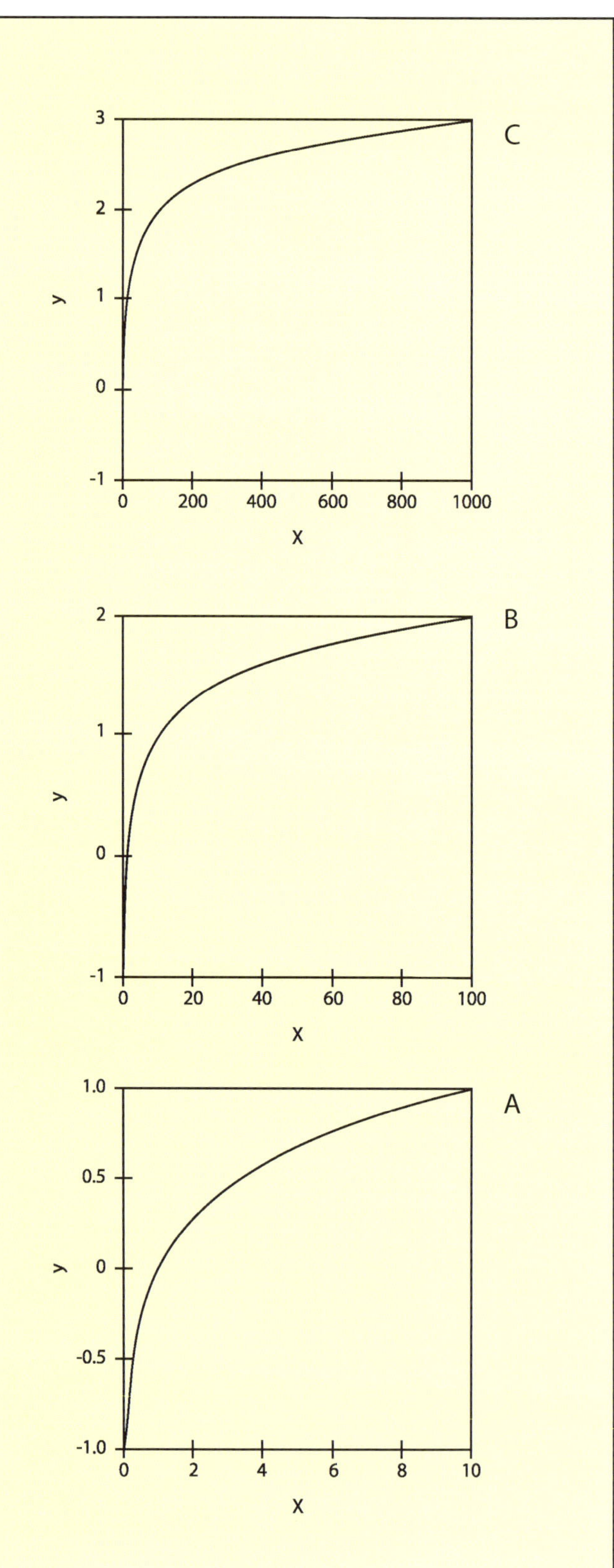


**Figure 1.** These are graphs of $y = \log x$ on different scales. Notice that $y$ is undefined at $x = 0$ and at all negative values of $x$. All values of $x$ greater than 1 have positive logs, all values of $x$ less than 1 have negative logs, and 1 has a log of 0. Each 10-fold increase or decrease in $x$ results in a corresponding increase or decrease in $y$ equal to 1. It's this last feature that makes it useful as a shorthand expression for large and small values. **A.** When $x = 10$, $y = 1$. **B.** When $x = 100$, $y = 2$. **C.** When $x = 1{,}000$, $y = 3$.

**Table 1.** Common logarithms for round, factor-of-ten values of *x*.

| *x* | Exponential Notation | $\log_{10} x$ |
|---|---|---|
| 0.00001 | $1 \times 10^{-5}$ | -5 |
| 0.0001 | $1 \times 10^{-4}$ | -4 |
| 0.001 | $1 \times 10^{-3}$ | -3 |
| 0.01 | $1 \times 10^{-2}$ | -2 |
| 0.1 | $1 \times 10^{-1}$ | -1 |
| 1 | $1 \times 10^{0}$ | 0 |
| 10 | $1 \times 10^{1}$ | 1 |
| 100 | $1 \times 10^{2}$ | 2 |
| 1,000 | $1 \times 10^{3}$ | 3 |
| 10,000 | $1 \times 10^{4}$ | 4 |
| 100,000 | $1 \times 10^{5}$ | 5 |

Until fairly recently, science textbooks always had tables of values of natural logarithms in the appendix. Any calculation you performed involving them meant turning to the back of the book, looking up an appropriate value in the table and placing it in your formula. Nowadays, every pocket scientific calculator has a button for calculating logarithms. However, be careful not to confuse the $\log_{10}$ ("log ten," also called the *base-10 logarithm* or the *common logarithm*) with the ln (pronounced "ell en," also called the *base-e logarithm* or the *natural logarithm*). There is a relationship between these two types of logarithms that you'll learn about in your higher math courses, but they are not the same.

## pH

In chemistry in particular, anytime you see the lower case p before a measured value it conveys that the expression is the **negative common logarithm** of the actual measured value. Examples in chemistry are pK (the negative of the log of an equilibrium constant), pε (the negative of the log of an "electron motive force") or, the most famous of all, pH. This symbol stands for the **negative of the common log of the hydrogen ion concentration**. *That* is a mouthful. However, it is easy to express mathematically ($-\log_{10}[H^+]$) and it's easy to understand. I'll show you how:

Let's begin with the hydrogen ion concentration of a strong base—a very low number, you will agree. (When there is a high concentration of free $OH^-$ ions in solution, the concentration of $H^+$ ions will be low.) If the hydrogen ion concentration of a sodium hydroxide solution is $1 \times 10^{-14}$ M, the log of the ion concentration is -14 and the pH is 14. Now, for a strong acid, the hydrogen ion concentration will be high. For example, a solution of hydrochloric acid may have a hydrogen ion concentration of 0.1 M. This is the same as $1 \times 10^{-1}$ M. The log of this concentration is -1, so the pH is 1. Because units are implied in the name pH, they are seldom used but are $-\log_{10}$ M.

If you are given a molar concentration of hydrogen ions, you can always calculate the pH by taking the negative log of that value ($-\log_{10} H^+$). If you are given the pH and asked for the molar $H^+$ concentration, you take the antilog ($10^x$ on your calculator) of the negative of the pH.

Here are a couple of practice calculations for you. First, find the pH of a 0.345 M $H^+$ solution. That is, calculate $-\log_{10}(0.345)$. Use your calculator instruction manual if you need to. You should get a pH of 0.462. (Taking a log has no effect on the number of significant digits reportable.) Now, return that pH to its $H^+$ ion concentration by making it negative and taking its antilog. Once again, your calculator's operation manual may come in handy. You should get a value that is the same as your starting value to at least three significant digits.

## pH of Familiar Environments

The strength of a solution in terms of its concentrations of acids and bases is usually expressed in pH units. Acids have pH values below 7 and bases have pH values above 7—the lower the pH, the stronger the acid, and the higher the pH, the stronger the base. Natural water will usually have pH values between 4 and 9. The pH of ocean water is about 8.2. The cytoplasm and living environments of most cells will be nearly neutral.

Most foods and drinks have pH values slightly below neutral (5-7), except for fruits which are usually acidic (mainly from citric acid) and range from about pH 4 down to pH 1.8 (lemons and limes). It is their acid that accounts for their sour taste. There are microbes that can live at any value of pH between 1 and 12. While pure water has a pH value of 7 by definition, distilled water contains dissolved carbon dioxide (carbonic acid, $H_2CO_3$), which tends to bring its pH down to around 4.

You can determine the pH of a nice, round number in your head, but without your calculator most people are not trained to calculate the logarithm of anything but an even factor of ten. Table 2 shows logs of the integers from 1 to 10, so you can see how they become fractional values. Try calculating the pH of a 0.2 M solution of HCl using your calculator. (Don't forget to change the sign. pH implies the *negative* of the log.) You should get an answer of 0.7 (0.69897 to one significant digit). Compare this pH to that of a 0.1 M solution (pH 1). When we add twice as much acid, the hydrogen ion concentration is doubled and the pH decreased. But because pH is a $\log_{10}$ value, it's not halved.

**Table 2.** Common logarithms of integer values of $x$.

| $x$ | $\log_{10} x$ |
|---|---|
| 1 | 0 |
| 2 | 0.301 |
| 3 | 0.477 |
| 4 | 0.602 |
| 5 | 0.699 |
| 6 | 0.778 |
| 7 | 0.845 |
| 8 | 0.903 |
| 9 | 0.954 |
| 10 | 1 |

**Exercises**

1) If you recall, pure water has a hydrogen ion concentration of $1 \times 10^{-7}$ M. What is the pH of this "neutral" solution?

2) If the pH of distilled water is 4, due to dissolved carbonic acid, what is the molar concentration of $H^+$ ions?

3) Calculate the pH of a dilute, 6M HCl solution.

4) Calculate the molar $H^+$ concentration of a sodium hydroxide solution that has a pH of 10.5.

# 60: Acids and Bases in Chemical Reactions

## Reactivity of Acids

Now that you know what acids *are*, let's talk about what they *do*. Reaction Box 1 shows a list of reactions in which acids participate. You will do well to memorize it. You already know the first reaction. The second looks much like the first. The third and fourth you will have seen in your laboratories.

## Reactivity of Bases

Likewise, we should learn a little about the reactivity of bases. The common reactions of bases are given in Reaction Box 2. The first one is an old friend by now. The second may be new to you, but is very practical in the chemistry laboratory and in the processing of metals. You don't need to memorize the third one because...well, it's a pain in the neck. Just remember that a base will react with a metal in water to make a mess—I mean, a *metal complex*.

**Reaction Box 1: Acids**

| | | | | | | | | |
|---|---|---|---|---|---|---|---|---|
| acid | + | base | → | salt | + | water | | |
| HCl | + | NaOH | → | NaCl | + | $H_2O$ | | |
| acid | + | metal oxide | → | salt | + | water | | |
| 2 HCl | + | $Na_2O$ | → | 2 NaCl | + | $H_2O$ | | |
| acid | + | metal | → | salt | + | hydrogen | | |
| 6 HCl | + | 2 Al (s) | → | 2 $AlCl_3$ | + | 3 $H_2$ | | |
| acid | + | carbonate | → | salt | + | $CO_2$ | + | water |
| 2 HCl | + | $Na_2CO_3$ | → | 2 NaCl | + | $CO_2$ | + | $H_2O$ |

**Reaction Box 2: Bases**

| | | | | | | | | |
|---|---|---|---|---|---|---|---|---|
| acid | + | base | | | → | salt | + | water |
| HCl | + | NaOH | | | → | NaCl | + | $H_2O$ |
| base | + | metal salt | | | → | metal hydroxide | + | salt |
| NaOH | + | $CuSO_4$ (aq) | | | → | $Cu(OH)_2$ | + | $Na_2SO_4$ |
| base | + | metal | + | water | → | complex salt | + | hydrogen |
| 2 NaOH | + | $Al^0$ | + | $H_2O$ | → | 2 $NaAl(OH)_4$ | + | 3 $H_2$ |

## Neutralization and Formation of Salts

The first reaction listed in each reaction box is that between an acid and a base to produce a salt and water:

$$HN + MOH \rightarrow N^+M^- + HOH$$

where N is usually a nonmetal and M is usually a metal. In general, a base can be described as a **metal hydroxide**. Both HN and MOH are ionic substances, NM is a salt, and HOH is, of course, water. Although two ionic substances are reacting here, water is molecular. It is not considered to be ionic (even though a small amount of ionization occurs), nor is pure water a good electrolyte.

The most common example of a neutralization reaction, and one that we have shown you several times, is the reaction of hydrochloric acid with sodium hydroxide, as follows:

$$HCl + NaOH \rightarrow NaCl + H_2O$$

Later, in the exercises, I'm going to ask you to predict the products of the reactions between acids and bases. You will always get the same two products—an ionic compound (a "salt") and water.

This completes an important aspect of your chemistry education. You are coming to know when a reaction will occur and are beginning to be able to predict their products.

## Writing Net Ionic Equations

The equations written above show all of the reacting substances and products as molecules. But these are ionic compounds, so they don't really exist as molecules in water. They are completely ionized. Sometimes it's helpful to show the actual entities that take part in the reaction. For this purpose, we might have chosen to write our example reaction in its *ionic* form as follows:

$$(H^+ + Cl^-) + (Na^+ + OH^-) \rightarrow (Na^+ + Cl^-) + H_2O$$

You may hear this called an **ionic equation** or a **total ionic equation**. Notice that $H_2O$ is not written in ionic form, because it no longer exists as ions. It is a molecular compound, so it would be inappropriate to show it as ions. (Recall that you can predict which compounds will be ionic and which will be molecular by comparing their electronegativities. Atoms having a difference in electronegativity values greater than or equal to 2.0 will be ionic. See Lesson 17 if you need a refresher.)

## Ionic Lazy-Boneses

The sodium and chloride ions in our example reaction really don't react with anything. At the beginning of the reaction they exist as separate, individual ions, and after the reaction they still exist in solution as separate, individual ions. Only the $H^+$ and $OH^-$ ions undergo a change of form. Sometimes we are concerned only with the real players in this game and choose to ignore these lazy **spectator ions**. For the purpose of focusing on the players, we may write the equation as a **net ionic equation**:

$$H^+ + OH^- \rightarrow H_2O$$

The equation you see here is the net ionic equation of any acid/base reaction where the acid is of the form "HN" and the base is of the form "MOH," where N is a nonmetal and M is a metal.

## Normality

Don't worry. When I say "normality" I'm not referring to your personality. When dealing with acids and bases, there is a special way to express concentration. By now you are quite familiar with the term molarity (M, moles per liter). When dealing with acids it's easy to make a mistake in thinking that acid strength is directly related to molar strength of the solution. You might think, for example, that 1 M hydrochloric acid is the same strength as a 1 M solution of sulfuric acid. This would be true were it not for the fact that some acids are **polyprotic,** meaning that each molecule donates more than one proton. Some examples of these are $H_2SO_4$, $H_3BO_3$ and $H_3PO_4$. Hydrochloric acid (HCl) donates only one proton per molecule. However, sulfuric acid donates two protons per molecule, and phosphoric acid donates three protons per molecule. We refer to these last two as **diprotic** and **triprotic** acids, respectively.

Chemists make a correction for these multiple proton donors by introducing a new term. A 1 M solution of $H_2SO_4$ is "normalized" for its two protons if we refer to it as a 2 **normal** solution. The term **normality** (abbr. **N**) refers to the proton-donating strength of an acid or the proton-accepting strength of a base. *You get the normality by simply multiplying the molarity by the number of protons the molecule donates or accepts.* For example, a 1 M solution of $H_2SO_4$ is 2 N because it donates two protons.

Likewise, many bases accept more than one proton. A 1 M solution of $Ca(OH)_2$ is 2 N because one mole of this compound can accept two moles of protons. Consider the effect of this new term on reaction chemistry. A 2 N solution of $H_2SO_4$ (1 M) will neutralize an equal volume of a 2 N KOH (2 M). Notice that the normalities are the same while the molarities are different. That's because the normalities are the expressions of the concentrations of the $H^+$ and $OH^-$ ions that are actually involved in the reaction. Anytime there is to be a balanced reaction between an acid and a base, the reaction must represent equivalent proportions of $H^+$ and $OH^-$ ions. The concentrations of those ions in solution are represented by their normalities.

I haven't told you anything new. If you understand a balanced equation like this one:

$$H_2SO_4 + 2\ KOH \rightarrow K_2SO_4 + 2\ H_2O$$

you know that it takes two moles of KOH to neutralize one mole of sulfuric acid. But if you have five different acid solutions on the shelf, and each of them is a 1 N solution, you don't have to write and balance equations if you want to know how much of each will be required to neutralize 100 mL of a 1 N KOH solution you can use 100 mL of any acid you choose. Any volume of a 1 N acid will neutralize that same volume of a 1 N base, and vice versa. If the normalities are equal, the volumes will also be equal.

You can solve any problem that incorporates the idea of normality if you simply learn to convert between N and M. Just remember that molarity is equal to moles of acid or base per liter, and normality is equal to moles of $H^+$ donated and moles of $OH^-$ accepted per liter.

**Exercises**

1) Complete the following reaction equations:

   a) $H_2SO_4 + Ca(OH)_2 \rightarrow$
   b) $HCl + Ba(OH)_2 \rightarrow$
   c) $H_3PO_4 + NaOH \rightarrow$
   d) $HNO_3 + Mg(OH)_2 \rightarrow$

2) Please write a total ionic equation for each of the reactions in 1.

3) Please write a net ionic equation for each of the reactions in 1.

4) Please write down the spectator ions in each of the reactions in 1.

5) Please give yourself an extra 20% for bonus points if you remembered to balance all of your equations without being told. (If you read ahead to see this message before doing the other exercises, give yourself the 20% anyway and be sure to balance your equations.)

6) Please give the molarity of a 0.5 N solution of each:

   a) KOH
   b) $H_2SO_4$
   c) $H_3PO_4$

7) Please give the normality of a 3 M solution of each:

   a) HI
   b) $Ca(OH)_2$
   c) $HNO_3$

8) How much of a 2 N solution of $H_2SO_4$ will be required to neutralize 10 mL of a 1 N solution of KOH?

9) Using the unit factor method, calculate the mL of 0.0300 N $H_2SO_4$ that will be required to neutralize 27.4 mg of $Ca(OH)_2$.

## 61: Chemical Equilibrium

We have had many opportunities to mention the concept of equilibrium. We ran across it most frequently when we spoke of phase changes. We also spoke of how equilibria change as temperature and pressure change. Any equilibrium is a balance among opposing forces. In a study of physics you will run across the concept of physical equilibrium. An example might be a seesaw with weights on both ends. The position of the seesaw will depend on the balancing point of those forces. We tend to think of physics and chemistry as separate disciplines, but this principle of equilibrium is one that runs through every discipline. Whether we speak of physical, chemical or biological equilibrium, this basic principle of balance permeates nature. It is the application of the principle of "central tendency" in mathematics and statistics. The idea of balance apparently gave our Creator great pleasure as it is critical to effectively every natural process. This is what makes it so important to every person—even those who don't know it.

The word *equilibrium* has come to take on a rather specific definition in chemistry. **Chemical equilibrium** is defined as *the balance point of a reversible chemical process.* Notice that word *balance* being used again. It implies that there are forces in opposition. Every chemical equilibrium is best understood if those basic influences that bring about the equilibrium are understood. I'm going to give you three different examples of chemical equilibrium, to show you how far-ranging the concept is.

First, the evaporation of water comes about because individual water molecules absorb heat, and that heat is converted to kinetic energy. The condensation of liquid water comes about because of the loss of that kinetic energy to the surroundings. So then, there are opposing processes that occur between the two phases that might be shown as:

$$H_2O\ (l) + heat \leftrightarrow H_2O\ (g)$$

Water reversibly evaporates.

The opposing arrows indicate that this reaction can proceed in either direction. That is, water may absorb heat and enter the gas phase or its gas may lose heat and become a liquid. This is called a **reversible** reaction. The reaction from left to the right is in opposition to the reaction that goes from right to left. In a closed container, at a constant temperature and pressure, a balance between the two phases will occur resulting in some amount of liquid water and some amount of gaseous water. If you'll recall, that's where our concept of vapor pressure came from.

Our second example of equilibrium is the dissolution of an ionic solute in water. The following reaction describes the ionization of the weak acid, hydrogen fluoride:

$$HF\ (aq) \leftrightarrow H^+ + F^-$$

HF "reversibly dissociates" to $H^+$ and $F^-$ ions.

Once again, an equilibrium exists between the concentration of HF molecules and the concentrations of its ions in solution. The truth is that both the dissociated and undissociated states are present in solution simultaneously. When the concentration of HF goes up, the concentrations of its ions must decrease accordingly. A balance exists between the two states and different things can happen that affect that balance.

A third example of equilibrium is the dissolution of a solute. A solid will enter solution until it reaches its saturation point at which it will reach an equilibrium with its dissolved molecules. That equilibrium might be expressed as:

$$C_{12}H_{22}O_{11}\ (s) \leftrightarrow C_{12}H_{22}O_{11}\ (aq)$$

Sucrose reversibly dissolves in water.

A balance exists between dissolved sucrose and undissolved sucrose. Once again, things can happen that will adjust the balance point.

None of these equilibria is static; every one is dynamic. While, at a distance, these systems may appear as though no changes are taking place, a closer look will prove them to be constantly busy. Water molecules continue to enter and leave the liquid phase as they lose and gain heat. HF molecules continually break apart into their ions and rejoin. Molecules of solid sucrose continually enter solution while dissolved molecules leave the solution to join the solid sucrose molecules on the bottom of the beaker. However, at equilibrium, the rate of each process is equal to that of its opposing process.

## Stress on Equilibrium

The basic principle of equilibrium was stated by the French chemist Henri Le Chatelier (1850 to 1936) when he said that *anytime a new stress is placed upon a reversible chemical process, the process will shift to adjust to that stress*. This is **Le Chatelier's Principle**.

Let's return to our first example of equilibrium to reconsider the evaporation of water:

$$H_2O\ (l) + \text{heat} \leftrightarrow H_2O\ (g)$$

We've said that, if liquid water is placed in a container and sealed, there will be a specific proportion of liquid water to gaseous water present at equilibrium. If we were to add additional heat to this system, we would expect a shift in the equilibrium to favor the formation of additional gas. The equilibrium expressed in our equation will have shifted to the right. If we were to cool down our little system, we would expect the equilibrium to shift to the left. In other words, a new equilibrium will be established after each application of every new stress. That equilibrium will be the best balance to minimize the opposing stresses.

Stresses that might shift an equilibrium include:

1) Concentration of one or more of the reactants or products – Addition of one substance on either side of the equation drives the reaction in the opposite direction. When the system is balanced, too much of one substance or the other will put it out of balance. The system will regain its balance by producing more of the substances on the other side.
2) Temperature – Temperature effects equilibrium by shifting it in the direction of the process that absorbs heat. In the evaporation/condensation of water, it is the evaporation that absorbs heat, so added heat would favor evaporation over condensation. At the new equilibrium a higher proportion of the water would be in the gas phase.
3) Pressure and volume – Pressure and volume affect reactions that involve gases. Higher pressures per amount of volume require that molecules be packed more efficiently. So then, an increase in pressure would drive a reaction in the direction of the fewest molecules per unit volume. For example, in the reaction:

$$2\ N_2 + O_2 \leftrightarrow 2\ N_2O$$

Higher pressure promotes greater efficiency of formation of the two molecules of $N_2O$ on the right, over the formation of the three molecules on the left (two molecules of $N_2$ and one of $O_2$). Higher pressure encourages greater compaction.

**Exercises**

In which direction (left or right) will the following reactions shift in response to the indicated change?

1) $2\ A + 3\ B \leftrightarrow 3\ C + D$
   Change: Increase in [C]

2) $H_2\ (g) + Cl_2\ (g) \leftrightarrow 2\ HCl\ (g) + \text{heat}$
   Change: Increase in temperature

3) $2\ SO_2\ (g) + O_2\ (g) \leftrightarrow 2\ SO_3\ (g)$
   Change: Increase in pressure

4) $N_2 + O_2 \leftrightarrow 2\ NO$
   Change: Increase in $[O_2]$

## 62: Quiz 10

1) Be able to determine whether or not to expect hydrogen bonding among the molecules of a compound when given its structure.
2) Know the general effect of hydrogen bonding on boiling point and vapor pressure.
3) Know how to write ionization equations for acids and bases.
4) Know how to tell whether a compound is an acid, a base or a salt by looking at its molecular formula.
5) Know how to calculate the pH of a solution given its $[H^+]$ or the converse.
6) Be able to complete and balance an equation for a neutralization reaction given the reactants.
7) Be able to write a total ionic equation and a net ionic equation for any neutralization reaction. Also be able to identify spectator ions.
8) Know how to calculate the normality of an acid or base when given its molarity, and *vice versa*. Be able to use concentrations expressed either way in stoichiometric calculations.
9) Be able to anticipate the direction of a shift in an equilibrium based on the type of stress placed on the system.
10) In a general way, be familiar with the three concepts of an acid and know the correct name to attach to each of the concepts.

Every organism has an optimal pH at which it operates best. Acidic soils (around pH 4) can produce some beautiful results.

# 63: Environmental Chemistry

## Environmental Complexity

You have been introduced to the word *homogeneous*, which is used to describe anything that is uniform in composition. Solutions are homogeneous because their particles are completely and randomly dispersed. When a system contains more than one phase, the phases are usually not well dispersed. If you sample a nonuniform system, you might get more of one phase on the first sample and get more of another phase in the second. These multiphase systems are referred to by the term *heterogeneous*.

True homogeneous solutions are practically a laboratory artifact (something invented). The waters in lakes, streams, underground aquifers and oceans do have some dissolved matter in them, but they also contain particles that are not completely dispersed. A sample from any natural environment will contain solids (which will include some amount of living and some amount of nonliving matter), liquids and gases. Solid matter may come into suspension and be held there for long periods of time by turbulence in the water. Generally, the smaller and less dense the particle, the easier it is to hold in suspension. Large and dense particles settle out quickly or require a great deal of energy input to keep them in suspension.

The easiest way to know the difference between suspended and dissolved matter is simply by sight. A suspension will look cloudy while a true solution will be clear. If there remains a question as to the presence of suspended matter, shine a light through the liquid. Dissolved molecules and atoms do not scatter light appreciably, but suspended matter does. You can actually see a beam of light that is shone through a suspension because the rays scatter out toward your eyes. If you shine a light through a solution, however, the beam will not be seen. For cases where the amount of suspended matter is too low to see directly, there is an instrument, called a nephelometer, that is designed for measuring scattered light. The amount of suspended matter can be estimated from the amount of light scattered.

Large particles tend to be denser than water and fall out of suspension quickly because the viscosity of water is not high enough to keep them in suspension. Molecular particles will remain suspended indefinitely. Even denser molecules will be maintained in suspension by **Brownian movement**—the random bumping of other molecules that keeps things stirred up on the molecular level. Between the size of large, suspended particles and the atom- or molecule-sized particles of solutions lie intermediate-sized particles called **colloids**. These are small enough to remain in suspension indefinitely due to the rapid motion of molecules beating them about, but they are not true solutions. Colloids will pass all but the finest of filters because they range in size from a few nanometers ($10^{-9}$ m) to a few micrometers ($10^{-6}$ m). According to their size, bacteria are in the range of large colloids.

Colloids are notorious for their complex surface properties. They may have a variety of fixed surface charges that attract dissolved ions around them. They may also have polar and nonpolar sites that adsorb (attract to their surfaces) different sorts of molecular substances. Their surfaces may be not only chemically complex, but physically complex as well. The many facets of their surfaces, their crevices and their craters give them extraordinarily high surface area relative to their size.

Of such is the study of environmental chemistry. One liter of lake water may contain 100 ppm (parts per million weight) of suspended matter, of which half is mineral and half is organic, 30 ppm of dissolved gases, 150 ppm of dissolved organic carbon, 25 ppm of mineral salts, 1,000,000 bacteria, 100,000 algal cells, 10,000 microscopic animals and the occasional fish. (See Figure 1.)

## Environmental Pollution

Pollution is an easy concept to talk about, but a hard one to define. One person looks at a city skyline

**Figure 1.** While solutions made up in the laboratory are usually simple, this glass of iced tea is more like the real world. It contains dissolved acids, plant degradation products, sugars and minerals (in the drinking water). It contains solids (tea grounds) and colloidal matter in suspension. It contains water in two different phases, as well as dissolved gases (from the air). There are bacteria and other microorganisms that live in iced tea and do quite well, thank you very much, despite the fact that sodium hypochlorite was added to stay their growth during underground transport to your house. During that transport, the water picked up some dissolved metals from the water pipes and some metal oxides from pipe corrosion along with the bacteria that aided that corrosion. I'd say that slice of lemon is a nice touch, wouldn't you?

and mourns over the marring of the native landscape while another sees the skyline itself as thing of beauty and a testimony to human creativity. Difficulties in definition aside, it is clear that humans have had negative effects on their living environment, and that those effects are, at times, devastating in terms of life, health and quality of life. "Negative effects" may range in severity and type from a distasteful odor, which is easy to track, to a cancer-related death which may be impossible to link to its source.

Human activity is not the only, or in most cases, the predominant source of negative effects. Nature itself can deal a hard blow by chemical means. Some of nature's chemical weapons include the noxious gases from volcanoes, alkaline water from ground sources, contamination of drinking water by heavy metals leaching from natural minerals, radon in basements (from the decay of uranium) and swamp fires caused by methane combustion.

It is common for chemistry texts to consider pollution as a problem in direct relationship with water. This is mainly because the regulation by the U.S. government of polluting activities has developed separately in water, air and soil, and the first sweeping measures by the government to regulate pollution were directed at water. However, in the real world, attempts to separate contaminants as purely problems of a single medium are artificial. There are constant interactions among these media so that polluting influences in one affect them all to some extent.

It is also common in textbooks to blame industrialization for the onset of severe pollution problems that have occurred over time. In truth, the most devastating pollution problems, related to infectious disease, happened before the industrial revolution and have been solved in industrialized nations by modern water treatment methods. These challenges still exist in undeveloped countries. In this country, the government (especially the Departments of Energy and Defense) has more polluted sites under its control than any other single entity. While the abundance of contaminating chemicals is largely due to industry and government agencies that produce them, their wide distribution results from their use and misuse by consumers.

**Sources.** Although there are as many different sources for pollution as there are polluting chemicals, they may be considered in two broad categories: **point-source** and **non-point-source** pollutants. Point-source pollutants are those that have an identifiable source such as a single factory that dumps harmful chemicals directly into a river. One example of such pollution is an accumulation of mercury in the sediments of the Great Lakes through the 1960's. Once the problem was identified, regulations were imposed on specific manufacturing facilities to halt the release of mercury. Dredging operations were used to remove contaminated sediments to prevent further injury.

Non-point-source pollutants are used broadly by consumers and may find their way into the environment in a variety of ways. An example might be the accumulation of phosphates in lakes, which causes an overgrowth of algae. The algae stink when they decay. There is not a single source for phosphates, but they come from a variety of fertilizers, detergents and cleaning solutions that are swept into receiving streams by rainfall or pass through waste treatment plants and end up in a receiving basin, such as a lake.

## Types of Pollution

Pollution is represented by several broad categories filled with subcategories. In the following paragraphs we briefly address the broad types so that you gain an understanding of where the most commonly referenced problems lie.

**Organic Chemicals.** Some of the more common organic pollutants are manufactured pesticides (DDT, dieldrin, trifluralin, and hundreds of others), wood treatment chemicals (largely pentachlorophenol and creosote), munitions (TNT and related compounds) and solvents (chlorinated solvents, alcohols and ketones). The most common pollutants from industry and government agencies are refined petroleum products—gasoline, jet fuel, diesel fuel, lubricating oils and heavy fuel oils. Most of the problems associated with organic chemicals arise from their toxicity and potential long-term contributions to cancer deaths.

**Human and Animal Waste.** A subcategory of organic pollution that deserves special attention is human and animal waste. These waste products are managed aggressively in the United States and most of the more developed countries using wastewater treatment systems. A treatment system may be something as simple as a holding pond on a farm or as complex as a facility that receives several millions of gallons of waste daily and subjects it to three different levels of treatment. The major problems associated with these wastes include the spread of infection from microbes and the high oxygen demand that starves air-breathing fish of their needed oxygen.

**By-products of Fuel Burning.** Fuels are another group of organic compounds, most of which are obtained from petroleum. When fuels are burned in air the major end products are carbon dioxide and water. There is some evidence that carbon dioxide can accumulate in the atmosphere over time and increase the retention of heat from the sun. This phenomenon, called **global warming,** or "**the greenhouse effect**" has received considerable attention from the world media, and some international initiatives have been taken to regulate it. However, the anticipated effects of global warming are speculative.

A known problem associated with fuel burning is that fuel contains sulfur compounds that are released as sulfites in the atmosphere. When these compounds are dissolved in rain they form acids which can increase erosion and affect wildlife. Stricter controls on emissions have reduced acid precipitation, but more economic ways of removing sulfur compounds, before and after combustion, are being sought.

**Nuclear Pollution.** While nuclear power plants do generate radioactive waste, that waste is managed carefully according to plans that were designed before the construction of those plants and were approved by the government. Most of the radioactive pollution problems that have resulted in contaminated property are related to the metals-processing facilities where nuclear source materials (especially for weaponry) were manufactured since the 1940's. These pollutants are in the form of mine tailings and slag to be found in soil. Although rare, accidents such as the reactor meltdown that took place in Chernobyl, Ukraine in 1986 are devastating in terms of their immediate and long-term effects. This accident, the worst of its kind to date, resulted in an estimated 8,000 dead, either directly during the disaster and cleanup, or indirectly from thyroid and other cancers.

Although the Chernobyl disaster was an avoidable tragedy (it resulted from an unauthorized test on an inoperable reactor), the increase in reliance on nuclear power comes with concerns of the increased likelihood of similar disasters. A close call in the U.S. at Three Mile Island in Pennsylvania in 1979 resulted in increased regulatory activity around nuclear power plants in the U.S.

**Heavy Metals.** Contamination by metals (especially the radioactive ones) is particularly difficult to deal with because they are not easily rendered nontoxic. They can potentially remain hazardous indefinitely. This is in contrast to organic chemicals that are naturally broken down by bacteria and fungi into nontoxic end products (such as carbon dioxide and water). Metals are notorious for creating a wide range of metabolic problems in living things, especially disorders of the central nervous system (brain and spinal cord) where they tend to accumulate.

**Heat.** Heat is a simple form of pollution. While not a chemical itself, heat affects the chemical environment and results in chemical problems. For example, oxygen is less soluble in warm water than in cold water, so the presence of excessive heat yields the water less livable to oxygen-requiring animal life. Excess heat may, by itself, render water unusable for some forms of life. However, cases can also be cited of certain animals that inhabit artificially warmed waters that could not otherwise exist there. This may have negative or positive ecological outcomes. For example, there are manatees that inhabit the warm streams coming from power plants in Florida. On the other hand, wildly growing zebra mussels plug the outgoing streams of industrial plants. This illustrates the point that anytime a change takes place in the environment, the ecological arrangement (the specific collection of animals, plants and microbes and their activities) will change in response to it. These changes prove complex and impossible to predict.

## Chemical Purification Processes

Chemical processes are not by any means the only ways of reducing the impact of pollution, but there are several chemical methods to consider. Disinfectant chemicals are used to remove infective agents. Activated carbon is often used to remove compounds that reduce the value of water because of unappealing tastes or odors. Trace amounts of toxins may also be removed from air or water by adsorption on activated carbon. Hydrogen sulfide can be removed by wet chemical "scrubbers." These are liquid systems that are used to chemically remove and/or destroy noxious chemicals from gases. Strong oxidizing chemicals may be used to react with toxic organic chemicals.

Apart from pollutants, naturally occurring salts may also contribute to water use problems. For example, naturally occurring calcium and magnesium ions are responsible for **hard water**. These ions combine with soaps to form insoluble scum. When heated, insoluble salts form that cause "scale" inside boilers and in the fabric of clothing during washing. This scale plugs hot water lines and makes heating elements less efficient. A **cation exchange** process is used to remove these divalent cations from water and replace them with sodium ions from sodium chloride in household water softeners. On an industrial scale, calcium and magnesium ions are precipitated.

Sea water is approximately 3.6% sodium chloride. This is fine for saltwater plants and animals, but not so good for people. A few places in the world have insufficient fresh water to serve the population, so sea water must be **desalinated** for potable (drinkable) water. This was historically accomplished by distillation, but, more recently, by **reverse osmosis** filtration. We have introduced osmosis as the tendency for water to pass through a semipermeable membrane to dilute a more concentrated saline solution. This process is spontaneous. Reverse osmosis, as the name implies, goes against the spontaneous passage of water into a saline solution and uses energy to force it out of the salt through a membrane—a form of molecular filtration. Most desalinization plants rely on this costly process to remove salt from sea water for drinking and for industrial purposes.

## Biochemical Purification Processes

Microbes have been used since before recognition of microbes as existing entities for the removal of chemical pollutants. Composting is an example of a microbiological process for the stabilization of manure. In recent years, the wide-ranging catalytic capabilities of bacteria have been exploited for removing chemicals that are toxic to humans. Many processes have been refined that are quite capable of removing even high concentrations of severely toxic chemicals from soil, water and air. Bacteria have been isolated and identified that have the ability to utilize toxins from virtually every class of organic chemicals as sources of carbon and energy. Although metals can't be further broken down to nontoxic chemicals as organic chemicals can, bacteria have been shown to accumulate, modify and immobilize them. This is a fast-developing field of study, and thousands of operations are ongoing around the country that take advantage of these marvelous little creatures.

**Figure 2.** This looks like a big white box, and it is. All steel with a nice coat of white paint on the outside. What's special about this big white box is what goes on inside. Water contaminated with toxic pollutants is pumped in one end of the box and water, pure enough to drink comes out the other. The purification that takes place in the box is done by harmless bacteria. While the pollutants are toxic to you and me, they are like candy to these organisms. God put bacteria on this earth that can consume just about anything--even toxins we have invented ourselves. We simply build a home for these organisms and give them things that they need to help them grow. They do the rest for free.

# 64: Organic Chemistry and Biochemistry

## Organic Chemistry

Organic chemistry is the study of the chemistry of carbon. Why does carbon chemistry deserve so much special attention? Look in any index of chemicals and compare the number of carbon compounds with the number of compounds of any other element. What you'll find is that carbon's love for bonding with other carbon atoms and its four-bond requirement allows it to form chains, branched chains and rings of all sizes from a single carbon atom to hundreds of thousands of carbon atoms. What's more, all of these structures may be substituted with halogens or with sulfur-, nitrogen-, phosphorus- or oxygen-containing groups. These groups have different effects on the structures and chemical activities of their carbon skeletons.

The field of organic chemistry is dedicated to the understanding and exploitation of these wide-ranging chemical properties. Organic chemistry gives way to several large fields of research, discovery and practice. These include petroleum chemistry, polymer (largely plastics) and textile chemistry, and biochemistry—the chemistry of living things. On the last several pages of this Lesson I've drawn many carbon-based structures for you. These are not arbitrarily selected, but each group and each compound within the group is chosen either (1) because it's famous or (2) it illustrates something specific about carbon chemistry. The instruction that goes along with each set of compounds is provided in its caption. Please read them all carefully.

Carbon is another of those one-of-a-kind, hand-picked natural treasures that helps us to realize that our being here was no accident. There is not another element that can take carbon's place in bringing life into existence. If there were 26 different elements that could serve as the basis for life, 107 different molecules that could do what water does, and other kinds of bonding that could stand in for hydrogen bonding in living things, we might be fooled into thinking that evolution had a fighting chance. There is only one carbon. Even the other elements in its group fail as substitutes.

## The Biochemistry Bridge

Much more biochemistry will be provided in ***The Spectrum*** course on biology. However, it's important that you realize that there is a logical connection between chemistry and biology. The natural bridge is formed by the field of biochemistry, which explains biological processes in chemical terms. You may learn some biology terms, but you'll never *understand* biology until you understand the chemical processes that make biology work. You now have a basic understanding of chemistry terminology, atoms, chemical bonds, reactions and the energetics that make reactions go. You also understand how the energy from one reaction can give up energy to drive other reactions.

While there is a continual thought process that takes us from chemistry to biology, in a larger sense it's like going from one world to another. Just imagine the difference between a world without life and a world with life. This jump isn't easily made. It's a jump from molecules made of two to eight atoms to a world where molecules are made of tens and hundreds of thousands of atoms. It's a jump from a place where a molecule is as purposeless as a rock on Mars to one where it has an integral function among tens of thousands of others which cannot function without it.

On the chemistry end of things, biochemistry is no different than the chemistry of other molecules. Atoms still do what they do because of the basic forces that exist among them. Sugars link together by the same chemical bonding influences that link carbon dioxide together. Those sugars make long chains with the seemingly ordinary polymer structure to make **polysaccharides**. Likewise, amino acids link together in long chains to form **peptide** sequences. Long-chain fatty acids are grouped together by a phosphate-based structure to form a fat molecule.

Somewhere in between amino acids and a functional protein, however, a transformation occurs. That transformation takes us from a nonfunctional string of amino acids to a functional protein—one piece in an immense and complex puzzle that, when it is complete, creates a picture that we call life. Life is not chemicals. The chemicals are the individual pieces of shapely, inanimate cardboard on which the picture is painted. One piece of the puzzle tells no story. It doesn't enlighten. At best it reminds us that there is a bigger picture without which it is misplaced and valueless.

All of those pieces jumbled together into a box wouldn't approach the orderliness of life. Chaotic chemicals will remain chaotic or minimally organized under their forces. No amount of mindless mixing would cause these pieces to congeal into a picture puzzle. Even if they did link randomly, no picture would appear.

As we enter the realm of biology, we'll remember that there is continuity between the world of chemistry and the world of biology–both obey the laws of physics. But we'll also remember that there is a gap in nature. Nothing lies between chemistry and biology in terms of complexity and organization unless it is biochemicals which were formed by a living system, then decomposed to some intermediate form. A minimally organized cell is unspeakably more organized than the most complex chemical.

A fully formed biochemical is integral to something that is of a higher order. Returning to our analogy of the puzzle, life doesn't come about from random arrangements of big molecules anymore than a picture puzzle comes about from random assortments of shredded cardboard at the recycler. It starts with a photograph or a painting. The painting is printed on a large piece of cardboard, then the cardboard is cut into pieces that fit nicely together. If this were not true, the pieces would no sooner make a picture than any other unguided, random arrangement of inanimate matter. Our Creator knew, before life was breathed into existence, that the pieces would come together to make something that transcended the pieces themselves.

| Amino Acid | 3-Letter Abbr. |
|---|---|
| Alanine | Ala |
| Arginine | Arg |
| Asparagine | Asn |
| Aspartic Acid | Asp |
| Cysteine | Cys |
| Glutamic Acid | Glu |
| Glutamine | Gln |
| Glycine | Gly |
| Histidine | His |
| Isoleucine | Ile |
| Leucine | Leu |
| Lysine | Lys |
| Methionine | Met |
| Phenylalanine | Phe |
| Proline | Pro |
| Serine | Ser |
| Threonine | Thr |
| Tryptophan | Trp |
| Tyrosine | Tyr |
| Valine | Val |

Yes, a protein is a series of amino acids in sequence. But it may also be a three-dimensionally folded **enzyme**. It fits together with one other, select, molecule that it was designed to act upon like a lock matches perfectly to a specific key. By doing so, it **regulates** a complex biological process—perhaps it triggers the production of a dozen other enzymes like itself, each of which regulates another process. It is but a single piece in a complex puzzle.

Life was **presupposed** in the structure of this and every other protein, but it was also presupposed in the nature of water. It was presupposed in the structure of carbon and the elements, and in the nature of chemical bonds. It was presupposed in the nature of atoms.

So we now move on to biochemistry—a simple, conceptual leap from chemistry, but a bridge over the gorge that exists between the simplicity of nonlife and the immense complexity of life.

## Biochemicals and Processes

The major biochemicals include **proteins** (polymers of **amino acids**, see Table), **complex carbohydrates** (polymers of **sugars**), **fats** (two- and three-unit groups—**dimers** and **trimers**—of **fatty acids**) and nucleic acids (polymers of **nucleic acid "bases"**).

Proteins are of two types, **structural proteins**, such as **collagen** and **elastin**, that make up the fibrous structures that hold animals together, and **enzymes**—organic biocatalysts that speed up chemical reactions. The increase in reaction rate due to an enzyme can be huge (such as a factor of a hundred million). The presence or absence of the enzyme can decide whether or not a reaction occurs for practical purposes, so biological regulation is dependent upon the timely formation and then decomposition (or "**degradation**") of the enzymes that control those processes.

Fats and carbohydrates are used for storage of energy. Carbohydrates are readily broken down into sugars for the short-term needs of cells, while the more reduced structures of fat molecules are available as potent energy producers. Fats produce much more heat upon degradation because there is much more energy stored in them.

In addition to their energy storage function, fats and sugars also have structural functions like proteins do. Specifically, a group of lipid molecules, called phospholipids, make up the cell membranes of every living cell. Sugars link up with lipids (in **glycolipids** and **lipopolysaccharides**) and with proteins (in **glycoproteins**) to form protective cell walls around the cell membranes of many organisms.

A nucleic acid molecule contains the genetic code of an organism in its sequence of nucleic acid bases. Nucleic acids are of two broad types—**deoxyribose nucleic acid** (DNA) and **ribose nucleic acid** (RNA). DNA molecules make up the **genomes** (genetic codes) of all living organisms.

A section of genetic code is read by an RNA transcriptase enzyme and a negative of the code is stored in a **messenger RNA** molecule. The code is transferred to a **ribosome** where the code is used to make proteins from its information and from the amino acids that are brought to it by small **transfer RNA** molecules. This may sound like a pure miracle, and in a limited sense of the word, it might be. However, the making and breaking of chemical bonds to form RNA, the action of ribosomes in piecing together proteins from the code read from the RNA structure, the diffusion of amino acid molecules to the site where the protein is being constructed, and the assembly of the protein molecule are all physical and chemical processes. All the players are made up of atoms; they are bound together by covalent bonds, ionic bonds and hydrogen bonds; they take place under the same types of attractions and repulsions that drive every other chemical process; and they absorb and give off energy in predictable amounts. But, oh, we have just begun to explore the complexity of the process in the simplest of terms.

As you can see, biochemistry does not consist only of a set of chemicals; it's a set of processes. Instead of speaking of simple chemical reactions, we'll be talking about chemical reaction **pathways** that involve several reactions (sometimes dozens) in concert (see Figure). Some of the more significant overall processes include:

1) growth—increasing cell mass by adding more molecular units to cellular structure
2) reproduction of cells—their walls and membranes, and their genomes (the DNA which contains their genetic codes), and duplication of their contents
3) fueling reactions—using food molecules as a source of energy by two major chemical oxidation pathways, and generation of a special fuel exchange molecule called adenosine triphosphate (ATP)
4) regulation—accepting stimuli from the environment and changing the course of the cell's actions to compensate for changing conditions
5) maintenance—repairing damaged structures by either patches or destruction of old material to be replaced with new material
6) storage—the joining of fuel molecules make them compact form for use during times when food is less abundant
7) uptake/excretion—selective acceptance or rejection of molecules that come to the cell membrane for either inclusion into or exclusion from the cell's contents
8) containment, partitioning and organization—division of the cell's contents into cell compartments for organizational purposes
9) motility—movement of the cell in response to stresses imposed on it by its surroundings

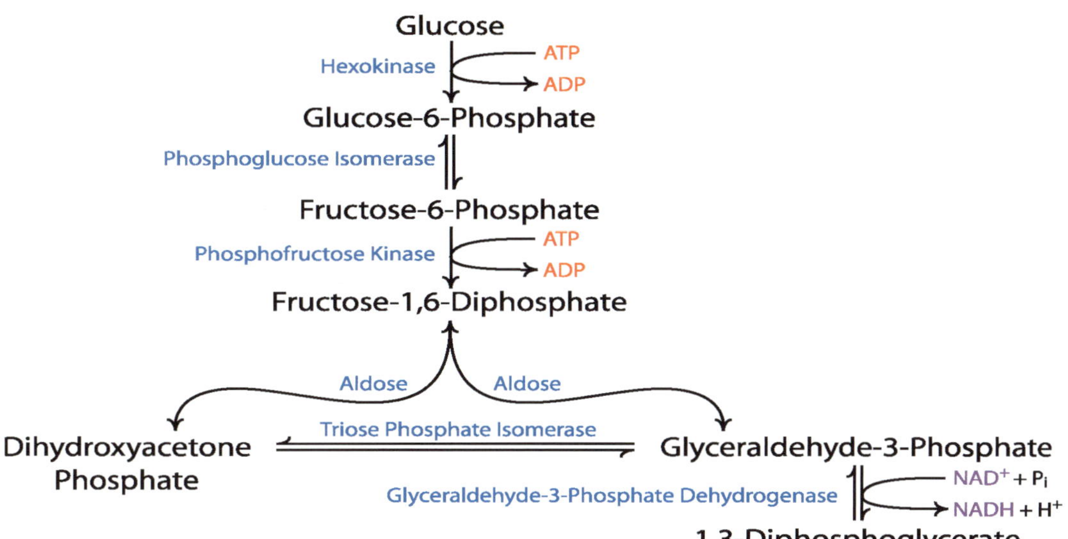


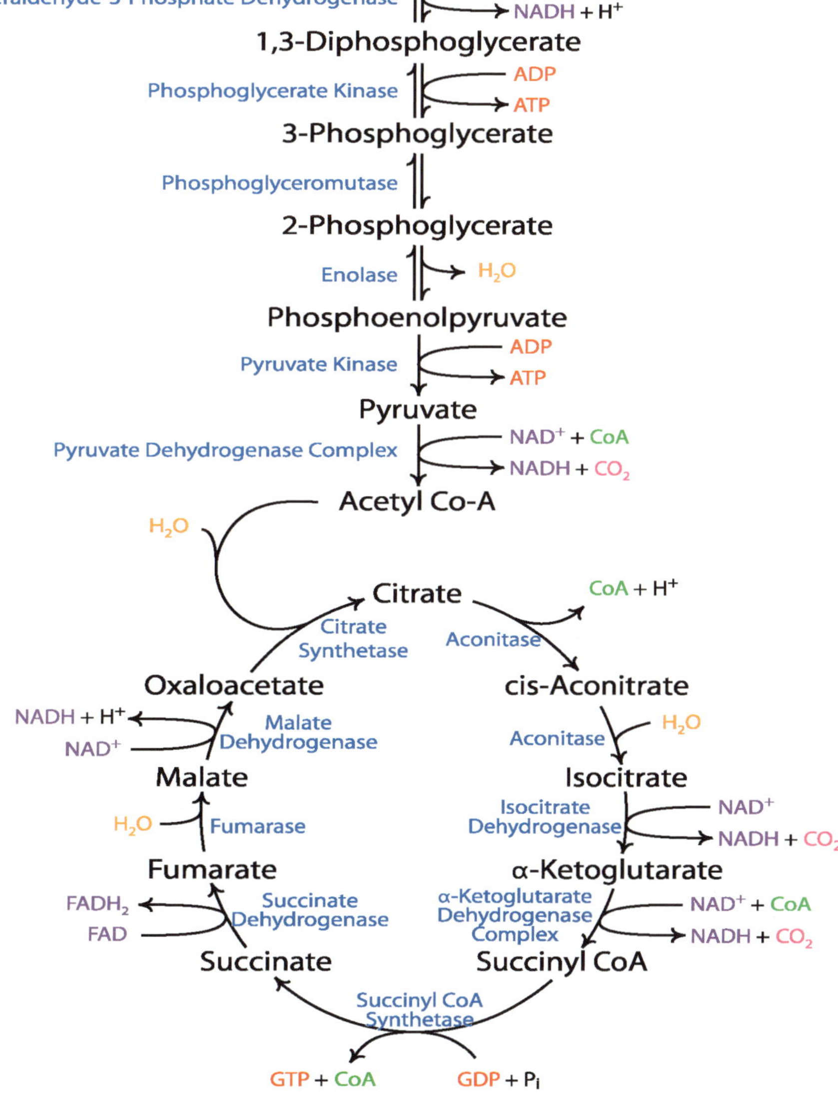


**Figure.** This diagram illustrates two of the most fundamental reaction pathways in all of biology. By fundamental I mean that they are nearly universal--nearly all organisms possess them. They represent the use of glucose to generate ATP and related compounds for energy (in red). They also generate adenine nucleotides (NADH and relatives, shown in purple) which are used by cells as "reducing power" for synthesis (building things). The two pathways involve 20 different enzymes (the names of which appear in blue), most of which act very specifically on the chemical for which they were designed. These two processes are named *glycolysis* (down to the pyruvate molecule) and the *citric acid cycle* (the circular pathway). Welcome out of chemistry and into biology. It's a different world, but it operates off of the same natural laws as the rest of creation.

Diagram design after Streyer, 1981

10) communication—the sending of biochemical/physical signals from cell to cell and over larger distances to coordinate actions (Hormones are chemicals that act in this way.)

Of course, each of these processes is immensely complex in and of itself, but consider a place where all of these processes occur simultaneously. Welcome to the *simplest* of all living things. Welcome out of chemistry and into the world of the living—biology.

The functions of biomolecules are related to their structures. People don't put wood down for a driveway. They use asphalt or concrete. Similarly, lipids are chosen for biological membranes because of their properties. They are hydrophobic, preventing the needless loss of water and ions from the cell. They are fluid, allowing movement of items about the membrane, but disallowing passage across the membrane. The same point could be made for every chemical in the cell. Rarely could another chemical be called in to substitute for the one that is used in a biological system. If there is a good alternative, it's probably being used by some organism.

## Biochemistry Research Areas

Drugs are usually either concentrated and purified natural chemicals or minor chemical modifications of a natural substance from a plant, animal or microbe. When a drug company discovers a new drug with biological activity, they **derivatize** it by attaching different **functional groups** (such as methyl, amino, carboxyl, alcohol or ketone) to it. This gives the drug a range of properties to be further explored and exploited. For example, they may add an atom to it that makes it more difficult to break down and thus allows it to stay in the human body longer.

When bacteria and fungi eat pollutants, they do so because they produce enzymes which are good scavengers of food of all sorts from their surroundings. Their ability to increase the rate at which harmful chemicals are destroyed is based on the activities of their enzymes and their ability to grow and reproduce when they are presented with an abundance of these chemicals to support them. Understanding the processes (process rates, energy use, growth rates, by-products, end products and conditions) by which microbes do these things are topics of biochemistry. To know what an organism *is* and what it *does* might be biology, but to know *how* it does it is biochemistry.

So biochemistry is a discipline that has an impact on biomedical research, genetic engineering, veterinary practice, medicine, environmental cleanup, and even computer logic, programming and semiconductor research. As a biochemist/microbiologist I have been involved in water purification, soil science, space exploration, biomedical research, viral research, organic synthesis and bacterial physiology research. Biochemistry is one of the hottest areas of science today and promises to be for as long as The Lord delays His return.

**Exercises**

1) Cover the names of the chemicals shown in the Figures and see if you can place names on them.

2) Provide the names of the biopolymers that are made from these monomers:

   a) fatty acids
   b) nucleic acid bases
   c) amino acids
   d) sugars

3) What characteristics would be useful for collagen to possess based on what you know about its function? In the Teacher's Helper we tell you some of its actual properties.

4) Ribosomes are necessary structures in all living things that manufacture proteins. The ribosomes are different in humans and in bacteria in ways that have been chemically described. Suggest one way in which we might use this knowledge.

## A Structural Tour of Organic Chemistry

**Figure 1.** The sequence of structures shown on this page is provided to show you how an organic chemist might abbreviate structures for convenience.

**1A.** Here you see two different views of methane. The first one is flat. All of the hydrogen atoms look as though they lie in the same plane. But you know that's not the way methane looks in reality. It really looks more like the second structure. The two hydrogen atoms that have dark, solid bonds are sticking out toward you, and the two that have hatched bonds are sticking into the paper, away from you. This is one way of showing that structures are three dimensional rather than flat.

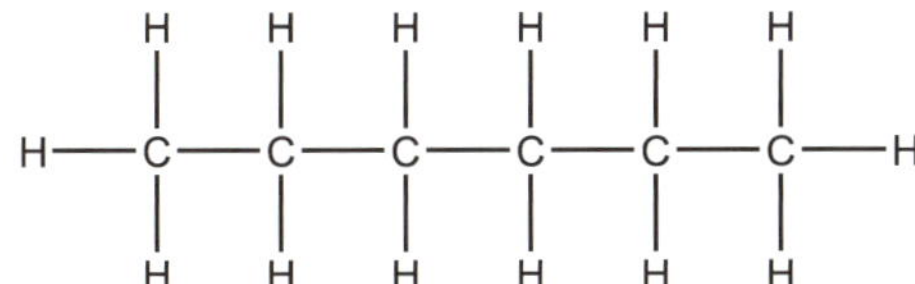

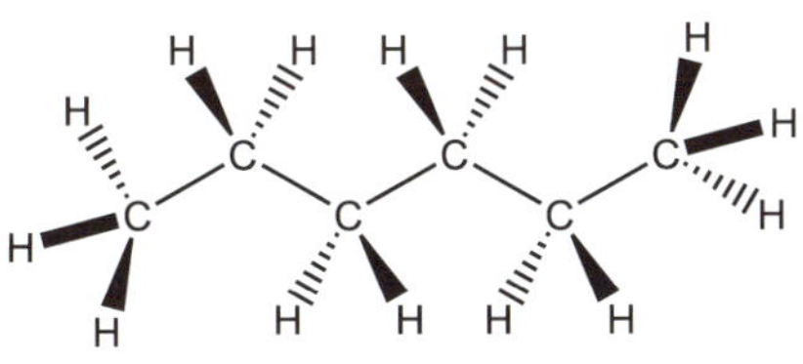

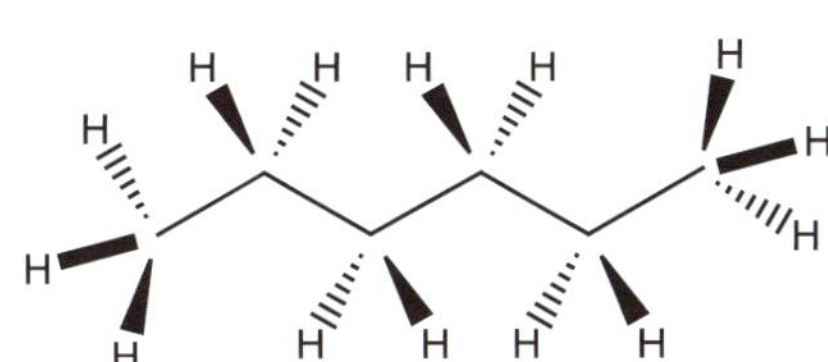

**1B.** If you look at the 3-D structure of hexane (the six-carbon, straight-chain hydrocarbon), the C atoms don't line up nicely as shown in the first drawing, but instead they buckle, making a zigzag line as shown in the second one. Chemists get tired of showing all of those C's every time they draw a molecule. If we can just agree that every time we see a bend in a chain that there is a carbon atom at the corner, we can give up writing them. That's what we do in the third view.

While we are in the mood for simplicity...we all know that carbon always forms four bonds. Can't we just assume that anytime a bond is missing from a carbon atom that it belongs to hydrogen? Then we don't have to draw all those H's, either. If it doesn't belong to hydrogen, we'll show the symbol of the actual element that fills that spot. The last view of the structure at the bottom left is the abbreviated form we'll use from now on. See how simple the structure gets when we can all agree?

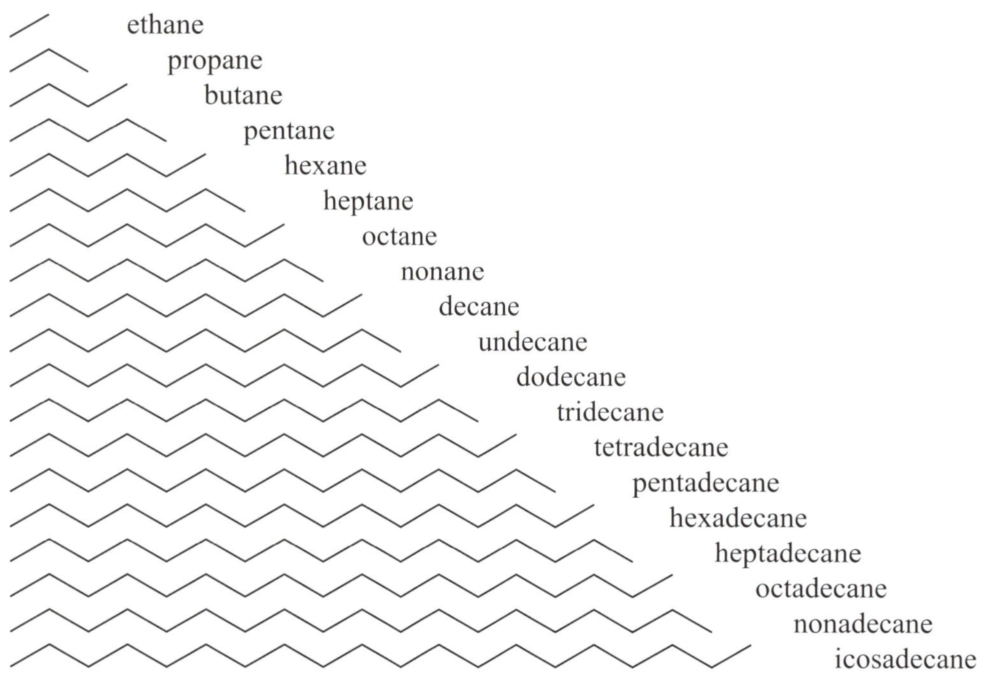


**Figure 2.** Here are all of the "**linear alkane**" molecules (straight-chain hydrocarbons with all hydrogen atoms surrounding the carbon chain) starting with $C_2$ all the way up to $C_{20}$. ($C_1$ is missing because it can't be drawn in this fashion. You have to have two carbon atoms to show a bond between them. Methane is the $C_1$ compound that fits in this series.) Notice how they are named in sequence using the Greek prefixes once you get above $C_4$ (butane). Notice that they all end with -ane.

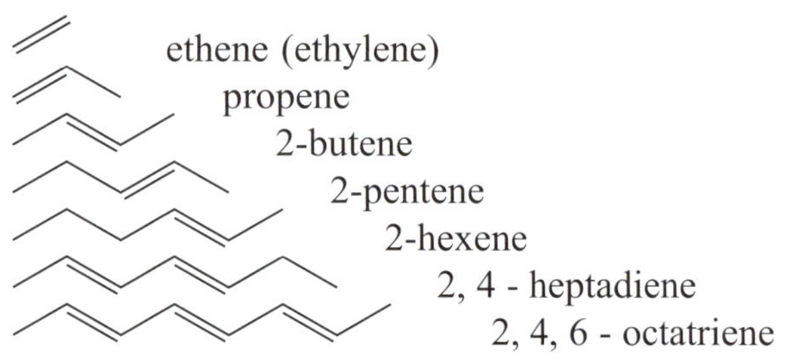


**Figure 3.** Of course, you know already that carbon can form double bonds as well as it can form single bonds. That just leaves one less bond for each of the double-bonded carbon atoms to share with hydrogen. These are all linear **alkenes** because each has at least one double bond. To specify where that double bond is located, you can tell the number of the first carbon atom it attaches to. We always count from the end of the molecule that is closest to the first double bond when numbering carbon atoms. A diene has two double bonds, and a triene has three. Some of these compounds have "trivial names" shown in parentheses.

ethyne (acetylene)
propyne
2-butyne

**Figure 4.** Carbon can also form a triple bond. These **alkynes** are named just like the alkenes.

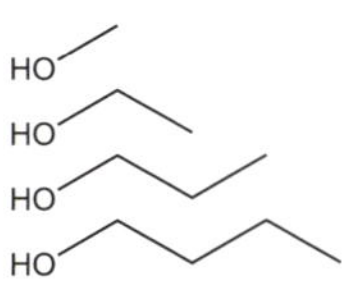


methanol
ethanol
propanol
butanol

**Figure 5.** **Alcohols** are named like the other compounds, but you have to drop the "e" off the end and add "ol." A number preceding the name tells to which carbon the –OH group is attached. Notice that the OH in the first structure (methanol) has a single carbon atom (as in methane) bound to an oxygen atom. If the –OH group is on the second carbon in the chain rather than the first, it's an "iso" alcohol. Isopropyl alcohol ("rubbing alcohol") is 2-propanol.

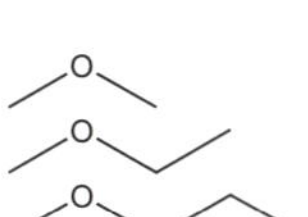


methyl ether
methylethyl ether
methylpropyl ether

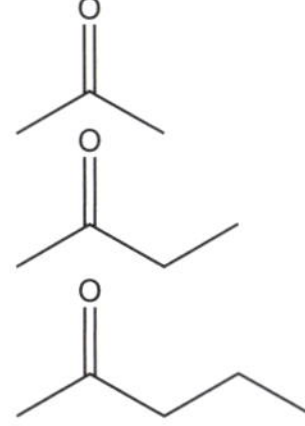


methyl ketone

methylethyl ketone

methylpropyl ketone

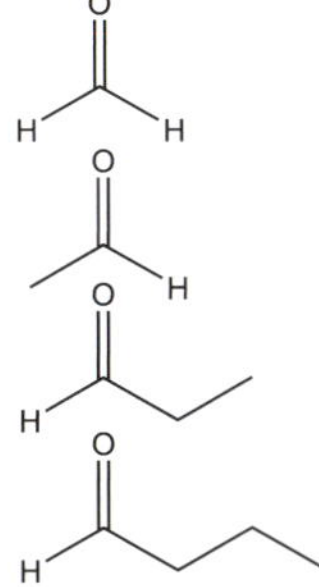


methanal (formaldehyde)

ethanal (acetaldehyde)

propanal (propionaldehyde)

butanal (butyraldehyde)

methylamine
ethylamine
propylamine
butylamine

methylpropylamine

diethylamine

triethylamine

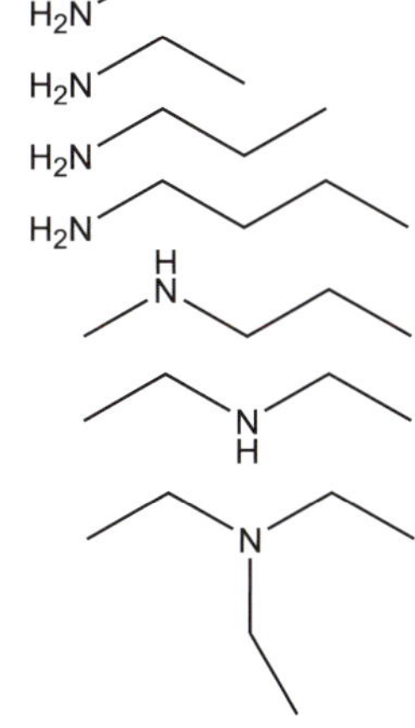


chloromethane
bromoethane
iodopropane

Cl
Br
I

2 - chloropentane

Cl

**Figure 6.** The amino compounds have an $NH_2$ group in them that makes them basic. There are also halogenated compounds that come from placing one or more halogens in place of a hydrogen on a hydrocarbon chain.

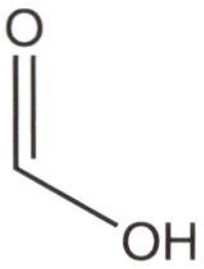

methanoic acid (formic acid)

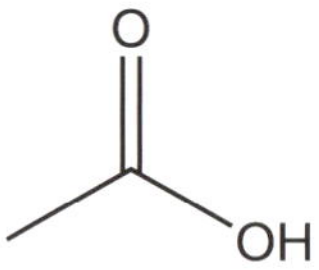

ethanoic acid (acetic acid)

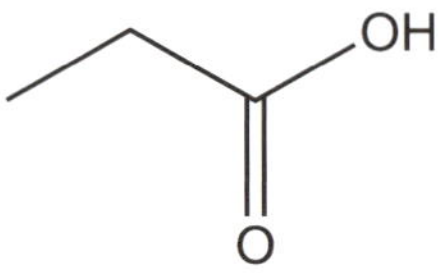

propanoic acid (propionic acid)

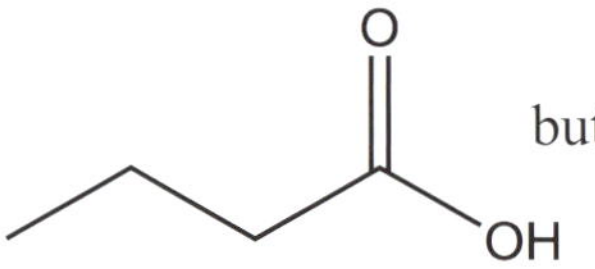

butanoic (butyric acid)

**Figure 7.** Of course, if there are organic bases, there must be organic acids. Those shown are carboxylic acids. Each has a COOH group on it. Acetic is (in vinegar) probably the most famous. Citric acid and ascorbic acid (vitamin C) are also famous carboxylic acids.

methanoate (formate)

ethanoate (acetate)

propanoate (propionate)

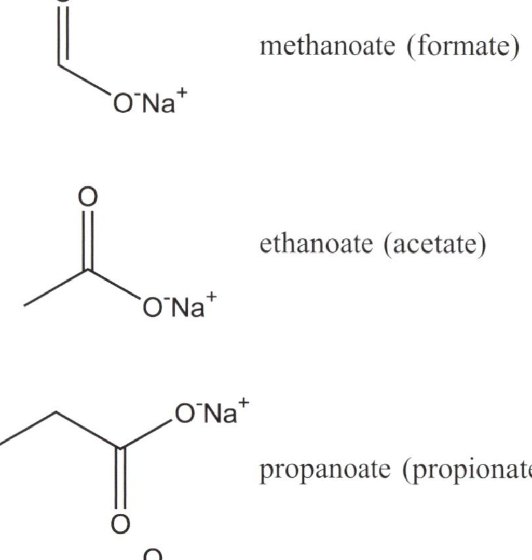

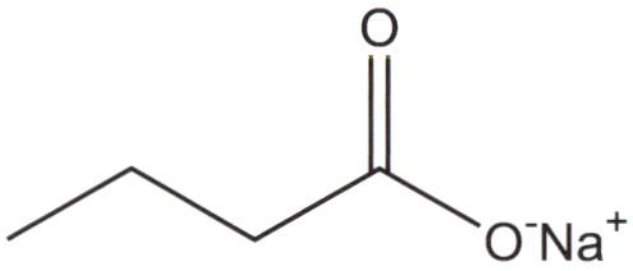

butanoate (butyrate)

**Figure 8.** When a carboxylic acid is neutralized by a base, you get a salt. If the base is sodium hydroxide, you get these sodium salts. Notice how their "-oic acid" endings are replaced by "-oate." Each of these compound names might (more formally) appear preceded by the word *sodium*.

**Figure 9.** Just when you think you know organic chemistry, carbon chains pull a fast one on you. The chain chases its tail and makes a cyclic compound. Here are a few examples. The most common of these rings (because of its greater stability) is cyclohexane. It occurs in two different geometries--affectionately the "chair" and the "boat" formations. All of these compounds are present in petroleum fuels. There are also double bonded cyclic compounds like those at the bottom right.

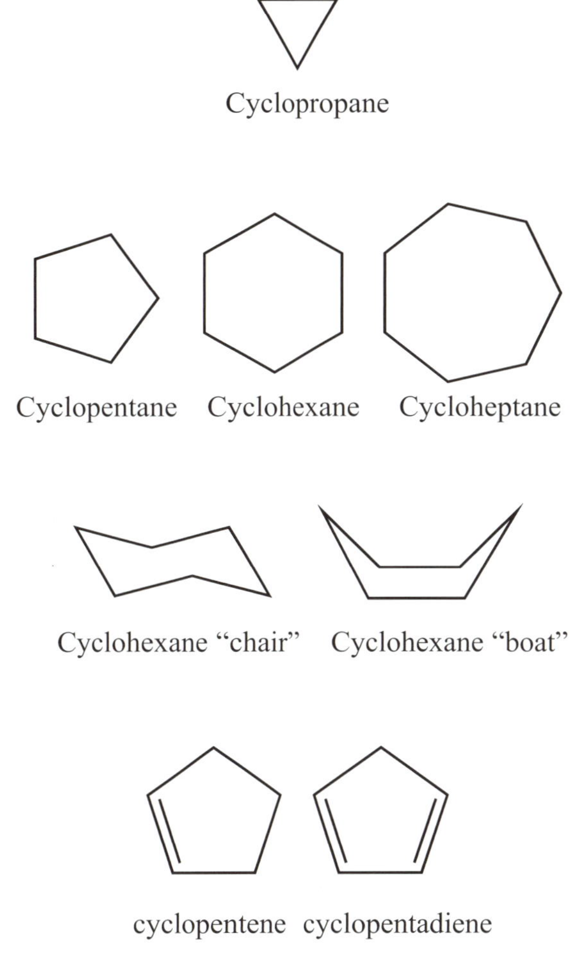


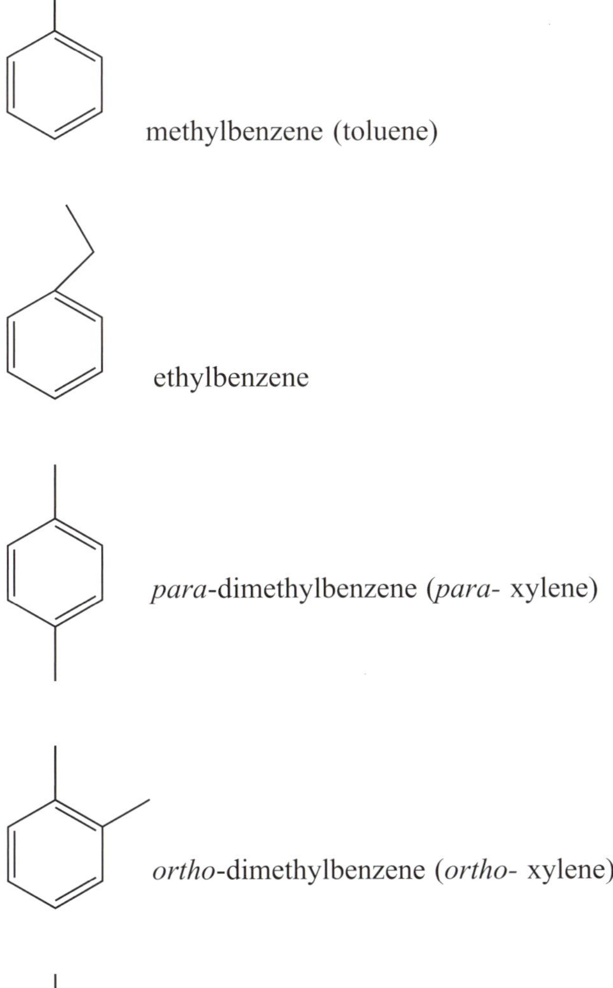


**Figure 10.** A six-membered ring can be strengthened by the presence of double bonds. You have already become acquainted with a benzene ring ($C_6H_6$), which has three double bonds that "resonate" to give the molecule uniform strength around its ring. Those benzene molecules can be "substituted" with carbon atoms and carbon chains. Here are some of the best-known "alkylbenzenes." They are among the most common "aromatic" (ring) compounds in gasoline.

**Figure 11.** Hydrocarbon chains are not the only thing a benzene ring can be substituted with. Here is a hall of fame for substituted benzene compounds. These are the work-horses of synthetic organic chemistry. If you learn how these compounds might be treated to manipulate their structures, you can make most any kind of structure. That's what the field of synthetic organic chemistry is all about.

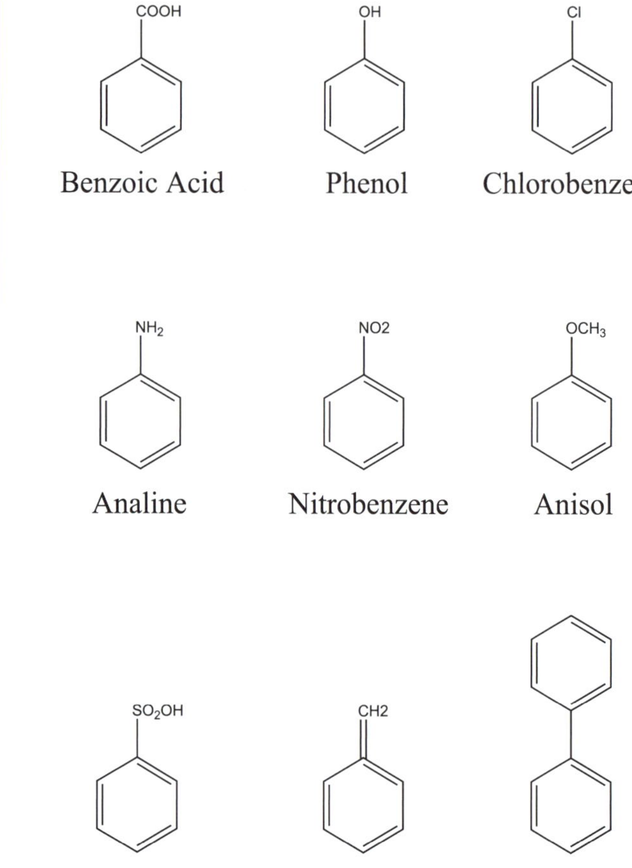


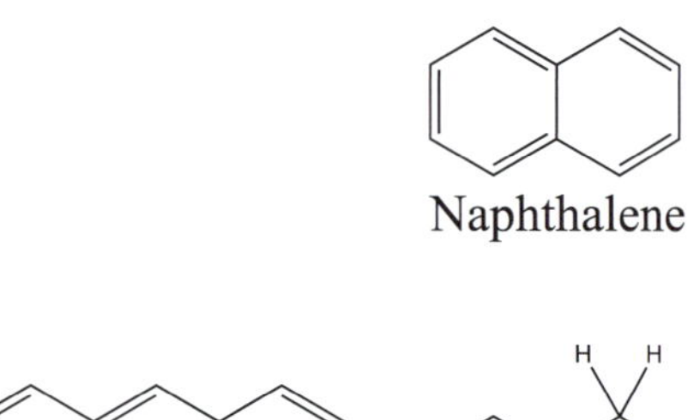


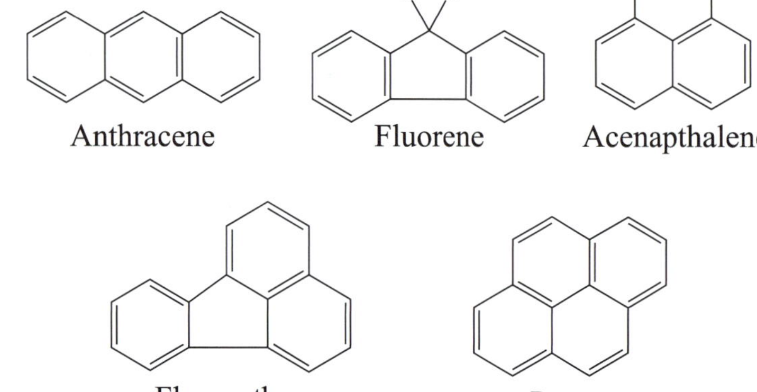


**Figure 12.** Benzene rings can also be joined together into large complex molecules. Molecules like these having multiple ring structures are referred to as polynuclear aromatic hydrocarbons--PAH's, or PNA's. Many of these are cancer-causing agents. They are found naturally in tar and creosote.

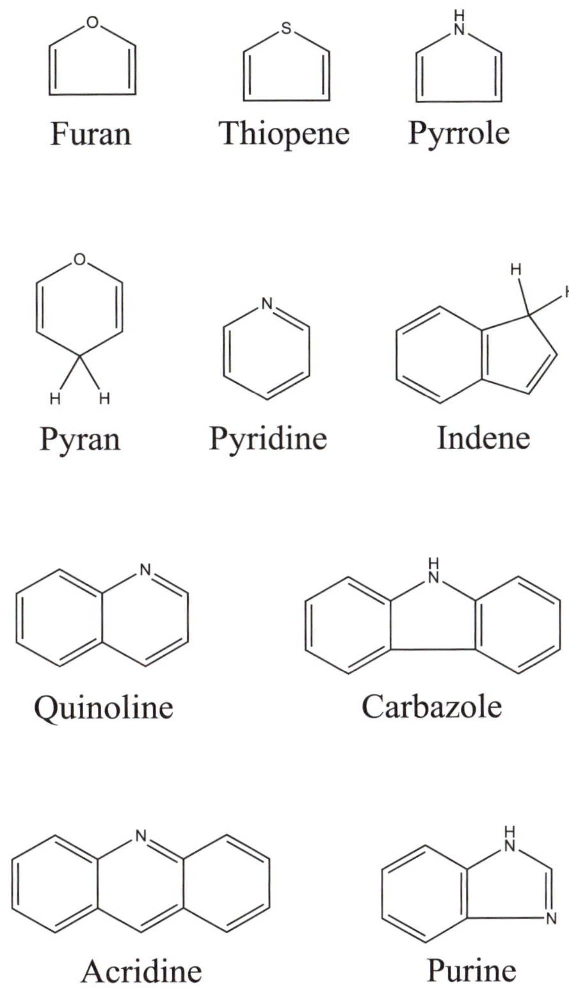


**Figure 13.** We end our pictorial tour of organic chemistry by showing you structures of several so-called heterocyclic (from *heteros*, meaning "other" and *cyclic*, meaning "round") compounds. They are odd cyclic structures. Of course, every compound in this chapter has its own unique chemical and physical properties. Just imagine the vast assortment of chemicals you invent by combining different ones. Of course, you'll have to take a college course in organic chemistry or read some books on your own to learn how.

Don't try to memorize all these structures unless you plan on being a professional chemist. If so, you can go ahead and get started. It'll save you the effort later.

## 65: Chemistry and Evolution: An Essay

In the study of chemistry, you may have noticed the lack of focus on evolutionary theory. You might consider this intriguing.

It isn't that chemistry has no place in an attempt to support the theory. In fact, one of the greatly touted successes in the support of chemical evolution was a study in biochemistry. That was the study by Stanley Miller in 1952 that yielded amino acids from a proposed "primordial soup." In fact, chemical evolution is absolutely crucial to Darwinian evolution, for without development of biomolecules through *chemical* evolution, *Darwinian* evolution of living things is a non-issue. No evolution can take place where there is no complex arrangement of molecules in existence that can give rise to life. That is, evolution isn't evolution until something exists for it to act on.

Why then do chemistry texts bypass discussions of the chemical mechanisms of evolution? I propose that it is because there is little to discuss. The attempts that have been made to show the reproducible and stable production of living things, or even reasonable precursors to living things, were thwarted in the years subsequent to the highly publicized Miller experiment. We are not substantially closer to spontaneously forming a living being of the most modest sort from nonliving molecules than we were fifty years ago.

Yes, I am aware that the Miller experiment has been repeated using a number of different "soup" recipes and that different molecules can be synthesized in this way. But we know too much about the complexities of cells to be held captive by the fairy tale that a flask full of monomers will ever produce anything that can reproduce itself. (Reproduction is, after all, the very *material* definition of life.)

Attempts at making chemical data fit the evolutionary model have failed. Calculations on the probabilities of forming substantial concentrations of biopolymers (proteins and nucleic acids) render evolution exceedingly unlikely. Calculations of the decay rates of these molecules under conditions in which they might form render evolution all but impossible.

The only thing that has been left undone to sew up the case against chemical evolution is to provide sufficient time for the failed attempts to become a matter of public embarrassment. After all if we achieve through intelligence that which we declare requires no intellect, then what have we proven even if we should generate life? If we succeed, we have proven only that life can be formed by intelligence, not that it can be achieved without it. On the other hand, if we never achieve it, at what point in our laboratory studies will we all agree that it isn't going to happen? Are there criteria for failure to prove that the theory is dead? Or do we simply insist that it is true despite the lack of proof and wait forever for it to happen?

The inputs of energy that would be necessary to drive evolution are the very inputs that prevent it. Evolution is like a johnboat pushed into the Mississippi River in New Orleans ending up at the headwaters in Minnesota. Against the boat runs the mighty current of entropy—the tendency of all organized systems to decay. In its favor, it has only the occasional wind of energy which blows discontinuously, and mainly from west to east. It has no sail formed by organized thinking. If it had, it would be of no assistance—there is no intelligence to guide it. The boat will be seen in the Gulf of Mexico until it finds its way into the world's ocean where it will perish in the depths.

The voyage north to Minnesota is not impossible, it's just unspeakably unlikely. If I find a boat in central Minnesota, I'll never assume it came there on its own from New Orleans. It's much more plausible to believe that some intelligence put it there.

You will hear about chemical evolution in biology class, and that same old experiment done in the mid-1950's will be cited. But you won't read much factual information about evolution. Especially not in a chemistry text where such information is most desperately needed if evolution is to be believable.

**Figure 1.** This boat is not on its way to Minnesota. It'll never make it there on its own. Not in a billion years--neither it nor a billion other boats just like it.

**Figure 2.** The journey from simple chemistry (left) to the chemistry of life (right) is far more difficult than a 2000-mile trip up the Mississippi River. The current that moves against complex chemistry is a stronger, more permanent influence than the strength and direction of the river. That influence is called entropy--the tendency of all organized systems to come undone.

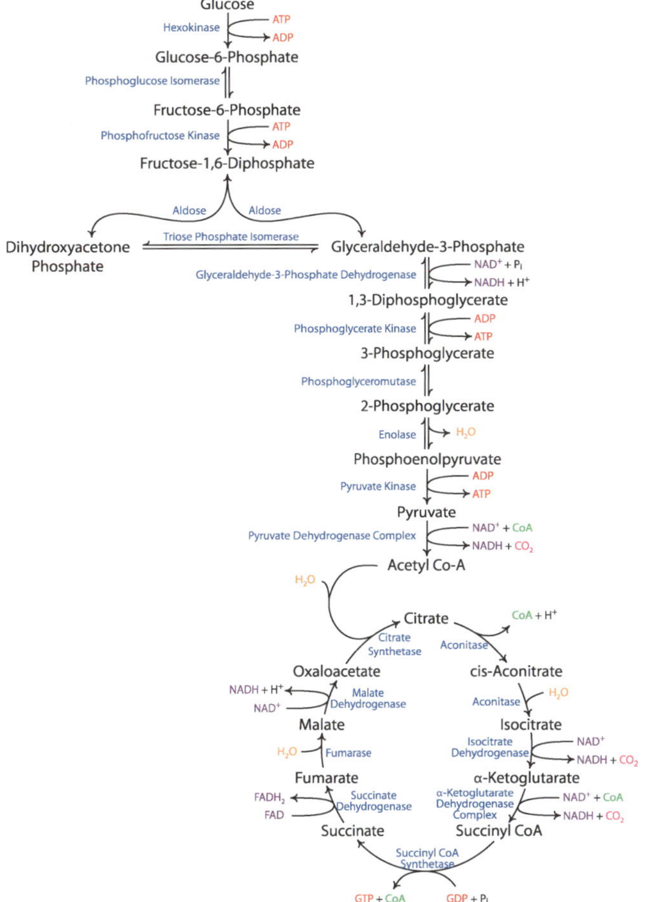


The authors wish to personally congratulate all who finished this course. If you did well, you should seriously consider pursuing a career in the sciences and/or perhaps teaching your own children so that they might carry the torch into the next generation. The world needs outstanding scientists, and we think you'd be some of the best. The world also needs more Godly people who have a solid technical foundation for their faith. These will be the ones who carry our society to its next level of scientific achievement. Seek after God with all your heart and He'll be found by you.

## Method 1: The Syringe

### General Information

A syringe has several functions in laboratory and medical settings. It is most often thought of as a device for injecting patients with medications, but the same features that make it useful for measuring and injecting medications make it useful for other functions as well.

First of all, a syringe allows you to take up a liquid by suction. When you pull on the plunger the cavity inside the syringe has a pressure that is lower than atmospheric pressure ("vacuum"), so liquid will enter to fill the demand of that lower pressure. For this reason, the syringe is used in any setting where suction is required. I keep syringes around the house to start siphons, remove water from the lawn mower's gasoline tank, meter oral medicines (a different syringe, of course) and for anything else that requires suction.

Second, a general-use syringe is a measuring device of moderate precision. It is more precise than a beaker or a flask, about the same as a graduated cylinder, but usually less precise than a finely graduated pipet. (You can purchase syringes that are as precise as any pipet and that measure down to low microliter quantities with extreme precision. These are often used in analytical chemistry.)

Third, while most measuring instruments rely on gravity to deliver their liquid, the syringe forces it out under variable pressure controlled by your own hand. This makes it useful for delivering even viscous fluids, or where resistance ("backpressure") is expected. (I use them on occasion for delivering glue, caulk and sealants. After all, a caulking gun is just a form of a syringe. See Figure 1.) There are also syringe pumps that are designed for mechanically depressing the syringe plunger so the syringe delivers its liquid in a controlled manner over time (Figure 2). For example, using a syringe pump, you can deliver 100 cc (or $cm^3$) of fluid over a 12-hour period at a constant rate.

Syringes can be purchased that are glass, having Teflon plungers, so they are chemically inert. They can be made "gas-tight" to use with gaseous materials without losses. They are readily available in volumes from a few microliters to around 100 mL. They may be disposable, single-use devices that cost a few dollars, or glass/Teflon/stainless steel devices that cost hundreds of dollars and may be used indefinitely. Some are sterilizable, so they can be reused after they are exposed to infectious agents.

### Syringe Methods

**The "No-Headspace" Method.** There are two methods for using syringes. The first is to do what most people do—draw the liquid in and expel it. This method is fast and easy, but it is not entirely without problems. The first problem is that syringes, by design, draw in a little air with the fluid (Figure 3). This air bubble, once inside the syringe, can be expelled before the liquid is measured if you proceed as follows:

1) Draw in more fluid than you will need.
2) Point the syringe tip upward and draw in some air. This will also draw in the small amount of liquid that remains lodged in the tip.
3) If the air bubble is stuck to the plunger, tap on the syringe barrel to dislodge it so it will float up to the tip of the syringe and burst.

**Figure 1.** A caulking gun is a type of syringe. Instead of putting pressure directly on the "plunger" you put pressure on the lever, and that pressure is mechanically transferred to the plunger. The tube of caulk is like the barrel of the syringe, except that, in this case, it comes pre-filled.

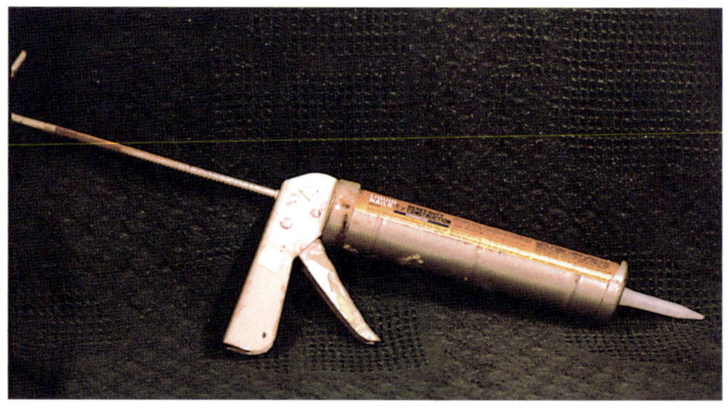

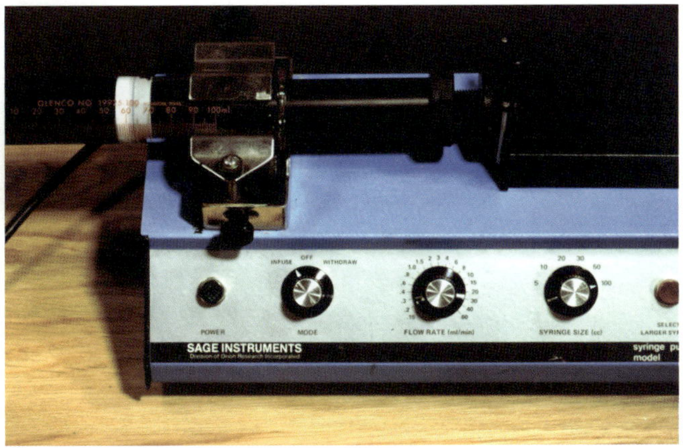


**Figure 2.** This syringe pump is designed to deliver any fluid in the syringe at a uniform, set rate.

4) Hold the syringe level and expel the air by pushing lightly on the plunger until all the air is expelled and the liquid is just at the tip. (See Figure 4.)

Once the air is expelled, it's no longer necessary to hold the syringe upside down. You can push out the excess liquid and measure just the amount you need using the graduations on the barrel.

The tip of the plunger is pointed so that its shape conforms to the shape of the barrel when the liquid is expelled. When measuring the liquid, measure by the rim of the plunger where it meets the barrel. Don't use the point. (See Figure 3.)

When the liquid is expelled from the barrel, a small amount of liquid will remain in the tip. This liquid is not included in the measured volume, so there is no need to work to get it out of there. However, if you are rinsing out the syringe to reuse it, remember that this small amount of fluid remains from your previous measurement. Some procedures require that your liquids be of high purity. You'll have to be sure this fluid is removed and your syringe is rinsed before it is reused.

**The Headspace Method.** While the no-headspace method works well for some applications, there are times when expelling liquids straight up is not acceptable. I've spent many hours dispensing strong acids, caustic solutions, damaging solvents and radioactive fluids with syringes. If great precision of measurement is needed, you can't live with the air bubble and you can't squirt it into the air. In these cases, the headspace method can be the solution to many problems. This is what you do:

1) Draw some air into the syringe. When using a 5-cc syringe, you might consider drawing half a cc or more.
2) Draw the fluid into the syringe and hold it straight up and down as you pull in just the right amount of fluid. Determine the amount of fluid you've

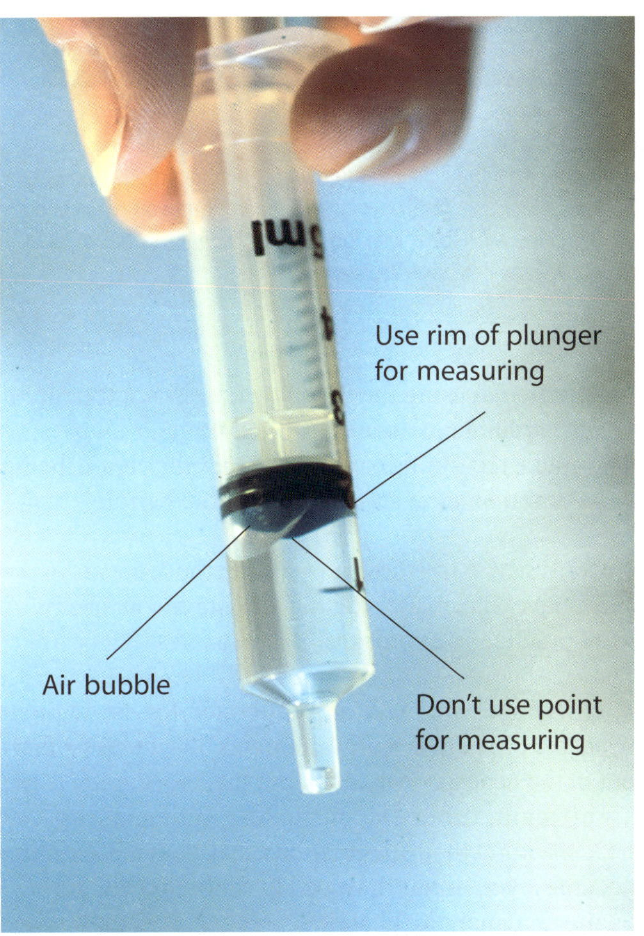


**Figure 3.** Syringes always draw in a little air with the fluid the first time they are filled.

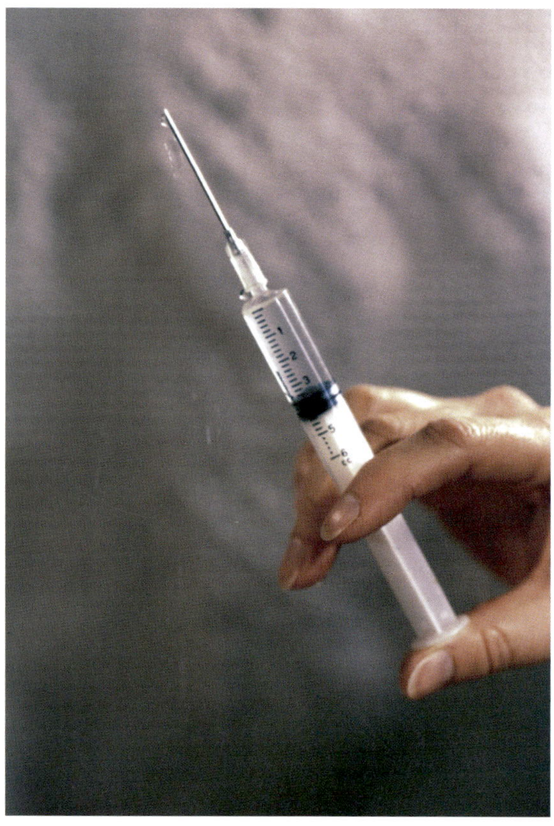

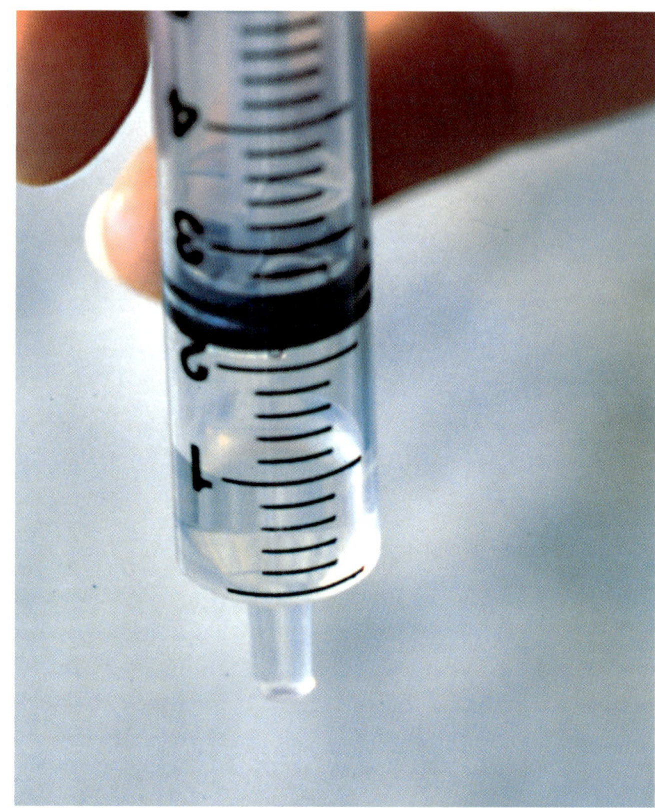

**Figure 4.** If the fluid you are using is not hazardous and not valuable, you can always draw in more than you need and expel the air bubble with the excess. Nine times out of ten, this is how it is done in practice.

pulled in using the fluid level itself, not the rim of the plunger. (Figure 5.)

3) With the syringe upright, expel the liquid being sure that none stays behind in the barrel.

**Note:** When using the headspace method, the air in the headspace can make volume measurement tricky. A little practice is in order.

Syringes are made and calibrated for use without headspace. There is a tiny amount of inaccuracy inherent in the headspace method, but that inaccuracy is not significant for most purposes. When using the headspace method in situations that demand extreme accuracy, you might first need to calibrate your syringe against an analytical balance.

**Figure 5.** (Above) Using the headspace method, you use the fluid level rather than the plunger to gauge the amount of fluid in the syringe. (Below) Syringes come in incredible variety for different purposes. Below you'll find (from top to bottom) an all-glass and Teflon 100-cc syringe for use with reactive materials; the same in a 10-cc variety; a "gas-tight" syringe for use with gases; a sterilizable, glass syringe; a small-volume (5 microliter) syringe for analytical chemistry using sensitive instruments; and a disposable "tuberculin" syringe. They range in price from a few cents to several hundred dollars.

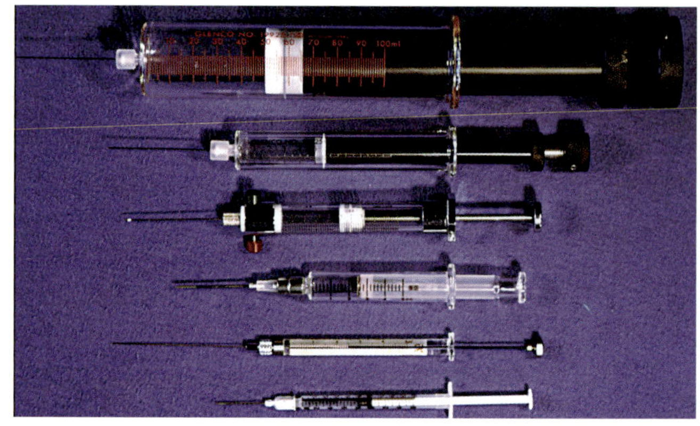

## Method 2: Choose Your Weapon

Measurements are such an integral part of science that you can hardly think of doing anything in the laboratory without some basic knowledge of measuring devices. Generally when we think of the variety of measuring tools the vast majority are for measuring liquids—solids are nearly always measured on an analytical balance. Because we can dispense with the discussion of the measurement of solids in such few words, we'll tackle it first.

### Measurement of Solids

Because solids come in such a variety of forms, mass is the common standard for their measurement. Few other methods are ever used. Since the 1980's, analytical balances have become exceptionally easy to use. Most models have but a single button to push unless some specialized feature is being used. The technician places a receiving vessel of some kind (weigh boat, aluminum dish or piece of glazed paper) on the balance and pushes the "tare" button. This causes the balance to read 0. Then the solid substance is added to the receiving vessel until the appropriate amount is reached. Because balances have become so convenient to use, many workers prefer to measure out liquids this way as well.

The old triple-beam balances that I used when I was in school are now all but obsolete. They are as costly as a simple digital, electronic balance today, and they are much more difficult to use and maintain. While they are not entirely without educational value, using one is like using an old slide rule instead of a calculator. There is little reason to use one.

The primary choice from among analytical balances is based on (1) the precision of the balance, (2) the capacity of the balance and (3) the price. An inexpensive ($100) balance may be purchased that has a capacity of perhaps a kilogram and a precision of ± 0.1 g. At slightly higher prices, balances can be purchased that have similar precision but have much higher capacities. Expensive balances (a *few thousand* dollars) have capacities in the tens of grams but have precision on the order of a few micrograms. One of my favorite activities with people who visit my lab is to pluck a hair and weigh a 1-cm segment of it on the sensitive analytical balance. So far, centimeter for centimeter, my wife has the heaviest hair of all my visitors. My balance weighs with a precision of 0.00001 g (10 micrograms). (See the photographs in Lesson 3.)

### Measurement of Liquids

The number of devices for measuring liquids is practically endless. Different subdisciplines of science sometimes have measuring devices that are made specifically for them. Sometimes out of dissatisfaction with the range of available devices, we make our own. However, there are a few old lab standards that every scientist is familiar with. You should learn the basic tools and how to select from among them.

### Glass or Plastic?

These days most measuring devices come in both glass and plastic varieties. How do you decide which to use? Glass is needed in cases where there is concern that plastic might react with the chemicals in use. Few materials are as inert as glass. Glass is also clear and may be made to exacting standards of optical clarity for use with instruments that are demanding of such clarity. The raw material for glass (sand) is abundant, although glass is still somewhat more expensive than comparable plastic.

Plastic, however, has several advantages to its use. Plastics can be made virtually indestructible for practical purposes, so its better to use it when you can to avoid the expense and hazards of breakage. The sound of breaking glass is all too common around a professional science laboratory. Rarely does it result in injury, but occasionally a glass container of a hazardous material is broken and lots of distress follows.

With plastics, you get different advantages from different products. Some (e.g., crystalline polystyrene) are transparent while others are translucent or even opaque (polyethylene and fluorocarbon polymers). Some plasticware devices (esp. vinyl plastics and poly-

Figure 1. Nearly every kind of laboratoryware comes in glass and plastic varieties. The greatest problem with glassware is that it is breakable. Different plastic devices have different limitations and different prices associated with them.

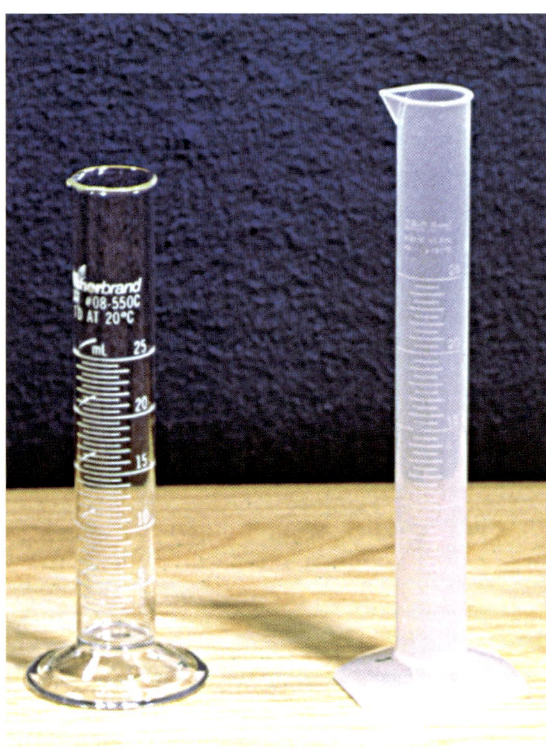


styrene) are so inexpensive that they are made for single use. Others are expensive because of their coveted properties. (For example, fluorocarbons are extremely chemical resistant.)

## Kinds of Devices

There are two broad categories of measuring devices for liquid: measuring/delivery tools and vessels (for simply containing liquids).

## Measuring/Delivery Tools

**Droppers.** The simplest of all devices used for containing and measuring liquids is perhaps a **medicine dropper**. Most are not even graduated. Nevertheless, they may be used as measuring devices that deliver volumes with great accuracy in increments of a single drop. Because every style of dropper makes drops of different sizes, each dropper must be calibrated to determine the volume dispensed with each drop. Or, more often, the user will determine the approximate number of drops in a milliliter of liquid so the dropper can be used to dispense fractional milliliters or (at most) a few milliliters with rough precision. Akin to the dropper is the "**transfer pipet**" or "**Pasteur pipet**" which is a simple dropper-like glass or plastic device—sometimes graduated.

Figure 2. Droppers and transfer pipets are most useful when you don't need to measure a precise quantity of a small volume of liquid. They will typically hold a milliliter or two. These turn out to be among the most used devices in the laboratory.

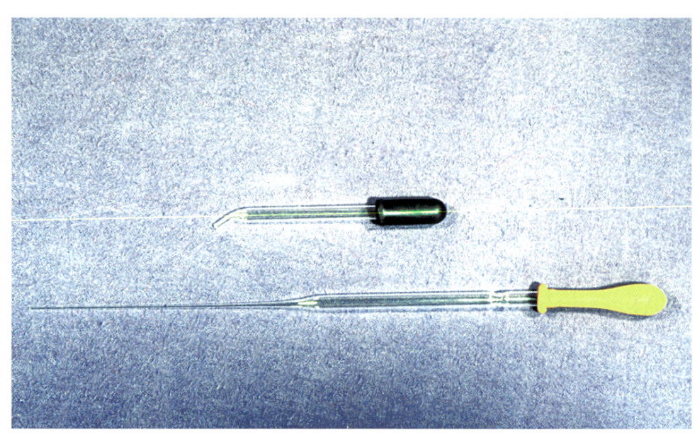

**Pipets.** Besides transfer pipets, most pipets are graduated finely for high-precision measuring of small volumes. In addition to these small-volume, high-precision devices, there are **volumetric pipets** (Figure 3) that show only a single graduation for measuring a single, larger volume, but they do so with great precision. These are often used in making up analytical standards by which other measurements will be calibrated.

Figure 3. The volumetric pipet is a most accurate way to dispense one specific volume of liquid.

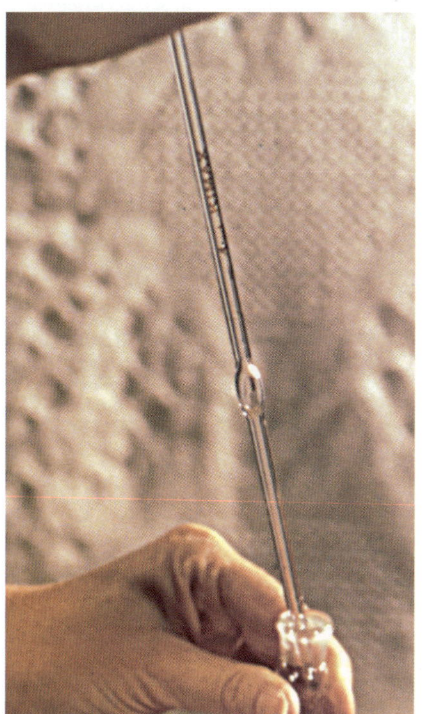

**Micropipets.** The ability to mold glass in high precision has given scientists the ability to manufacture small-volume devices with high precision. **Glass capillary tubes** can deliver single-volume quantities of liquids in microliter quantities. If the use of plastics is suitable, **micropipettors** have also been made that deliver a select amount of fluid with high precision, even down to single-microliter quantities (Figure 4).

**Syringes.** As described in Method 1, syringes are made for measuring volumes from a few microliters up to about 100 milliliters. The precision of larger syringes is not on par with comparable volumetric pipets, but are generally more precise than the large-volume devices discussed later. The unique and often useful trait of syringes is the ability to deliver the liquid they contain under pressure. See Method 1 for additional information and use instructions. (See photographs in Method 1.)

**Graduated Cylinders.** The graduated cylinder is most useful for measuring volumes on the order of a few tens to a hundred milliliters (Figure 5). For a few large-volume applications, I have graduated cylinders that hold volumes up to four liters. Rarely do I use the smallest of my cylinders which measures out 5 mL with pretty high precision. It's too tedious. Pipets are usually more useful in this volume range. Cylinders measure volumes generally with mid-range precision, comparable to syringes, because their diameters tend to be about the same. You might even think of a syringe as a graduated cylinder with a plunger.

**Burettes.** Developed for applications in analytical chemistry, the burette is a device that is designed for measuring variable volumes of a few tens of milliliters with high precision. (See Figure 6.) The real utility of the burette is in cases where the precise amount of liquid to be used is not known ahead of time. It is usually used with a color indicator that indicates when the volume of liquid dispensed is the right volume. It allows the worker to calculate, after the fact, how

Figure 4. Small variable volumes can be dispensed using a micropipettor. They can be purchased to cover a wide range of volumes from a few milliliters down to a few microliters. When used with care they can be extremely accurate and precise.

**Figure 5.** Graduated cylinders come in a wide variety of sizes from a few milliliters up to several liters. The most useful ones are in the range of fifty to a hundred milliliters.

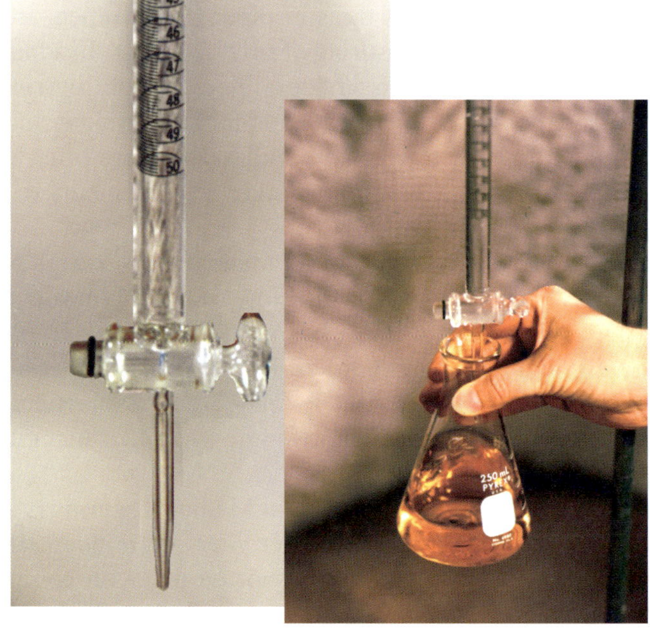

**Figure 6.** The burette is used in analytical chemistry using color-change indicators to determine when the proper volume has been dispensed. That volume can be determined after the fact by the difference between beginning and ending volumes on the graduated scale.

much liquid has been "**titrated.**" This will become your nearest companion during your first semester of analytical chemistry in college. The photograph shown is of the tip of a burette that is almost a meter long.

**Volumetric Flasks.** Volumetric flasks are available in a wide range of volumes (Figure 7). Although they are included among the mid-range tools, they may hold as little as 10 mL or as much as several liters. They have single volume marks etched on their very thin necks. This permits the measurement of even fairly large volumes with the precision of a smaller device.

**Figure 7.** These are volumetric flasks that are used to do precise dilutions. The ones shown range in capacity from 10 to 1000 mL.

## Vessels

Often times high precision and high volume don't go together. When they do, large volumetric flasks are typically used. (See previous.) The other tools in this section are typically low-precision devices. Typical precision might be ± 5% if they are even graduated at all.

**Figure 8.** Beakers are a laboratory icon. They don't measure with high precision, but they are easy to use for quick tests. If you need to store something a reagent bottle (left) is a better choice.

**Beakers.** Beakers are general-use glassware for holding volumes on the order of a few tens to several hundred milliliters. They are made to contain volumes as small as a few milliliters to several liters, but they lose their usefulness at the extremes. Because of their wide open mouths, they are used as reaction and mixing vessels, but because they lack covers they are usually for short-term use. For long-term storage or use, various bottles and jars are available with or without crude graduations.

**Flasks.** There are many types of flasks designed for different purposes, but the workhorse is the **Erlenmeyer flask**. This flask is designed with angled walls for applications where some swirling will take place to mix liquids. This turns out to be most of the time. Erlenmeyer flasks may come threaded for a screw-cap or with a "ground glass" fitting for insertion of a ground glass stopper. They are also suitable for rubber or silicone stoppers. Because they are flat on the bottom, they sit on the countertop and fit into the sockets of a shaker table. With stoppers and glass tubing you can use flasks to make all kinds of reaction vessels. Other flasks have round bottoms, stopcock closures, "baffled" bottoms (to increase aeration when they mix) and all kinds of built-in specialty features.

**Side-Arms.** Because of its usage, one particular type of Erlenmeyer-style flask is worth mentioning. Its use is so specialized that it deserves its own heading. The **sidearm flask**, or **vacuum flask**, has a hollow arm at

Figure 9. Erlenmeyer flasks have slanted walls that make them useful for swirling liquids without spilling.

the neck for connection of a vacuum hose. This is particularly useful for drawing a vacuum on a filter funnel to hurry along a slow filtration.

**Test Tubes.** Test tubes are useful for doing lots of repetitive work or for doing simple tests. They hold small volumes and can be placed in racks by the dozens. This allows you to try a lot of different experimental variables at the same time. (Figure 10)

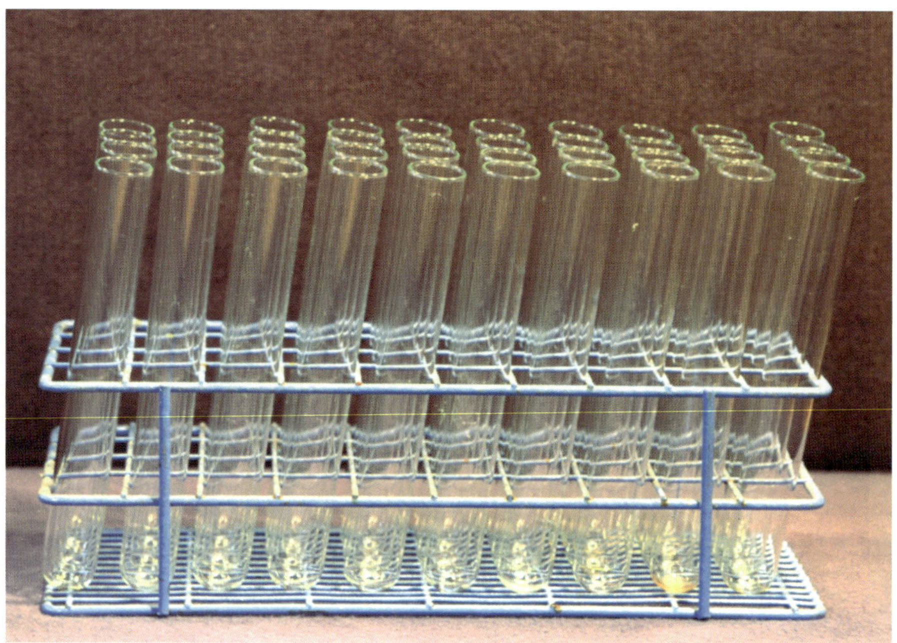

Figure 10. Test tubes are convenient for doing large numbers of small-volume tests.

# Method 3: Filtration

Filtration is an everyday occurrence in just about any laboratory. It has far more uses than you might even expect. Sure, it's used to remove solids from liquids, but did you know that filtration can also be used to improve separations between liquids? Did you know that it can also be used to remove gases from liquids? These days, special "in-line" filters can remove bacteria-sized particles from liquid or gas streams. Ultrafiltration and reverse osmosis filtration can even be used to remove dissolved solute molecules from liquids.

It all begins with understanding basic filtration with a paper filter. The paper filter is a series of fine fibers pressed together. When a liquid containing a solid is applied to the surface of the paper, gravity pulls the liquid through the fiber mesh. Particles are too large to pass through the network, so they get trapped. Liquids work their way among the fibers to the other side of the paper. Then they drip down the funnel into the receiving vessel. That's filtration in a nutshell.

## Selection of Supplies

There are more types of filters available through scientific supply houses than we could even list on this page. Selection of filters is based on (1) what substances are involved and (2) what pore size is required to remove the fraction of the material to be removed. A "fluted" filter (one that has been folded, accordion-style as shown in Figure 2) won't stick to the walls of the funnel. That can improve filtration efficiency in cases where the filter might tend to get plugged up with particles.

Selection of the appropriate funnel also depends on the type of work you are doing. A glass funnel is required if you are using organic solvents that attack plastic. Special funnels are made for high-temperature work and for filtering under vacuum. Specialty filter products sometimes require specialty "funnels" (some of which no longer look like funnels).

The **receiving vessel** or **collection vessel** is the vessel that will collect your **filtrate**—the material that passes through the filter. This vessel is nearly always made of glass so that we can look in on the progress of the filtration and on the clarity of the filtrate. Since the funnel and filter are placed on top of it, the weight of glass (rather than plastic) is helpful in keeping it from getting knocked over. A filter flask is used if we want to apply vacuum to the underside of the filter and hurry the filtration along.

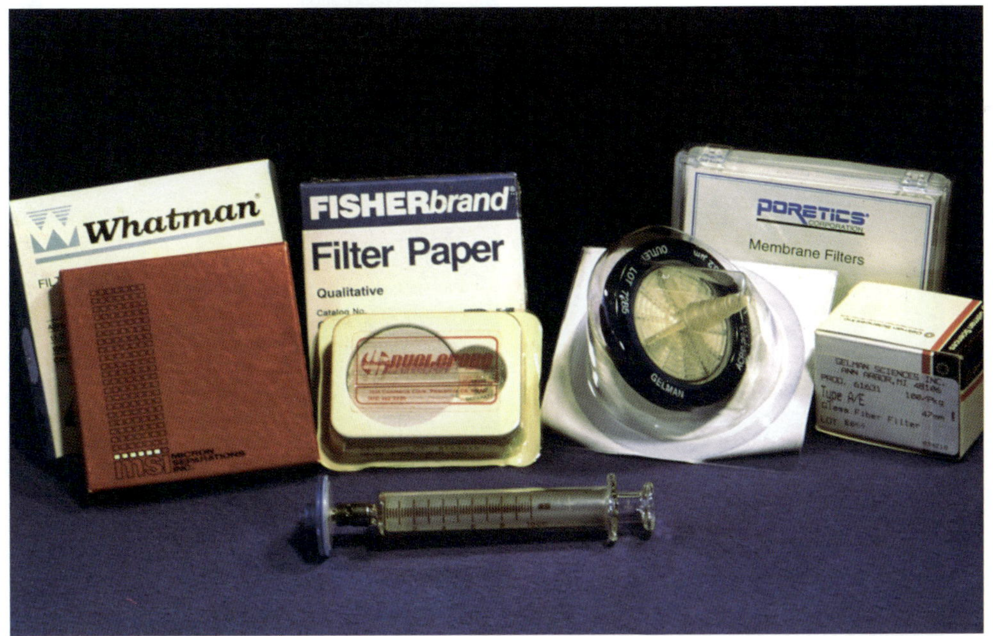


**Figure 1.** This is a sampling of the large variety of filters that are available through scientific suppliers. They include paper, glass fiber, membrane, in-line (air and liquid), and syringe filters. There are many others.

**Figure 2.** This is the filtration apparatus you will be using throughout your lab work. Notice that the filter paper on the tabletop (and the one in the funnel) is "fluted." That means it's folded to provide greater effective surface area for contact with the liquid.

**Figure 3.** Here are two views of an apparatus that is commonly used for filtering suspensions under vacuum. The "Buchner funnel" seals against the neck of the vacuum flask so suction can be drawn on the flask. The bottom of the funnel is flat so a filter paper disk can fit snugly against it.

## Quantitative Filtration

A complete discussion of quantitative chemistry is provided in Method 5 called *Quantitative Transfer*. Please read that method before returning to your lab work.

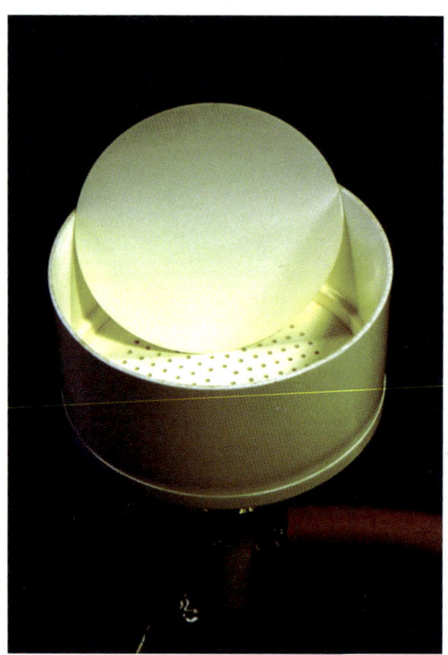

## Method 4: Serial Dilution

The word *serial* has nothing to do with breakfast. It has to do with things done in a *series*. In this case we're carrying out a series of dilutions. Each dilution in the series represents the same degree of dilution compared to the one before it and after it. For example, in a tenfold dilution series, every dilution step represents a factor-of-ten dilution.

Here is an example of a case where a tenfold serial dilution might be needed. Suppose a sample comes into your lab from an outside source. You know it has a particular chemical in it, but you don't know the concentration of that chemical. You have an instrument that can measure it, but only if it is present in a certain range of concentrations. You might do a tenfold serial dilution of the sample and put each of the diluted samples into the instrument to see which one is in the right concentration range.

To do this you need:

1) an original, undiluted, liquid sample
2) a set of "dilution blanks" (vessels, usually test tubes, vials or beakers, all holding the same amount of the dilution liquid). Dilution blanks contain a carefully selected volume so that, when a portion of the sample is added, the desired dilution is achieved.
3) a device (usually a pipet) to move a portion of the undiluted sample into the first dilution blank and a portion of each diluted sample into the next dilution blank
4) a means of mixing the sample after each dilution

Here is one procedure for serially diluting an undiluted sample in 10-fold dilution steps:

1) Prepare dilution blanks by adding 9 mL of distilled water to each of a series of test tubes. The number of dilution blanks should equal the number of 10-fold dilutions you wish to do. Label the blanks in sequence in a way that indicates that each blank represents a 10-fold lower concentration than the one before it. [If the undiluted sample represents a relative concentration of $10^0$ (or 1 X), the first dilution represents a relative concentration of $10^{-1}$ (1/10 X), the second dilution $10^{-2}$ (1/100 X), and so on.]
2) Homogenize (mix, stir, blend, etc.) the undiluted sample. Any solids in the sample must be completely suspended and dispersed when a sample is taken out of it.
3) Withdraw 1 mL of the homogenized sample in a pipet and expel all of it in the first dilution blank.
4) Thoroughly mix this diluted sample. This may be done by repeatedly drawing sample into the pipet and expelling it beneath the surface of the diluted sample. If the sample contains solids, some mechanical mixing may be necessary. (Special devices called "vortex mixers" are made for this purpose. See the Figure.)
5) Rinse the pipet in the diluted sample to assure that the transfer of sample material is complete and to prepare for the next dilution step. (In procedures where precise dilutions are critical, a new or clean/dry pipet should be used for each transfer.)
6) Perform all other dilutions down the series in the same way (steps 2-5), being sure to rinse the pipet thoroughly in the diluted sample with each transfer. This will prevent carryover of a portion of a higher-concentration sample to a lower-concentration sample.

This procedure will provide any number of 10-fold dilutions of the sample. The concentration of each dilution is 10-fold less than the concentration of the dilution before it. If it is the $10^{-5}$ dilution, then the concentration is that of the original sample $\times$ $10^{-5}$. In this way, the calculation of the concentration at any dilution level in a 10-fold series is kept very simple.

The other most common dilution series used is a twofold series. Of course, every step in a serial twofold dilution is half the concentration of the step before it. To do this, you would follow the same procedure as before, but the volume in the dilution blank would be equal to the volume transferred. For example, you might choose to use 5-mL dilution blanks and to pipet 5-mL sample portions. If larger volumes were required, you might use 50-mL dilution blanks (prepared in beakers) and move 50-mL sample portions using your graduated cylinder.

If the concentration of the undiluted sample is 1 X, the concentration of the first dilution is ½ X, and the concentration of the next is ¼ X. To find the concentration of any sample in the series, just multiply the concentration of the original sample by the dilution level. For example, if the original concentration was 20 g/L, then the concentration of the ¼ X sample is 20 g/L × ¼ = 5 g/L.

**Figure.** A vortex mixer is a device that is used for thoroughly mixing a sample in a test tube.

## Method 5: Quantitative Transfer

The word *quantitative* has to do with the *quantity* of something. By *quantitative filtration* we mean doing your filtration in such a way that the quantities of all materials are saved—nothing is wasted. Let's say you have a sample that contains soil in water. You might filter this suspension to remove the soil from the liquid. If you don't care about saving all the liquid, you might just take that damp filter and throw it away when you're done with it.

Compare that with a case where you need to know exactly how much of a solute is contained in a liquid. As part of your purification of the solution you have to pass it through a filter. You can't just throw away the filter, because, if you do, you're throwing away the portion of the solution that's trapped in the paper itself. Instead, you must rinse the filter several times with the solvent to remove any trapped solute. Depending on how serious you are about getting every last bit of the solute, you might rinse the filter five or even ten times. If you want to know exactly how much of it you are losing, you might do an experiment to see what portion of the solute can be recovered from a filter by rinsing.

A quantitative transfer is one in which *all* of the important substances are transferred *completely* for practical purposes. Any significant losses of materials are *quantified*—that is, a value is placed on the loss. If you are careful to rinse your filter correctly so that you leave none of your important materials behind, you should measure the volume of the filtrate and be sure it is homogeneous. Then, if you accidentally spill some, you can measure how much is left after the spill and know exactly what portion of the original quantity has been recovered.

A college course that is required of all chemistry majors is called *Quantitative Analysis*. In this course a student is given many exercises in measuring and tracking quantities of substances as he carries out various procedures. It's important that you begin now to develop good habits of paying attention to what's happening to all the substances that are involved in your experiments. It's okay to discard some things, but you want to be sure that you are in control of what you keep and what you discard. The only way to do that is to give thought during your procedures as to whether you are giving up things you don't want to lose. A good chemist controls and accounts for those losses.

Here is a summary of the important points made about quantitative transfers:

1) Be careful not to spill or splash during transfers.
2) Pay careful attention to volumes of added materials to be sure they are accurate.
3) Record any losses of important materials that result from spills and consider the effects of these losses on your results.
4) Rinse glassware and filters to be sure that important materials are not left behind after transfers. Collect all rinse liquids. (The first rinse liquids contain the most solute, so they are the most important ones.)

So then, if you are told to transfer material quantitatively from a vial to a beaker, what is meant? It means be sure you transfer every bit of it. Usually you will dissolve the entire contents of the vial in water or some other solvent and transfer the entire solution to the beaker being careful not to lose any by dripping or splashing. Then you will rinse the vial several times with the solvent and transfer the rinsates to the beaker as well. This is how you can be absolutely sure you are not losing materials in your transfers.

## Method 6: Droppers

You wouldn't think that something as simple as a medicine dropper would require any instructions to go along with it. Unfortunately, droppers may be misused in a laboratory setting in ways that can bring about failure of an experiment or misleading results. Also, a dropper is such a simple device that you may not realize its full usefulness as an accurate means of measuring liquids.

### The Dropper as a Measuring Device

Even a dropper with no graduations may be used as a measuring device if you count the drops of fluid as they are delivered. For example, when you finish reading this entire method, get out your graduated cylinder and see how many drops of liquid are required to fill it to the 10-mL mark using your medicine dropper. Yes, I know it will be tedious counting well over a hundred drops, and yes, if you make a mistake you will have to start over. But once you have finished, that dropper will be useful for measuring small quantities for the rest of your natural life.

If you don't drop the dropper on a concrete floor or place it in your back pocket and sit on it, this simple device will never change—its drops will always be of the same size. This means that if you divide the 10 mL by the number of drops that went into the cylinder, you will get a fixed, decimal number of milliliters per drop. That will be the same number of milliliters per drop that you will always get from that dropper until the day you either drop it or drop dead.

### Priming the Dropper

When you draw your dropper full of water, the first drop from the tip is often a dud. It may contain air or it may not fully form before it drips. This is why the dropper should be primed before calibrating it. You'll also need to prime it before each time you use it in a quantitative way. It's really not difficult and takes no significant amount of time, but it will significantly improve your precision.

Here is the principle for using a dropper to measure a liquid: *Every liquid has a certain surface tension, so it will always form the same size of drop from the same glass dropper tip.* However, the person controlling the dropper must give care not to introduce errors into this measurement by artificially affecting the sizes of the drops. This will happen if you:

1) move the dropper while it is dropping fluid. (A drop may release before it reaches its full size under the pull of gravity.)
2) expel air from the tip of the dropper as though it were part of a drop. (An air bubble may cause a drop to release before it is full grown.)
3) touch the drop to the receiving vessel (the graduated cylinder), causing the drop to cling to the surface and leave the dropper too soon.
4) hold the dropper at inconsistent angles. (Which angle you use doesn't matter as long as it is always the same. One simple way to keep it consistent is to always hold the dropper so that its tip is vertical. See the Figure.)

The following are instructions for priming and calibrating your dropper. You will need: room temperature distilled water, a thermometer, a dropper, a 50-mL beaker and a graduated cylinder. Please follow these instructions:

1) Lay your dropper on your work surface (don't hold it in your hand) and leave it there while you allow a beaker of distilled water (40 mL or so) to come to room temperature.
2) Inside the back cover of your lab workbook, write the following information:

Medicine Dropper Calibration

Date: ____________________

By: ____________________________________

Room Temperature: __________

**Figure.** The angle at which the dropper is held should always be approximately the same. The easiest way is to hold the tip approximately vertical.

Number of whole drops required to fill graduated cylinder to 10-mL mark:

__________ drops/10 mL

__________ drops/mL

__________ mL/drop

__________ mL/drop

3) Fill in the date and your name, and, when the distilled water in the beaker has reached a constant temperature, record the temperature.
4) Place your distilled water near the graduated cylinder.
5) Pick up the dropper near the rubber bulb. Do not handle the glass tip. You want the tip to remain as near room temperature as possible.

**Note**: Obviously, there is no way to draw water into the dropper and expel it without touching the rubber bulb. However, holding the bulb warms it and makes it more difficult to control the flow of liquid from the dropper. I don't grab the bulb until it's necessary.

6) Squeeze the bulb to expel the air. Immerse the tip of the dropper in the water, then draw the dropper full.
7) Prime the dropper by lightly squeezing the bulb to expel a couple of drops of water back into the beaker. Without releasing the bulb or letting air return into the tip of the dropper, move the dropper over the graduated cylinder and begin expelling and counting drops. Be sure you count every drop that is expelled into the cylinder. (Even a single miscounted drop will affect your results for as long as you use the dropper.) Stop placing drops into the cylinder before you add the last of the liquid. (If you run out of water during your last drop, you will be introducing error into your volume.) The goal is to have every drop that enters the graduated cylinder be uniform in size. That can happen only if you don't have partial drops.
8) Record the number of drops you added to the cylinder before refilling the dropper.
9) Repeat steps 8 and 9 as many times as necessary to fill the graduated cylinder to the 10-mL mark.
10) Sum the numbers of drops to get the total and do the math to fill in the blanks in the back of your workbook.

The size of a drop of water has to do with the size of the surface area and material that it clings to as it forms. If the dropper has a fine tip, the drop will be smaller than if it has a blunt tip. Every dropper has its own characteristic drop size. You will need to go through this exercise with every new dropper you want to use in measuring liquids. Also, every liquid has its own characteristic surface tension, so each will have its own drop size. You must recalibrate your dropper for use with every different liquid.

## Bibliography and Credits

Davis, B.D., R. Dulbecco, H. Eisen, H.S. Ginsberg, W.B. Wood, Jr. and M. McCarty. 1973. *Microbiology.* 2nd Ed. Harper & Row, Publishers, Hagerstown, MD.

Hein, M. and S. Arena. *Foundations of College Chemistry.* 10th Ed. Brooks/Cole Publishing Company. Pacific Grove, CA.

Lehrman, R. and C. Swartz. 1965. *Foundations of Physics.* Holt, Rinehart and Winston, Inc., New York, NY.

Masterton, W.L., and E.J. Slowinski. 1977. *Chemical Principles.* W.B. Saunders Company, W. Washington Square, Philadelphia, PA.

Morrison, R.T. and R.N. Boyd. 1973. *Organic Chemistry.* 3rd Ed. Allyn and Bacon, Inc., Boston, MA.

Perry, R.H. and D. Green. 1984. *Perry's Chemical Engineers' Handbook.* 6th Ed. McGraw Hill Professional Publising, New York, NY.

Sittig, M. 1985. *Handbook of Toxic and Hazardous Chemicals and Carcinogens.* 2nd Ed., Noyes Publications, Park Ridge, NJ.

Skoog, D.A. and D.M. West. 1976. *Fundamentals of Analytical Chemistry.* 3rd Ed. Holt, Rinehart and Winston, Inc., New York, NY.

Skoog, D.A. and D.M. West. 1980. *Principles of Instrumental Analysis.* 2nd Ed. Saunders College, West Washington Square, Philadelphia, PA.

Streyer, L. 1981. *Biochemistry.* 2nd Ed. W.H. Freeman and Company, San Francisco, CA.

Stumm, W. and J.J. Morgan. 1981. *Aquatic Chemistry: An Introduction Emphasizing Chemical Equilibria in Natural Waters.* 2nd Ed., John Wiley and Sons, New York, NY.

Thaxton, C.B., W.L. Bradley and R.L. Olsen. 1984. *The Mystery of Life's Origin: Reassessing Current Theories.* Philosophical Library, New York, NY.

Weast, R.C., M.J. Astle and W.H. Beyer (Eds.). 1985. *CRC Handbook of Chemistry and Physics.* 66th Ed. CRC Press, Inc., Boca Raton, FL.

Wetzel, R.G. 1983. *Limnology.* 2nd Ed., Saunders College Publishing, Philadelphia, PA.

Wilson, J.D. 1977. *Physics Concepts and Applications.* D.C. Heath and Company, Lexington, MA.

Windholz, M., S. Budavari, R.F. Blumetti and E.S. Otterbein (Eds.). 1983. *The Merck Index.* 10th Ed. Merck & Co., Inc., Rahway, NJ.

Professional photographs, pp. 60, 85, 106, 113, 121, 139, 145, 151, 168, 173, 179, 197, 209, 212, 230, 249 © Corel Corporation and used under license.

Photographs, pp. 113, 145 ("happy cat" and "hungry cat") retouched by Durell Dobbins.

All other photographs © Beginnings Publishing House by Durell Dobbins.

Illustration, pg. 98, by Donald Clark.

All other illustrations by Durell Dobbins.

Text editing by Gloria Dobbins; additional text editing by Therisa Pennington and Shannon Gottke.

Glossary by Therisa Pennington and Durell Dobbins.

Indexing by Shannon Gottke and Durell Dobbins.

Text layout and design by Durell Dobbins with assistance from Shannon Gottke and Therisa Pennington.

Printed by Sexton Printing, Inc., W. St. Paul, MN.

## Glossary of Terms

**absolute** – pure and unmixed

**absolute zero** – zero point on the Kelvin temperature scale; theoretical temperature at which molecular motion ceases

**accuracy** – refers to the extent to which a measurement gives a true value of the quantity being measured

**acid** – any substance that raises the hydrogen ion concentration of pure water, that "donates" a proton or that "accepts" a pair of electrons in a chemical reaction

**acid anhydride** – non-metal oxide that reacts with water and is acidic in water solution

**activity series** – list of elements (e.g., metals or halogens) in the periodic table in order of their readiness to participate in a chemical reaction

**actual yield** – the actual amount of product obtained from a reaction

**adhesive force** – intermolecular force that causes a liquid to be attracted to a solid surface

**alloy** – a homogeneous mixture formed by melting metals to their liquid states and fusing them with other metals

**amorphous** – solid matter that does not exist in structured patterns; noncrystalline

**anhydrous salt** – ionic compound which remains when the water has been removed from a hydrate

**anion** – atom or molecule bearing a negative charge, having gained one or more electrons above its neutral condition

**Arrhenius concept** – idea that an acid is a substance which produces an overabundance of $H^+$ ions when added to water, and that a base is a substance which ionizes to produce $OH^-$ ions

**atom** – the smallest particle of an element which has all the properties of that element

**atomic mass (atomic weight)** – a number which represents the mass of an atom of an element as it relates to the mass of another that is accepted as the standard (e.g., $^{12}C = 12$). It is expressed in atomic mass units (AMU).

**atomic mass unit (AMU)** – unit of measure of atomic mass, equal to 1/12 the mass of the most abundant isotope of carbon ($^{12}C$).

**atomic number** – represents the number of protons in the nucleus of an atom, and, therefore, the number of electrons in that atom when in its neutral state.

**Avogadro's law** – states that, at a constant temperature and pressure, the same molar quantity of two different pure gases will occupy the same amount of space (e.g., one mole of oxygen occupies the same volume as one mole of hydrogen).

**axis** – central line in a geometric figure that forms a reference for symmetrical arrangement of parts

**base** – any substance that raises the hydroxide ion concentration of pure water, that "accepts" a proton or donates a pair of electrons in a chemical reaction

**basic anhydride** – compound that forms a base when added to water (the metal oxides are basic anhydrides)

**bent molecule** – molecule having all atoms lying on a single imaginary plane and forming a single angle (e.g., $H_2O$)

**binary compound** – compound consisting of only two elements

**biochemical** – molecule from among the largest and most complex of all molecules, in or resulting from living things (polysaccharides, nucleic acids, proteins, lipids, etc.)

**biology** – the study of living things, with regard to their structures, functions, origins, development and behavior

**boiling point** – temperature at which, under intensive heating, the rate of cooling by evaporation is equal to the rate of heating; temperature at which the vapor pressure of the liquid is equal to the pressure of the atmosphere above the liquid

**boiling point elevation** – colligative property of solutions describing an increase in the boiling point of a solvent that occurs when a solute is added to it

**bond** – means by which atoms or groups of atoms in molecules are linked

**bond angle** – angle formed by imaginary line segments connecting bonded atoms

**bond length** – distance between the two nuclei of bonded atoms

**Boyle's Law** – states that the pressure and the volume of the air are inversely related by a constant value

**Brönsted-Lowry concept** – broadened Arrhenius' definition of acids and bases by suggesting that the process of donating a proton can take place outside of a water solution; also known as the "proton donor" concept

**Brownian movement** – random bumping of molecules which maintains dense molecules in suspension

**bulk liquid** – large part or pool of liquid, as opposed to any smaller fraction of such liquid

**calibrate** – to graduate, correct, or check the scale of a measuring utensil (as, to calibrate a thermometer)

**calorie** – unit of energy that equals the amount of heat necessary to raise the temperature of one gram of water by one degree centigrade (equals to 4.184 joules)

**capillary action (capillarity)** – movement of a liquid through a small tube (capillary) contrary to the force of gravity due to its attraction to the capillary walls

**catalyst** – A substance that changes the rate of a chemical reaction without participating in that reaction as a reactant

**cation** – atom or molecule which bears a positive charge having lost one or more electrons relative to its neutral condition

**cation exchange** – process used to remove divalent cations ("hardness") from water and replace them with sodium (monovalent) ions from sodium chloride; such a process is used in household water softeners

**Celsius** – temperature scale on which there are 100 degrees between the freezing and boiling points of water (zero and 100 degrees, respectively) and having its zero point defined as the freezing point of water

**centimeter** – one hundredth of a meter

**centroid** – center of a mass

**Charles' Law** – states that the volume of a gas is directly proportional to its temperature, at constant pressure

**chemical change** – change in chemical composition that results in physical changes as well; one in which one or more substances is either consumed or formed

**chemical equation** – expression which describes all the reactants and products of a chemical reaction in their correct molar proportions

**chemical equilibrium** – balance point of a reversible chemical process

**chemical formula** – symbolic representation of the composition of a compound such that the constituents are shown in their exact proportions

**chemical property** – characteristic of a substance that may be determined by its reaction with others

**chemistry** – the science that studies the properties of substances and the changes that take place within and among those substances as they interact

**coefficient** – a number or symbol put before a variable in an algebraic expression by which that variable is considered to be multiplied

**cohesive force** – force which causes molecules to cling to one another

**collagen** – fibrous protein that is the major component of connective tissue, bone, cartilage, blood vessels, skin, and teeth; present in nearly all organs and notable for having a high tensile strength

**colligative property** – characteristic of solutions that is the same regardless of the kind of solute under consideration but which changes according to the concentration of solute particles

**colloid** – small, insoluble, nondiffusible particles that remain in suspension, often having complex surface properties and a high surface area to volume ratio

**combination reaction (synthesis reaction)** – a chemical reaction in which chemical entities combine to form a single substance

**combined gas law** – states that, for a specific amount of any gas, PV=kT; that is to say, the product of the pressure and volume is related to temperature by a constant value

**common name** – name used to refer to certain familiar chemicals as opposed to using their standardized names (e.g., "water" as opposed to "dihydrogen monoxide").

**complex carbohydrate** – polymer of sugars

**compound** – substance composed of two or more elements combined in fixed proportions having distinct characteristics; the components lose their original, individual characteristics

**concentrate (adj.)** – describing the condition of a large amount of solute relative to the amount of solvent

**concentrate (verb)** – to increase the concentration of a solute by removal of solvent as by evaporation, distillation or reverse osmosis filtration; opposite of dilution (verb)

**concentration gradient** – condition of a solution in which the concentration of solute in one location is greater than that in another location within the same solution; increase or decrease in solute concentration over a distance

**conjugate acid** – according to the Brönsted-Lowry concept, the proton donating portion of a conjugate acid-base pair

**conjugate base** – according to the Brönsted-Lowry concept, the remaining portion of a conjugate acid-base pair once the conjugate acid has dissociated

**covalent bond** – a link between two atoms produced by the sharing of two electrons in a single molecular electron cloud

**critical pressure** – equilibrium pressure of a substance at critical temperature

**critical temperature** – highest temperature at which the molecules of a substance may be compressed into the liquid phase

**crystalline** – type of solid (such as salt and sugar crystals) in which the atoms are arranged in regular, repeated geometric patterns

**cubic centimeter** – volume of a cube having a measure of one centimeter in each dimension; equal to one milliliter

**Dalton's atomic theory** – theory on the nature of makeup and behavior of matter proposed by English scientist John Dalton

**Dalton's law of partial pressures** – states that in any container of gases, the total pressure of gas in the container is equal to the sum of the partial pressures of the individual component gases

**decomposition reaction** – a chemical reaction in which a substance disintegrates into two or more products different from the reacting substance

**degradation** – decomposition

**deliquescent** – chemicals that are capable of absorbing moisture to the point of their own dissolution

**density** – a ratio of the mass of a substance to its volume

**deoxyribose nucleic acid (DNA)** – component of all living matter and that which contains the genetic code; long-chain polymer of deoxyribonucleotides

**derivatize** – modify the structure of a molecule by replacement of one or more atoms

**desalination** – removal of salt from sea water so that it is suitable for drinking

**diatomic molecule** – molecule formed by the combination of two atoms of the same element; the seven diatomic molecules are hydrogen, nitrogen, oxygen, fluorine, chlorine, bromine, and iodine

**dietary calorie** – unit of energy, used on food labels, that equals the amount of heat required to raise the temperature of one kilogram of water by one degree centigrade (also called *large calorie*)

**diffusion** – process by which the molecules of a substance will randomly disperse with those of another by random collisions among their molecules

**dihydrogen dioxide** – standardized name for hydrogen peroxide

**dihydrogen monoxide** – standardized name for water

**dilute (adj.)** – describing the condition of a small amount of solute relative to the amount of solvent

**dilute (verb)** – to reduce the concentration of solute by addition of solvent; opposite of concentration (verb)

**dimensionless** – mathematical expression in which the units in the numerator and denominator cancel leaving no units in the result (i.e. mol/mol or g/g)

**dimer** – two-unit group

**dipole** – term applied to a molecule having opposite charges in different locations

**diprotic** – describing acids that have two ionizable protons

**dissociation** – process of breaking down a compound into simpler component parts, most often used of the separation of two ions of an ionic compound when placed in water

**distillation** – a process which separates a liquid from a mixture by vaporizing it, then condensing it, so that it can be collected as a more refined substance

**double displacement (double replacement) reaction** – reaction in which one element from each of two reacting compounds change positions, each displacing the other, to form two new products

**dynamic** – pertaining to constant motion or tending toward change; opposite of static

**elastic** – describing the ability of gas molecules to spring back upon collision without a loss of energy; frictionless

**elastin** – protein component of elastic tissues which, particularly when moist, can stretch to several times their original length and then return to their original shape and size again once the applied tension is released

**electrolyte** – a compound which, when added to water, dissociates into ions and promotes electrical conductivity in the solution

**electron** – fundamental particles of an atom, having a negative charge and a very low mass and moving about in a realm about the nucleus

**electronegativity** – property of an atom which is increased by its affinity for the electrons in a chemical bond; opposite of the tendency to ionize

**electrostatic separation** – separation of matter from a mixture by applying an electrostatic charge as the means of removing substances having the opposite charge

**element** – fundamental substance composed of atoms, and having a distinct number of protons in its nucleus; cannot be broken down into simpler substances except by nuclear disintegration

**empirical formula** – symbolic representation of the composition of a compound showing the smallest whole-number ratio of the elements present in the compound

**endothermic** – having to do with a chemical reaction in which heat is absorbed

**energy** – the capacity of a body to do work and to overcome inertia

**energy, kinetic** – the energy that relates to motion (as opposed to potential energy)

**energy, potential** – the energy that results from the positioning of a body as it relates to the influence of a force (as opposed to energy of motion or kinetic energy)

**enzyme** – biochemical catalyst

**equilibrium** – state of dynamic balance at which forward and reverse reactions occur at equal rates

**equilibrium constant ($K_{eq}$)** – number in an equilibrium system which places a condition on the concentration of reactants and products

**exothermic** – having to do with a chemical reaction in which heat is released

**Fahrenheit** – temperature scale on which water freezes at 32° and boils at 212°.

**fats** – two-and three-unit esters which are made from glycerol and a long-chain carboxylic acid ("fatty acid")

**fatty acid** – long-chain carboxylic acid; a component of a lipid (fat) molecule

**filtration** – process for separating solids from liquids in mixtures by passing the liquid phase through a fine barrier through which the solid cannot pass

**fine heterogeneity** – kind of variation in the composition of a mixture that appears uniform overall, but, on closer inspection, is not completely dispersed

**force** – that which initiates motion or causes a change in, or cessation of, motion

**freezing point** – temperature at which a substance's liquid solidifies or its solid liquefies (they are one and the same)

**freezing point lowering (freezing point depression)** – colligative property of solutions describing a decrease in the freezing point of a solvent that occurs when a solute is added to it

**functional group** – group of atoms that, when added to an organic molecule, give it a particular desired function

**gas** – matter made up of molecules that are very mobile and energetic, and capable of expanding in all directions so that they have neither definite shape nor definite volume

**Gay-Lussac's Law** – states that, at a constant volume, the temperature and pressure of a set amount of gas are related by a constant

**genome** – genetic code

**global warming** – warming of the atmosphere due to gradual accumulation of carbon dioxide and other "greenhouse gases" in the atmosphere that leads to increased retention of heat from the sun

**glycolipid** – sugar-containing lipid which is a derivative of sphingosine

**glycoprotein** – protein in which one or more sugar chains are attached to one of its amino acids

**gram** – in the metric system, the basic unit for measuring mass and weight; mass of one cubic centimeter of pure water at its greatest density (4° Celsius)

**gram atomic weight** – one mole of an entity; the quantity of an element that has a weight which equals the element's atomic weight (e.g., since the atomic weight of sodium is 23 AMU, the gram atomic weight of sodium is a quantity of sodium weighing 23 grams)

**gross heterogeneity** – kind of variation in the composition of a mixture involving large, distinct types of matter such as lumps of coal mixed with scraps of iron

**group** – any of the vertical columns in the periodic table dividing elements according to the numbers of electrons in their orbitals

**half reaction** – either of a pair of reactions which, when summed form a complete oxidation reduction reaction

**halide** – a compound which consists of a halogen and another element (i.e. fluoride, chloride, bromide, and iodide are halides.)

**hard water** – water having high concentrations of naturally occurring calcium and magnesium ions which tend to precipitate in water pipes and form insoluble scale when combining with soap molecules

**headspace** – gaseous space above any liquids or solids in a sealed container

**heat of fusion** – amount of heat required to change one gram of a solid to its liquid at its freezing point

**heterogeneous system** – a mixture which does not have uniform composition and may include more than one phase

**homogeneous system** – characteristic of a mixture when its composition is uniform and it possesses the same physical and chemical properties throughout

**hydrate** – ionic compound which combines in definite proportions with water by electrostatic attraction

**hydrogen bond** – bond formed when a hydrogen atom in one molecule is attracted to a strongly electronegative atom in a neighboring molecule

**hydronium ion** – positive hydrogen ($H^+$) ion; free proton

**hydrophilic** – substance having an attraction to water

**hydrophobic** – substance that repels water

**hydroxide ion** – that portion of a water molecule that remains after dissociation of the hydronium ion; $OH^-$

**ideal gas law** – PV=nRT; a relationship between pressure, volume, temperature and molar quantity of gas where R is the ideal gas law constant

**impermeable** – membrane which does not allow the passage of a particular substance

**infinite** – quantity which is always designated as being greater than any other quantity, whether it be large (positively infinite) or small (negatively infinite)

**insoluble** – property of a substance that will not allow it to completely dissolve or enter into a solution with another substance

**interface** – physical boundary between phases in a mixture

**intermolecular force** – force that affects interactions among molecules

**ion** – an atom or group of atoms which has either a positive or negative electric charge; positively charged ions (cations) are formed when neutral atoms or molecules lose one or more valence electrons, and negatively charged ions (anions) are formed when an atom or molecule gains one or more electrons

**ionic bond** – bond in which electrostatic forces hold together positive and negative ions resulting from the transfer of one or more electrons from one element to another

**ionic equation (total ionic equation)** – chemical equation showing a reaction among ionic substances in which the ionic compounds are shown dissociated into their anions and cations

**isotope** – one of a number of atoms having the same number of protons in their nuclei, but having different numbers of neutrons and therefore having different masses; (e.g., $^{235}U$, $^{238}U$ and $^{239}U$ as isotopes of uranium)

**isotope notation** – designation of an isotope consisting of the symbol of the element, a left superscript showing the mass number of the isotope and a left subscript showing its atomic number

**joule** – SI unit of energy that is a measure of the amount of work done when one Newton is displaced one meter in the direction of the force (equals 0.239 calorie)

**Kelvin scale** – temperature scale on which there are 100 degrees separating the freezing and boiling points of water, and measuring in degrees from absolute zero (–273.15 °C).

**kinetic-molecular theory (KMT)** – theory describing the nature of gases

**law of constant (definite) composition** – states that chemical compounds are composed of specific fixed proportions of elements by mass

**law of multiple proportions** – states that when two or more elements combine to form more than one compound, the product will contain the original elements in a proportion that is defined by a ratio of integers.

**Le Chatelier's principle** – basic principle of equilibrium that anytime a disturbance is placed upon an equilibrium the equilibrium will shift to counteract the change

**Lewis concept** – idea which broadened the existing definition of acids under the Brönsted-Lowry concept by removing the restriction that a species must have an ionizable hydrogen atom; defines an acid as an electron pair acceptor and a base as an electron pair donor

**Lewis structure** – diagram consisting of the symbol of an element, or of the elements in a compound, and using dots to represent valence electrons

**limiting reagent (limiting reactant)** – reactant in a chemical process that is present in a quantity below that theoretically required to carry out a balanced chemical reaction

**linear molecule** – molecule in which the nuclei of all atoms can be thought of as falling on a straight line

**lipopolysaccharide** – unusual molecule in the outer membrane of certain bacteria which consists of three regions: "lipid A", "core oligosaccharide," and "O side chain"

**liquid** – substance that has a definite volume but not a definite shape so that it flows easily, unlike a solid, but expands minimally, unlike a gas

**liter** – metric system basic unit for capacity; equal to the volume of one kilogram of water

**logarithm** – for a given value, the exponent to which a base value (such as 10) is raised to equal that value

**magnetic separation** – process in which a magnet is used for removing matter from a mixture

**mass** – quantity of matter of a body in relation to inertia; determined as the quotient of its weight divided by the acceleration due to gravity

**mass percent composition (or percent composition)** – percentage contribution of each element to the total molar mass

**matter** – material; that which occupies space, has mass and may be perceived by the senses

**melting point** – see **freezing point**

**membrane filter** – one used for filtering microscopic particles; its pores range in size from several micrometers (microns) down to a fraction of a micrometer

**meniscus** – curvature in the upper surface of a liquid as the result of adhesion to the walls of the container

**mercury (or U-tube) manometer** – device consisting of mercury in a glass tube bent in the shape of a "U" and used to measure gas pressure

**messenger RNA (m-RNA)** – type of RNA that serves as the template for the synthesis of proteins

**metal hydroxide** – alternate term for *base* (in the acid-base sense)

**meter** – the SI unit for length, equal to about 39.37 inches

**milliliter** – one thousandth of a liter

**millimeter** – one thousandth of a meter

**mixture** – a substance which contains two or more elements in varying proportions; differs from a compound in that the individual components retain their original properties

**molality** – unit of measurement of concentration, expressed as the number of moles of solute divided by the number of kilograms of solvent

**molar heat of vaporization** – quantity of heat necessary to vaporize one mole of a liquid at its boiling point

**molar mass (gram molecular weight, gram formula weight)** – mass in grams of one mole of an entity

**molar ratio** – the proportion of the moles of one substance to a corresponding number of moles of another

**molar stoichiometry** – stoichiometry involving molar, rather than mass, ratios

**molarity** – unit of measurement of concentration, expressed as the number of moles of solute divided by the number of liters of solution

**mole** – chemical mass unit containing $6.022 \times 10^{23}$ molecules, atoms or ions and equal to the gram formula weight

**molecular mass (molecular weight, formula weight)** – number that equals the sum of the atomic weights of all the atoms in a molecule

**molecule** – smallest particle of an element or compound that can exist separately and still retain all the original properties of the element or compound

**nanometer** – one billionth of a meter

**negative common logarithm** – negative representation of a base 10 logarithm, denoted by a small p before measured value (as in pH or pK)

**net ionic equation** – chemical equation for a reaction among ionic substances from which the spectator ions have been removed

**neutral solution** – solution in which the concentrations of hydrogen and hydroxide ions are in balance

**neutron** – fundamental particles within the nucleus of an atom, having an atomic mass of 1 AMU and no electric charge

**noble gas configuration** – stable structure in which electrons are shared so that they fill the outer valence shell of both atoms, obeying the octet rule; also called an inert gas structure

**nomenclature** – the system of names used by a particular person or group for a certain type of activity or purpose

**non-point-source pollutant** – pollutant that cannot be tied to a single source

**nonpolar** – molecule characterized by equal sharing of electrons, with no distribution of charges (e.g., $H_2$ and $O_2$)

**normal boiling point** – boiling point at one atmosphere of pressure

**normality** – concentration of an acid or base obtained by multiplying the molar concentration by the number of protons donated or accepted by the acid or base (respectively) under consideration

**nucleic acid** – groups of acids that occur in organic nuclear material and which are composed of phosphoric acid combined with a carbohydrate and a nucleic acid "base"

**nucleic acid bases** – set of molecules that are combined into a nucleic acid molecule that form the genetic code of living cells; guanine, adenine, thymine, cytosine or (in RNA) uracil

**nucleus** – the dense region at the core of an atom which contains all of the atom's positive charge and most of its mass

**octet rule** – atoms tend to obtain, by extracting or by sharing electrons, eight electrons in their valence shells, thus obtaining complete *s* and *p* sublevels in their highest principal energy level

**organic chemistry** – study of the chemistry of carbon

**osmosis** – diffusion of a substance through a semipermeable membrane into a solution of lower concentration so that the concentrations on opposite sides of the membrane tend to be equalized

**osmotic pressure** – pressure inside cells caused by osmosis

**osmotic pressure elevation** – colligative property of solutions describing an increase in osmotic pressure that occurs when a solute is added

**oxidation** – process by which atoms or ions lose electrons

**oxidation number** – number that represents the electrical charge condition of an element in a covalent compound as though the bond were ionic

**oxidation state** – state of an element represented by the oxidation number

**oxidation-reduction reaction (or redox reaction)** – A reaction in which one of the reactants is oxidized (loses one or more electrons) and the other is reduced (gains one or more electrons)

**paper filter** – filter usually made of cellulose fibers in a crisscrossing network

**partial pressure** – portion of gas pressure exerted by a single component gas in a mixture

**pathway** – complex sequence of chemical reactions

**peptide** – sequence of amino acids not constituting a protein

**percent yield** – percent of the theoretical yield of products actually resulting from a chemical reaction; mathematically equal to the ratio of the actual yield to the theoretical yield, multiplied by 100

**periodic table of the elements** – systematic arrangement of the symbols of chemical elements by their atomic numbers as determined by the periodic law; groups (vertical columns) align elements with related valence electron configurations and periods (horizontal rows) that represent the highest principal energy level occupied by electrons.

**permeable** – describing a membrane that is open to the passage of a particular substance

**phase** – portion of a mixture, either solid, liquid, or gas, that is in a state of matter that is different from that of other portions of a heterogeneous system; as, ice is a phase of water

**phase equilibrium diagram** – diagram relating pressure to temperature and demonstrating the conditions under which a substance may exist as either a solid, a liquid, or a gas, and also the circumstances under which equilibrium occurs for the phases various phases

**physical change** – change in a physical property, but not in chemical composition

**physical property** – characteristic of a substance that may be determined from observation of that substance in isolation from other substances

**physics** – the most fundamental of the three main branches of science; deals with forces, energy and matter, their properties and their interactions

**planar molecule** – molecules whose atoms all lie on a single plane; may be biangular (bent, like water), trigonal (like boron trifluoride), or tetragonal (like ethylene)

**point-source pollutant** – pollutant that has an identifiable source

**polar** – describing a molecule in which there is a distribution of both positive and negative charge; this is characteristic of bonds atoms dissimilar in their electronegativities (e.g., HCl, HF)

**polyatomic ion** – ion consisting of more than one atom and acting as a single entity

**polyprotic** – describing acids that have more than one ionizable proton

**polysaccharide** – group of carbohydrates that can break down into more than three monosaccharides

**precipitation reaction** – reaction in which substances of low solubility are formed, bringing about the formation of a precipitate

**precision** – the extent to which repeated measurements agree with one another; a term used to convey how much variation may be expected in a given measurement

**presupposed** – that which is implied or required as a preceding condition

**product** – any substance that results from a chemical reaction

**protein** – polymer consisting of a chain of various amino acids

**proton** – fundamental particles in the nucleus of an atom, having a positive charge and an atomic mass of 1 AMU

**proton acceptor** – Brönsted-Lowry base

**proton donor** – Brönsted-Lowry acid

**pyramidal molecule** – tetrahedral molecule with no atom at its centroid, but with atoms at each vertex (e.g., ammonia, $NH_3$)

**rate of dissolution** – rate at which a solute dissociates (is "solvated") in a solvent

**reactant** – any of the beginning substances participating in a chemical reaction to form a product

**reduction** – The process by which an atom or ion gains electrons

**regulate** – to adjust and govern the function of a system according to a particular standard

**resonance** – movement of electrons from one atom to another equivalent atom within a molecule having a stabilizing effect on the structure of the molecule

**reverse osmosis** – process that goes against the spontaneous passage of water into a saline solution and uses energy to force it out of the salt through a membrane

**reversible reaction** – a reaction that may proceed in either direction according to variations in temperature, volume, pressure, or quantity of the substances involved in the reaction

**ribose nucleic acid (RNA)** – molecules present in the cytoplasm of all cells that serve as templates for the synthesis of proteins

**ribosome** – organelle responsible for protein synthesis

**saturated** – property of a solute that has reached its equilibrium solubility in a given solvent

**scientific notation** – a shorthand used in science and mathematics to convey either very large or very small numbers as powers of ten; consists of a decimal number having a single digit to the left of the decimal point, multiplied by ten to the appropriate power.

**semipermeable** – describing a membrane that is permeable in relation to the passage of solvent molecules, but impermeable in relation to solute molecules

**separatory funnel** – device used for recovering individual liquids from mixtures which separate into phases based on differences in density; liquids having the highest density will gravitate to the bottom of the funnel and can then be recovered by draining them from the funnel, one phase at a time, through a stopcock

**SI unit** – unit of measure designated by the International System of Units (SI) to represent the single standard unit for each type of quantity (For instance, the SI unit for length is the meter, mass is the kilogram, etc.)

**significant figures** – sum of all meaningful digits in a measurement; all those that can be determined without doubt plus one estimated digit

**single displacement reaction** – reaction in which one element in a compound is displaced by another more active element

**solid** – state of matter in which there is a definite shape and volume, firm resistance to pressure and restriction in molecular motion so that molecules tend to stay in a fixed position with relation to each other

**soluble** – capable of being dissolved in another substance

**solute** – in a solution, the substance which is dissolved (as opposed to the solvent)

**solution** – a homogeneous mixture that is formed by dissolving one or more substances, whether solid, liquid, or gaseous, in another substance, and in which they are dispersed uniformly throughout

**solvate** – dissolve

**solvent** – the substance (usually a liquid) which dissolves another substance to make a solution

**specific heat** – ratio of the amount of heat needed to raise the temperature of a unit mass of a substance one degree to the quantity of heat needed to raise the temperature of the same mass of water by one degree

**spectator ion** – ion dissociated from an ionic reactant which does not, itself, participate in the reaction

**standard** – that which has been established as a means by which to compare or measure extent, quality, quantity, capacity, etc.

**standard temperature and pressure (STP)** – reference conditions for reporting characteristics of gases equal to zero degrees Celsius and one atmosphere of pressure

**state of matter** – condition in which matter may exist: solid, liquid, or gas

**static** – characteristic of that which is immobile, at rest, and not advancing or progressing; not undergoing change

**stoichiometry** – the branch of chemistry that deals with masses of elements in combination with each other

**structural formula** – diagram showing the bonding arrangement of atoms in a molecule

**structural protein** – protein that constitutes part of the structure of a living being, having the function of support and/or cohesion

**sublimation** – process of direct vaporization of a solid without its passing through a liquid phase

**supersaturated** – concentration of a solute above its equilibrium solubility

**surface tension** – resistance of a liquid surface to penetration due to intermolecular attraction among its molecules

**tetrahedral molecule** – molecule having an atom at its centroid bound to four atoms or unshared electron pairs that form four corners of a tetrahedron (a geometric figure having four triangular faces, four axes and four vertices). If the tetrahedron is regular, all four faces are equilateral triangles, and the bond angles are 109.5 degrees (e.g., methane, $CH_4$).

**theoretical yield** – the maximum amount of product that is theoretically possible if the limiting reactant is completely consumed during its formation

**torr** – SI unit of gas pressure; equivalent to mmHg

**transfer RNA** – type of RNA which carries individual amino acids to ribosomes to be added to a peptide

**trimer** – three-unit group

**triple point** – single point at which, with regard to both temperature and pressure, the solid, liquid, and gaseous states of a substance exist in equilibrium

**triprotic** – describing acids that have three ionizable protons

**type I binary compound** – binary compound containing a metal that forms only one kind of cation

**type II binary compound** – binary compound containing a metal that can form more than one kind of a cation

**type III binary compound** – binary compound containing only nonmetals

**ultrafiltration** – process of filtering molecules through a fine membrane, usually with the aid of applied pressure

**unit-factor method** – method for organizing and solving problems that involves the use of factors that are equivalent to unity for converting units of expression

**unsaturated** – property of a solute that is less than saturated and able to dissolve to a greater extent in its solvent

**valence number** – the property of an element measured by the number of hydrogen atoms one atom of the element can either combine with (if negative) or replace (if positive); for instance, oxygen has a valence of two because one atom combines with two hydrogen atoms.

**valence shell electron pair repulsion (VSEPR)** – repulsion of electron clouds for one another bringing about a change in the geometry of the overall molecule

**vapor pressure curve** – an expression of vapor pressure as a function of the substance and the temperature

**vapor pressure lowering (vapor pressure depression)** – colligative property of solutions describing a decrease in the vapor pressure (the pressure exerted by the vaporization of the solvent) above the solution that occurs when a solute is added to it

**vertex** – the point of an angle in a geometric figure

**volatile** – describing substances that evaporate readily

**volume fraction** – unit of measurement of concentration used in cases where both the solvent and solute are liquids, or when both are gases; expressed as the volume of one substance divided by the volume of the other (e.g., L/L or mL/mL)

**volume percent** – volume fraction times 100

**water of hydration (water of crystallization)** – the water that combines with an ionic compound to form a hydrate

**weight** – measure of the force by which an object is attracted to the earth; expressed as a product of the body's mass and the acceleration due to gravity

**weight percent (mass percent)** – percent by weight of one substance in a mixture

**weighted average** – method of calculating averages in which individual quantities are "weighted" (multiplied) by their corresponding proportions

**work** – the effect of energy upon a body resulting in motion of that body; represented as a product of force and the amount of distance over which it is applied

# Index

## A



## B

## C

## D



## E

## F



## G



## H

I



K



L

## O



## P

## Q



## R



## S

## W



## X



## Y



## Z